MANAGEMENT OF TECHNOLOGICAL CHANGE

ETM WILEY SERIES IN ENGINEERING & TECHNOLOGY MANAGEMENT

Series Editor: Dundar F. Kocaolu, Portland State University

PROJECT MANAGEMENT IN MANUFACTURING AND HIGH TECHNOLOGY OPERATIONS
Adedeji B. Badiru, University of Oklahoma

MANAGERIAL DECISIONS UNDER UNCERTAINTY: AN INTRODUCTION TO THE ANALYSIS OF DECISION MAKING
Bruce F. Baird, University of Utah

INTEGRATING INNOVATION AND TECHNOLOGY MANAGEMENT
Johnson A. Edosomwan, IBM Corporation

CASES IN ENGINEERING ECONOMY
Ted Eschenbach, University of Missouri-Rolla

ENGINEERING ECONOMY FOR ENGINEERING MANAGERS: WITH COMPUTER APPLICATIONS
Turan Gönen, California State University-Sacramento

MANAGEMENT OF RESEARCH AND DEVELOPMENT ORGANIZATIONS
Ravinder K. Jain, U.S. Army Corps of Engineers
Harry C. Triandis, University of Illinois

STATISTICAL QUALITY CONTROL FOR MANUFACTURING MANAGERS
William S. Messina, IBM Corporation

KNOWLEDGE BASED RISK MANAGEMENT IN ENGINEERING: A CASE STUDY IN HUMAN-COMPUTER COOPERATIVE SYSTEMS
Kiyoshi Niwa, Hitachi, Ltd.

MANAGING TECHNOLOGY IN THE DECENTRALIZED FIRM
Albert H. Rubenstein, Northwestern University

MANAGEMENT OF TECHNOLOGICAL CHANGE
Yassin Sankar, Dalhousie University

PROFESSIONAL LIABILITY OF ARCHITECTS AND ENGINEERS
Harrison Streeter, University of Illinois at Urbana-Champaign

MANAGEMENT OF TECHNOLOGICAL CHANGE

YASSIN SANKAR
School of Business Administration
Dalhousie University
Halifax, Nova Scotia
Canada

A WILEY-INTERSCIENCE PUBLICATION
John Wiley & Sons, Inc.
NEW YORK / CHICHESTER / BRISBANE / TORONTO / SINGAPORE

Copyright © 1991 by John Wiley & Sons, Inc.

All rights reserved. Published simultaneously in Canada.

Reproduction or translation of any part of this work beyond that permitted by Section 107 or 108 of the 1976 United States Copyright Act without the permission of the copyright owner is unlawful. Requests for permission or further information should be addressed to the Permissions Department, John Wiley & Sons, Inc.

Library of Congress Cataloging in Publication Data:
Sankar, Yassin.
    Management of technological change / by Yassin Sankar.
       p.   cm. -- (Wiley series in engineering and technology management)
      "A Wiley-Interscience publication."
      Bibliography: p.
    1. Technological innovations--Management. I. Title. II. Series.
HD45.S275 1989
658.5'14--dc20
    ISBN 0-471-63147-7   89-8946
                                                                   CIP

Printed in the United States

10 9 8 7 6 5 4 3 2 1

To Mother Teresa M.C.
Ahillya, Romila, Shivana
Sri Sathya Sai Baba

"... Our time is distinguished by wonderful achievements in the fields of scientific understanding and technological application of those insights. Who would not be cheered by this? But let us not forget that knowledge and skills alone cannot lead humanity to a happy and dignified life. Humanity has every reason to place the proclaimers of high moral standards and values above the discoverers of the objective truth. What humanity owes to personalities like Buddha, Moses, and Jesus ranks for me higher than all the achievements of the inquiring and constructive mind."

ALBERT EINSTEIN

All royalties from the sale of this book will be donated to the Missionaries of Charity, founded by Mother Teresa.

Excerpts taken from the following:

"The Human Side of Robotics: How Workers React to the Robot," by Linda Argote, Paul S. Goodman, and David Schkade, in *Sloan Management Review*, Spring 1983, pp. 31-41.

Philosophy and Environmental Crisis, edited by William T. Blackstone. Copyright 1974. The University of Georgia Press.

Organizations: A Guide to Problems and Practice by John Child. Copyright 1984 by John Child. Reprinted by permission of Harper and Row, Publishers, Inc.

Motivation Theories and Practice by Chung, 1977, pp. 42, 44-53, and 56. Reprinted by permission of Publishing Horizons, Inc.

People and Performance by Peter F. Drucker. Copyright 1952 by Peter F. Drucker. Reprinted by permission of Harper & Row, Publishers, Inc.

Harvard Business Review. An excerpt from *How to Counter Alienation in the Plant* by Peter F. Drucker, issue Nov./Dec. 1972. Copyright 1972 by the President and Fellows of Harvard College; all right reserved.

Harvard Business Review. An excerpt from *New Templates for Today's Organizations* by Peter F. Drucker, issue Jan./Feb. 1974 by the President and Fellows of Harvard College; all rights reserved.

Innovation and Entrepreneurship by Peter F. Drucker. Copyright 1985 by Peter F. Drucker. Reprinted by permission of Harper & Row, Publishers, Inc.

Self-transedence as a human phenomenon by V. Frankl. *Journal Humanistic Psychology, 1966* Fall, Copyright 1966 by the Association for Humanistic Psychology. Reprinted by permission of Sage Publications, Inc.

"Matrix Organization Designs," by Jay R. Galbraith. February 1971, pp. 29-40. Reprinted from *Business Horizons*, February 1971. Copyright 1971 by the Foundation for the School of Business at Indiana University. Used with permission.

Management-Oriented Management Information Systems, 2e, by Jerome Kanter.. Copyright 1977, pp. 275-279. Adapted by permission of Prentice Hall, Inc., 1734 Englewood Cliffs, New Jersey.

MIS Concepts and Design, 2e, by Robert G. Murdick/John C. Munson. Copyright 1986, pp. 343-344. Reprinted by permission of Prentice Hall, Inc., 1734 Englewood Cliffs, New Jersey.

"The Scarcity Society" by William Ophuls. Copyright 1974 by *Harper's Magazine.* All rights reserved. Reprinted from the April issue by special permission.

In Search of Excellence by Thomas J. Peters and Robert H. Waterman. Copyright 1982 by Thomas J. Peters and Robert H. Waterman. Reprinted by permission of Harper & Row, Publishers, Inc.

The Man in the Assembly Line by C. R. Walker and Robert H. Guest, (Cambridge, Mass.: Howard University, 1952). Cited in Szilagyi and Wallace, *Organizational Behavior and Performance,* 3/e, p. 128, 1983.

*"If only if weren't for the people,
the goddamned people," said Finnerty,
"always getting tangled up with the
machinery. If it weren't for them,
earth would be an engineer's paradise."
from* Player Piano *by*

KURT VONNEGUT, JR.

This book [The Revolution of Hope] is born out of the conviction that we are at the crossroads: one road leads to a completely mechanized society with man as a helpless cog in the machine ... the other to a renaissance of humanism and hope—to a society that puts technique in the service of man's well being.

ERIC FROMM

Science and technology have made astonishing progress, but humanity is going on the downward path.... But science alone is not enough. There must be "discrimination" for utilizing the discoveries of science for right purposes. Science without discrimination, human existence without discipline, friendship without gratitude, music without melody, a society without morality and justice, cannot be of benefit to the people.

SRI SATHYA SAI BABA

Management of Technological Change

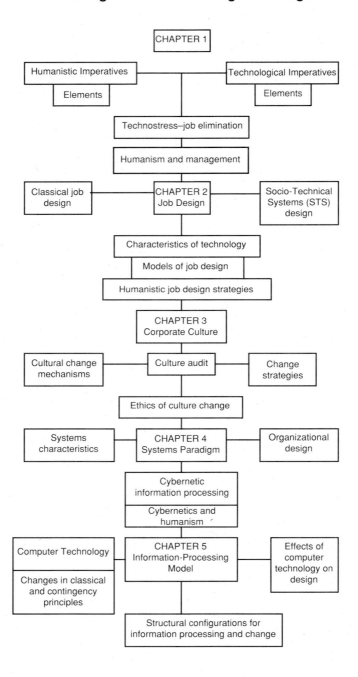

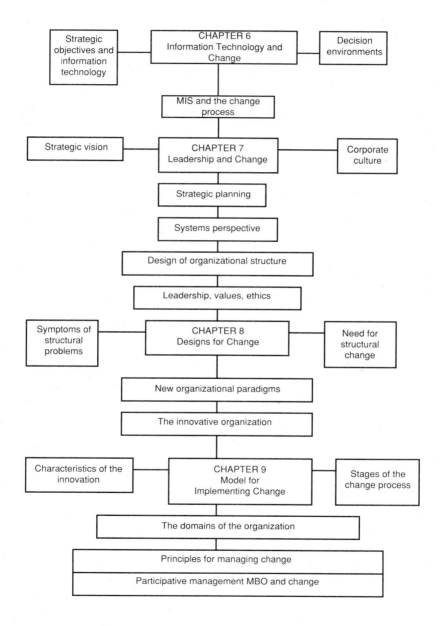

CONTENTS

Preface xvii

Acknowledgments xxv

1. Humanistic Imperatives and Technological Change 1

   1.1 Advance Organizer, 1
   1.2 Introduction, 2
   1.3 Technological Change, Work, and Progress: Peter Drucker, 3
   1.4 The Growing Phenomenon of Employee Alienation, 8
   1.5 Technological Change, Work, and Alienation: Karl Marx, 9
   1.6 Anatomy of Alienation, 13
   1.7 New Technologies: Lund and Hansen's Optimistic Perspective, 17
   1.8 Job Stress and Technology, 19
   1.9 Humanism and Technological Change, 22
   1.10 Industrial Humanism: A Contemporary Management Creed, 23
   1.11 Perspectives on Humanism, 24
   1.12 Humanism and Management, 33
   1.13 Conclusion, 40

2. Job Design and Technological Change   43

2.1 Advance Organizer, 43
2.2 The Technology of Job Design, 46
2.3 Some Positive Effects of Technology on Job Design, 49
2.4 Robotics: Some Effects of a New Technology, 50
2.5 Some Recommendations for Managers Introducing New Technologies: Sociotechnical Design, 52
2.6 Technology and Job Design, Slocum and Sims Model, 53
2.7 Job Design for New Technologies, 59
2.8 Job Design, Human Needs, and Technology: Humanistic Imperatives, 61
2.9 Two Models for Designing Jobs, Chung, 63
2.10 Some Job Design Action Steps for New Technologies, 69
2.11 A Scenario in Job Design, 70
2.12 Ethical Questions for Current Technologies, 72

3. Corporate Culture and Technological Change   75

3.1 Advance Organizer, 75
3.2 New Technologies: Some Perspectives, 76
3.3 Organizational Culture: Some Perspectives, 79
3.4 Types of Cultures: Three Profiles, 81
3.5 Strategies for the Management of Corporate Culture, 83
3.6 The Cultural Audit for New Technologies, 83
3.7 The Levels of Culture and Technological Change, 84
3.8 A Model of Organizational Culture and Change, 89
3.9 Changing Culture: Lorsch's Perspectives, 91
3.10 How to Assess Resistance to Culture Change, 93
3.11 Organizational Culture and Values, 94
3.12 Ethics of Cultural Change, 102
3.13 Merging Corporate Cultures, 103
3.14 The Ethics of Technological Change and Human Values: Mesthene, 105

3.15 The Technological Society: Its Value Orientations: Fromm (1968), 106
3.16 Industrial Engineering Ethics: A New Model, 107

4. A Systems Approach for Managing Technological Change   115

4.1 Advance Organizer, 115
4.2 Characteristics of the Sociotechnical Systems Theory, 118
4.3 The General Systems Concept, 120
4.4 The Organization as a System, 121
4.5 Characteristics of Open Organizational Systems, 121
4.6 The Manager and the Systems Approach, 125
4.7 A Systems Framework of the Organization, 126
4.8 Systems Differentiation, Change, and Innovation, 129
4.9 Negative Entropy and Change, 130
4.10 The Law of Requisite Variety and Change, 131
4.11 Progressive Segregation and Change, 132
4.12 Interdependence and Change, 133
4.13 Equilibrium and Change, 134
4.14 Cybernetics and Organizational Systems, 135
4.15 Communications, Change, and Innovation, 138
4.16 Cybernetic Principles for Management of Technology, 140
4.17 Implications for Management, 142
4.18 Leadership, Systems Thinking, and Change, 143
4.19 The Cybernetic Approach and Humanism, 145

5. An Information-Processing Model of Organizational Design for Technological Change   150

5.1 Advance Organizer, 150
5.2 Information Technology and Classical Design, 152
5.3 Information, Structure, and Change, 154
5.4 Power, Information Technology, and Change, 155

- 5.5 Some Design Principles, 156
- 5.6 A Change in Organizational Design, 158
- 5.7 The Impact of Computer Technology on Organizational Design, 161
- 5.8 The Computer-Based Information System and Organizational Design, 163
- 5.9 Information Technology, Centralization, and Control: Child's Perspectives, 164
- 5.10 The Information-Processing Model of Organizational Design, 166
- 5.11 An Information-Processing Model, 167
- 5.12 Changes in Classical and Contingency Design Principles, 169
- 5.13 Information Systems and Structure, 171
- 5.14 Information Pathologies and Organizational Design, 173
- 5.15 Information Imperatives in Organizational Design, 174
- 5.16 Design of Structural Configurations for Information Processing, 176
- 5.17 The Creative Individual in the New Corporate Design, 181

6. Information Technology and Change   187

- 6.1 Advance Organizer, 187
- 6.2 The Importance of Information Technology and Change, 188
- 6.3 Strategic Objectives and Information Technology: Child's Perspectives, 191
- 6.4 Control, Integration, and Design Changes, 193
- 6.5 Stability, Change, and Planning, 194
- 6.6 MIS and the Reduction of Uncertainty, 194
- 6.7 MIS, the Management Process, and Change, 195
- 6.8 Planning Change, 196
- 6.9 Environmental Scanning, 197
- 6.10 Information for Operations Control, 198
- 6.11 MIS, Feedback, and Change, 201

6.12 Decision Making, Change, and Information Technology, 202
6.13 MIS and the Management of Change, 206
6.14 The Bureaucratic Framework, MIS, and Change, 210
6.15 MIS, Integrated Data Bases and Change, 211
6.16 Information Technology and Organizational Change, 215
6.17 Some Factors Limiting MIS Applications in Change and Innovation, 218
6.18 Human Information Processing and Change, 221
6.19 Ecology, Technology and Environmental Monitoring, 226
6.20 Technology Assessment and the Environment, 230

7. Leadership, Strategic Planning, and Innovation 234

7.1 Advance Organizer, 234
7.2 Introduction, 235
7.3 Leadership and Innovation, 236
7.4 Functions of the Leader and Adaptive Coping, 238
7.5 Leadership, Strategic Planning, and Change, 239
7.6 Leadership and Strategic Decision Making, 239
7.7 Strategic Change and Innovation and Their Application at Northern Telecom, 242
7.8 Strategic Planning and Its Application at Northern Telecom, 244
7.9 Leadership and Organizational Culture: Schein's Perspective, 250
7.10 Leadership, Strategic Vision, and Change, 253
7.11 The Transformation Leader: Tichy's and Devanna's Perspectives, 255
7.12 Leadership, Structure, and Change, 260
7.13 Leadership, Humanistic Values, and Effectiveness, 266
7.14 Leadership, Values, and Ethics, 266
7.15 Values, Self-Esteem, and Leadership Excellence, 267
7.16 Values and Decision Making, 268

8. Designs for Change — 274

- 8.1 Advance Organizer, 274
- 8.2 The Need for a New Corporate Design, 275
- 8.3 Symptoms of Structural Problems in Conventional Organizational Design: Duncan's Perspective, 281
- 8.4 Reorganization: The Need for Structural Change: Child's Perspective, 283
- 8.5 A New Paradigm for Organizational Design: Trist's Design, 286
- 8.6 New Templates for Today's Organizations: Drucker's Views, 288
- 8.7 Some Attributes of the Innovative Organization: Peters and Waterman, 289
- 8.8 Organic-Mechanistic Systems and Innovation, 290
- 8.9 Temporary Problem-Solving Systems and Innovation: Bennis' Perspective, 291
- 8.10 The Innovative Organization: Thompson's Perspective, 292
- 8.11 The Adaptive Corporation: Toffler, 293
- 8.12 Collateral Organization and Innovation: Zand's Perspectives, 295
- 8.13 Matrix Organization: A Design for Change; Galbraith, 297
- 8.14 Adhocracy and Change: Mintzberg's Perspectives, 303
- 8.15 Organizational Design and Humanism: Argyris' Perspective, 305
- 8.16 Humanistic Organizational Design, 308

9. A Model for Implementing Changing — 312

- 9.1 Advance Organizer, 312
- 9.2 Introduction, 314
- 9.3 Conceptions of the Implementation Process, 315
- 9.4 A Model of the Implementation Process, 316

10. Summary and Conclusion   361

   10.1 Information Technologies: Dimensions of Technological Change, 361

   10.2 Concluding Comment: on the Individual in the Technological Society, 366

Index   371

PREFACE

The major focus of this text is on the dynamics of technological change. We live in a high-tech environment, in an era with complex changes in our values, basic assumptions, belief systems, expectations and needs. The impact of technological change on these elements is such that changes in our cognitive and ethical maps are being initiated invariably without any conscious, systematic, or integrated planning. Sometimes technological change impairs rather than serves the well-being of that entity for whom the change is designed, namely, the human being. Human social needs, values, the will to meaning and will to believe, the need for individuality and self-esteem, autonomy, self-actualization, and self-realization are all imperatives to be considered in the management of technological change. The potential benefits of technological change must not undermine these needs, values, and orientations. Our formula for managing change must incorporate these humanistic guidelines.

This text, while incorporating some conventional bits of information on the theme of technological change, also provides a different perspective on this subject. Technological change is generally managed within complex bureaucratic technological structures with efficiency as the primary goal. We recognize this dimension as essential as well, but do not believe it to be the dominant norm of a social system which subordinates all other human values, such as creative, experiential, attitudinal, aesthetic, ethical, and spiritual values—values that anchor one to the realm of human experience and to one's "will" to meaning.

Technology produce changes within the organization that must be considered for effective implementation of innovations. The management of change therefore emerges as the strategic function of the contemporary manager. Managerial excellence is associated with the use of humanistic imperatives in managing the process of change. Some of the changes produced by technology will impact job design, work flow, job stress, the elements of corporate culture, the organizational system, the information technology of the organization, the leadership style and strategic planning premises, the organizational design, and the value systems of managers and the organization.

Managers, to manage change effectively, must use humanistic strategies for designing jobs for new technologies, reorienting the corporate culture for change through culture audit, using the system's paradigm for managing systemic change, developing management information systems for managing the change process, designing the organization as an information-processing model for the imformation age, developing a conceptual model for implementing change, experimenting with a variety of organizational designs for change, and conducting an ethical audit for managing technological change. If a manager as a change agent attempts to blend technological and humanistic imperatives into a coherent strategy and framework for managing change, then industrial humanism will be the potent managerial creed for the new information society.

This book is written primarily for use as a framework for analyzing the dynamics of technological change by four main types of change analysts: the policy analysts who are regularly involved in the implementation of change programs, or the design of corporate strategy as an instrument of change; the manager who needs to identify the factors that facilitate and inhibit change in his/her management of the change process; the researcher who is searching for the foundations for a theory of change; the student of technological change, and organizational architects who design organizations for change. The student of change will thus be able to develop the conceptual and diagnostic skills as well as the systemic perspective for the initiation, facilitation, and management of change and innovation. The text addresses those who are interested not only in an intellectual appreciation of the complex architecture of change dynamics, but also those who are engaged in applied change or in the management of the change.

Some Major Themes in the Management of Technological Change

Technological change is a complex, dynamic, and creative process. The effective management of technological change is contingent on many conditions that we will briefly review.

The blending of technological and humanistic essentials will reduce the incidence of technostress and job alienation produced by technology. The elements of humanism, such as identity and integrity, compassion and empathy, individuality and potentiality, self-actualization, freedom, and autonomy, character and values, the will to meaning, and so on, must be incorporated into management for humanizing technology. Industrial humanism is the contemporary management creed for managing technological change (Chapter 1).

This management creed dictates that job design organized around classical industrial engineering methods must give way to a sociotechnical systems (STS) design that places the individual at the center of the design framework. The characteristics of technology must be tempered by a humanistic job design model. Job enlargement and enrichment strategies that emphasize such job dimensions as task variety, ability utilization, feedback, autonomy, goal setting, group management, man-paced control, etc., can enhance productivity and job satisfaction if extended to the management of new technologies (Chapter 2).

A corporate culture that has a high culture/value adjustment must be designed through a culture audit to promote the use of these job design strategies and to facilitate the adoption and diffusion of technical innovations. The strategic management of corporate culture in the management of new technologies must be a priority for the leader. The mechanisms for changing the elements of culture, such as values, beliefs, and basic assumptions, must be guided by the ethics of culture change (Chapter 3).

Managing the organization as a system and designing it as a system mean that it will be receptive to technological change. A systemic paradigm of the innovation process means that managers cannot view technical innovations as a simplistic linear process. Rather, the cybernetic open systems process should be reflected in the company's technological strategy (Chapter 4).

The design of the organization with the conventional rationale based on power and authority is no longer tenable. The information domain must be the design parameter for the organization of the information age. The information pathologies of the classical design and the information imperatives of the innovative organization call for a different organizational model. An information-processing model of organizational design with a variety of structural configurations to cope with technological change is needed (Chapter 5).

This new design based on the systemic paradigm, a dynamic organizational culture, new strategies for job design, and industrial humanism must be augmented by an information technology that manages the critical stages of the change process. The management information system based on integrated data-base systems, distributed data bases, and strategic information systems must conform to the structure of the organization, the variety of decision environments within the organization, and the corporate philosophy of the firm. Human information processing, with its dimensions of intuition, creative visualization, lateral thinking, and perception, is still the center of the decision-making process (Chapter 6).

The leader puts it all together with a strategic vision, a systemic perspective, and the design of organizational structure aligned with corporate strategy. Corporate strategy becomes the instrument of change, and innovation and corporate culture are managed strategically by the leader to facilitate the effective management of technological change (Chapter 7).

The designs for change must focus on the structural problems of the contemporary organization. The need for structural change is essential. The bureaucratic machine of the industrial era must be replaced by the adaptive corporation for the information age. New templates for today's organization make a variety of designs mandatory. The innovative organization, the ambidextrous organization, the matrix design, and adhocracy are some of the designs for technological change (Chapter 8).

A model for implementing such designs, change, and innovation is necessary to guide the process. The characteristics of the innovation, the nature and character of the domains of the organization (the behavioral, structural, management systems, technical, and process domains), and the strategies for changing the organization all indicate that implementation is a complex contingent process. The principles for managing change must be based on all of these variables in the implementation stage of the change process. Implementation dynamics point to the use of participative management, particularly management by objectives, as a vehicle for introducing technological change (Chapter 9).

The managers of technological change must attempt to blend the human and technical imperatives in an ethical culture. The ethical audit must monitor the changes in human needs and human values that technology affects. The question of identity, integrity, the dynamics of caring, and the will to meaning must be part of this ethical assessment. The role of technology in creating changes in creative, experiential, and attitudinal values must also be built into the moral audit. The meaning orientation provided by religion must be enhanced by technology, not subverted. By creating new opportunities, options, and choices, technology injects value premises that must be monitored into the decision-making process. In addition, the consequences of technology pose ethical dilemmas for corporate decision makers. These consequences raise in acute form all the traditional questions of beauty and ugliness, ends and means, good and evil, right and wrong. This makes technology a branch of moral philosophy. The urgency of these issues is reflected in the current attitude that a clean, safe environment is an inalienable human right for which technological change must be held accountable. Never in human history has such impetus for the resolution of these ethical issues been so compelling.

The management of technological change must enhance human dignity, identity, self-actualization, compassion, individuality, potentiality, freedom, autonomy, and the reverence for life. With this focus, a beautiful scenario emerges for humankind that reverses the questions posed by T. S. Eliot on our predicament:

"Where is the life we have lost in living
Where is the wisdom we have lost in knowledge
Where is the knowledge we have lost in information"
  May the Force be our partner in this joint venture.

                                        YASSIN SANKAR

Dalhousie University
Halifax,
Nova Scotia, Canada

ACKNOWLEDGMENTS

This work has been in the making for many years and is the product of many influences. One of the most influential is Professor John Walton of Johns Hopkins University, who was critical of the conceptual poverty in the field of organizational change and inspired me to devote time, effort, and resources to this field. There are also many researchers, excerpts of whose work have been cited verbatim in a number of sections in this book because of their style, comprehensiveness, and power of thinking. My sincere appreciation to Drucker, Marx, Walton, Maslow, Frankl, Fromm, Hackman and Oldham, Chung, Slocum and Simms, Bass, Schein, Sathe, Robey, Argyris, Galbraith, Knight, Mintzberg, Zand, Pastin, Schon, Scott, Mitchell and Birnbaum, Lund and Hansen, Duncan, Rankin V.P., Northern Telecom, Harrison, Kanter, Cadwallader, Child, Toffler, Elias and Merriam, and Davis and Associates.

There are several additional people I wish to thank for their help on this book: Peter Ng Yuen, a friend of two decades, who has consistently encouraged me in my academic efforts and has debated from a more pragmatic perspective; Dr. C. Ramcharran, whose advice was to focus on reality-based problems in the pursuit of excellence; Dr. C. David, who prescribed the infrastructure with which to view these reality-based problems; Nazir, Danny, Hannah, Gordon Karanja, members of the Caribbean Information Group, who provided a supportive atmosphere in which to engage in debate and acts of solidarity; Amrall Hasanally, Teddy Fakira, and Nazir Khan, for rap sessions after moments of academic saturation; the Gulati family, Kamal and Rajpaul, for their continuing warmth and hospitality, Ramesh and Savitri Panday who were available for consultation at a critical time; my sisters,

Shirine and Farida, and my brother, Farouk for always identifying with my career goals; Tally, Farah, Natasha, Royston, André, Muntaz Khan, Joseph Dzivence of Ghana, the McDooms, Aunt Twinnie, Uncle Bram, Fazil, Zorida, Raymond Kudrath, Bobby Khan, Barry, Dolly and Samantha, Drs. R. and R. Gupta, S. P. Gurdin and Florence, the Rawana dynasty, my brother Farouk, Isaac Saney for proofreading and comments, Radha, who consistently urged me toward excellence, humanism, and human dignity; and Sri Sathya Sai Baba, for his Divine message, inspiration, committment to excellence in human values and service to humanity with Love (Prema), Truth (Sathya), Nonviolence (Ahimsa), and Duty (Dharma).

Thanks also to my colleagues at Dalhousie and Saint Mary's Universities—Sandhu, Duffy, Patton, Laird, Schwind, Fitzgerald, Hari Das, Shripad Pendse and Y. Shafai who saw the potential contributions of this work, and to Dr. R. Storey, Director of the School of Business, who made it possible for me to meet the deadline by providing resources and facilities. The typing was done in an exceptional manner by Leslie Stockhausen, who also edited the text and made invaluable suggestions on style and format. Others included Teresa, who typed most of the manuscript, and Judy, Helen, Olga who assisted with segments. A special note of gratitude to Sumintra Naraine for proofreading the entire manuscript, to Frank Cerra, who was always available for consultation and progress reviews, and to Maggie Kennedy and the production staff for monitoring the progress of the final draft and Shelly Boutlier for assisting in segments on communications. Finally, I wish to express my appreciation to my wife, Ahillya, and my children, Romila and Shivana, for their patience and understanding as the deadline began to emerge, and to my mother, Massuman, who provided the foundation for all my academic efforts.

1
HUMANISTIC IMPERATIVES AND TECHNOLOGICAL CHANGE

1.1 ADVANCE ORGANIZER

The dimensions of work indicate that a humanistic perspective in managing technological change is crucial. The psychological, economic, social, and power dimensions noted by Peter Drucker are used to stress this theme of humanism. Another perspective is provided by Karl Marx; here, the theme is of alienation and dehumanization of work through technology and division of labor. An anatomy of alienation is developed from three major contemporary researchers, Walton, Blauner, and Argyris, which defines five major modes of alienation. Technostress and job stress from the viewpoint of Brod and Kets de Vries further reinforce this theme of the dysfunctional aspect of technology and change.

The positive features of technological change noted by Drucker are reinforced by current research by Lund and Hansen (1986) on the implementation of new technologies.

A humanistic perspective on the management of technological change is given major focus here. The theme of industrial humanism as a contemporary management creed is noted. The elements of humanism, such as compassion and empathy, identity and integrity, survival and transsurvival needs, the will to meaning, self-actualization, character values, individuality and potentiality, freedom and responsibility, and so on, are identified from major works in the field of humanistic psychology.

The imperatives of humanism conflict with those of technology and have the potential to produce stress. The humanistic perspective argues for actualization of the self and development of value potentialities of the individual. The development is facilitated by freedom and autonomy in the particular work setting. The worker brings to his (meaning his or her throughout this book) job situations certain hopes, expectations, and values. Emotions and perception are important components of the individual complex. The "inner world" of the worker is considered as important as external reality in determining productivity. The worker obtains his sense of identity through interpersonal relationships, and the better they are, the more responsive to the norms of the work group he will be. The employee is responsive to management to the extent that the supervisor can meet his needs for esteem and self-actualization, his will to meaning, and his character development for ethical excellence.

The technological imperatives conflict with the humanistic imperatives because technology requires specialization for efficiency—the fragmenting of jobs into tasks that utilize limited aptitudes, skills, abilities, and talents. Technical rationality must be observed in programming means and ends of the work flow for optimal performance. Planning to reduce uncertainty is critical in this programming of the work flow. Also critical to this process are (1) the predictability of decision outcomes and performance expectations, (2) control of the work flow through operations manual and programs, (3) stability of management systems to cope with complexity, and (4) order and conformity as major organizational values. Technology leads to the standardization of the work process through uniform application of job design principles. Next, formalization of the work process through rules, regulations, programs, and procedural specifications further increases standardization, which limits individual autonomy, discretion, and potential. The dysfunctional elements of these imperatives are noted by Scott et al. (1981), who contend that from the human standpoint, the individual characteristics of the worker must be separated as much as possible from the application of labor and intelligence to tasks.

The relationship between humanism and management is noted and strategies for managing technological change are linked with elements of humanism. The elements of humanism are extended into job design, organizational cultures, leadership, information technology, etc.

The framework in Figure 1.1 depicts the major themes for this chapter.

1.2 INTRODUCTION

This chapter provides the building blocks for our focus on the management of technological change. The need for more humanistic imperatives in the management of technological change must be our mandate for corporate and human excellence. Industrial humanism is our new management creed based on humanistic psychology, sociology, philosophy, and management. The elements of humanism are identified from a variety of perspectives to provide the foundation for humanizing technology. The managerial strategies that

1.3 TECHNOLOGICAL CHANGE, WORK, AND PROGRESS: PETER DRUCKER

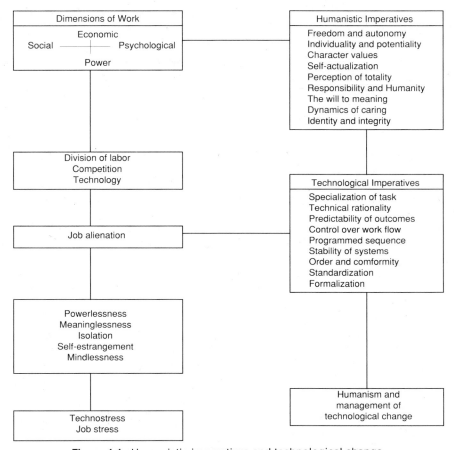

Figure 1.1. Humanistic imperatives and technological change.

incorporate these elements for managing change are outlined in this chapter and further expanded and developed in subsequent chapters.

We provide excerpts of the works of many researchers and scholars verbatim to illustrate the power of their thinking; these excerpts are punctuated by relevant commentaries. This provides a comprehensive focus on the theme of humanizing technology. The more pragmatic themes for the effective management of technological change are elaborated in subsequent chapters.

The first part of this chapter is organized around the works of Peter Drucker on the dimensions of works and the potential of technological change; the works of Karl Marx on the problem of alienation, work, and technology; the research by Walton on how to counter alienation in the plant; the modes of alienation as presented by Blauner; the positive effects of technological change as seen by Lund and Hansen; and technostress as discussed by Brod, Benson, and Ket de Vries.

The second part of this chapter stresses the theme of humanism. Major elements of humanism from the works of Fromm, Frankl, Herzberg, Maslow, Elias, and Merriam are outlined, with commentaries regarding their relevance for technological change. The cluster of elements include such powerful human and managerial attributes as compassion and empathy, identity and integrity, survival and transsurvival needs, the will to meaning, character values, self-actualization, freedom and autonomy, individuality and potentiality, responsibility and humanity, and the dynamics of caring and ethics. The imperatives of technology and those of humanism are summarized in Figure 1.5 and managerial strategies for change are outlined briefly, to be developed in detail in subsequent chapters.

1.3 TECHNOLOGICAL CHANGE, WORK, AND PROGRESS: PETER DRUCKER

1.3.1 The Nature of Work

Work has always been central to human consciousness. Man is not truly defined as the toolmaker, but by making tools; the systematic, purposeful and organized approach to work is specific and unique in human activity. What is needed to make work productive is quite different from what is needed to make the worker achieve. The worker must therefore be managed according to both the logic of the work and the dynamics of working. Personal satisfaction of the worker without productive work is failure, but so is productive work that destroys the worker's achievement. Neither is, in effect, tenable for very long.

Work is impersonal and objective. Work is a task. It is "something." To work, therefore, applies the rule that applies to objects. Work has a logic. It requires analysis, synthesis, and control.

Working is the activity of the worker; it is a human being's activity and an essential part of humanity. It does not have a logic. It has dynamics and dimensions (Drucker, 1977).

Working, according to Drucker, has at least five dimensions. In all of them the worker has to be achieving in order to be productive.

1.3.2 Machine Design and Human Design

The human being is not a machine and does not work like a machine. Machines work best if they do only one task, if they do it repetitively, and if they do the simplest possible task. Complex tasks are done best as a step-by-step series of simple tasks in which the work shifts from machine to machine, either by moving the work itself physically, as on the assembly line, or, as in modern computer-controlled machine tools, by bringing machines and tools in prearranged sequence to work, with the tool changing with each step of the process. Machines work best if run at the same speed, the same rhythm, and with a minimum of moving parts.

The human being is engineered quite differently. Altogether the human being is a very poorly designed machine tool. The human being excels, however, in coordination. A human excels in relating perception to action and works best if the entire person, muscles, senses, and mind is engaged by the work. Each individual has his or her own pattern of rhythms. Nothing, we now know, creates as much fatigue, as much resistance, as much anger, and as much resentment as the imposition of an alien speed, an alien rhythm, and an alien attention span, and above all, the imposition of one unvarying and uniform pattern of speed rhythm and attention span.

While work is, therefore, best laid out as uniform, working is best organized with a considerable degree of diversity. Working requires latitude to change speed, rhythm, and attention span fairly often. It requires fairly frequent changes in operating routines as well. What is good industrial engineering for work is exceedingly poor human engineering for the worker (Drucker, 1977).

1.3.3 Other Dimensions of Work

Work is an extension of personality. It is achievement. It is one of the ways in which a person defines himself or herself, measures his or her worth and humanity. So far, the task is still to make work serve the psychological need of humanity. Work is a social bond and a community bond. It largely determines status. It is also a means to satisfy our need for belonging to a group and for meaningful relationship to others.

Work is also a living. It emerges from the division of labour. The moment people cease to be self-sufficient and begin to exchange the fruits of their labour, work creates an economic bond that connects them. Work is living for workers. It is the foundation of their economic existence. But work also produces the capital for the economy. It produces the means by which an economy perpetuates itself, provides for the risks of economic activity and the resources of tomorrow, especially the resources needed to create tomorrow's jobs and with them the livelihood for tomorrow's workers.

In addition to the economic and the social dimension of work, there is also a power dimension. There is always a power relationship implicit in working within a group and especially in working within an organization. In any organization, no matter how small, there has to be a personal authority. The organization member's will is subordinated to an alien will. In an organization, jobs have to be designed, structured, and assigned. Work has to be done on schedule and in a prearranged sequence. In short, authority has to be exercised by someone (Drucker, 1977).

Anarchists, according to Drucker, are right in their assertion that "organization is alienation."

The dimensions of working, the physiological, the psychological, the social, the economic, and the power dimension, are separate. But they always exist together in the worker's relationship to work and job, fellow workers, and management. They have to be managed together. The basic fallacy of our traditional approaches has been to proclaim one of these dimensions to be *the* dimension.

Drucker (1977) states that Marx—and most other economists—saw that if the economic relationships could be changed, there would be no more alienation.

Marxism becomes bankrupt when it becomes apparent that the "expropriation" of the "exploiters" did not fundamentally change the workers' situation and their alienation because it did not change in any way any of the other dimensions.

Comment

Technological change affects all the dimensions of work identified by Drucker (Figure 1.2). At the physiological level, Drucker himself noted that nothing creates more anger and resistance than the imposition of an alien speed, rhythm, and attention span on the worker. Where technology fragments work, the total person, "his muscles, senses, mind" (Drucker, 1977), are not completely engaged by work, and alienation emerges. At the psychological level, work is the means by which a person defines himself or measures his worth and humanity. If a worker is merely an appendage to a machine, the sense of worth, humanity, and identity is compromised. The development of one's potential and actualization of one's creative values can be effectively curtailed by technological change if undue emphasis is placed on routinizing the elements of the task for predictability and control of work outcomes. With reference to the social dimension, technological change can also disrupt social networks and meaningful interpersonal relationships as well as impair the sense of identity that emerges from belonging to a group or work team. The economic dimension is still vital and may be augmented through technological change. "Work is living for workers," but the indivi-

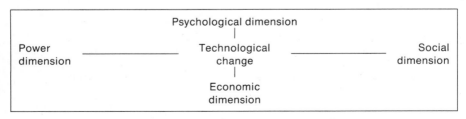

Figure 1.2. Technological change and the dimensions of work.

dual worker may perceive himself as a commodity, as an object where his labor power competes with machine power. As an object, he becomes alienated from his essence. In terms of the power dimension, authority is not based on the unique talents, capabilities, and faculties of the worker, namely, his expertise power, but on the technological imperatives of the machine. Technological change creates its own division of labor and its own degree of specialization that is requisite for optimal functioning of the machine, not the individual worker. The power base may shift to the technocrats, the systems designers, and the computer and software designers.

1.3.4 Technological Change

We are indeed in the early stages of a major technological transformation, one that is far more sweeping than the most ecstatic of the "futurologists" yet realize, greater even than Megatrends or Future Shock. For three centuries, advances in technology meant—as it does in mechanical processes—more speed, higher temperatures, higher pressures. Since the end of World War II, however, the model of technology has become the biological process: the events inside an organism. And in an organism, the processes are not organized around energy in the physicist's meaning of the term. They are organized around information (Drucker, 1986).

There is no doubt that high tech, whether in the form of computers or telecommunications, robots on the factory floor or office automation, biogenetics or bioengineering, is of immeasurable qualitative importance. High tech creates the vision for entrepreneurship and innovation. High tech is what logicians used to call the *actio cognoscenti,* the reason why we perceive and understand a phenomenon rather than the explanation of its emergence and the cause of its existence (Drucker, 1986).

1.3.5 Management as a Technology and an Innovation

As a "useful knowledge" technique, management is the same age as the other major areas of knowledge that underlie today's high-tech industries, whether electronics, solid-state physics, genetics, or immunology. Management is a vehicle of profound change in attitudes, values, and behavior. It is a technology: the application of knowledge to human work. What has made possible the emergence of the entrepreneurial economy, the new technology, are new applications of management, according to Drucker (1986),

Management is the new technology (rather than any specific new science or invention) that is crucial for the entrepreneurial economy. Innovation is the specific tool of entrepreneurs, the means by which they exploit change as an opportunity. It is capable of being presented as a discipline, capable of being learned, capable of being practiced. Management is the vehicle for articulating the knowledge and principles of successful innovation (Drucker, 1986).

1.4 THE GROWING PHENOMENON OF EMPLOYEE ALIENATION

Even the lyrics of the modern songs exemplify the alienation of modern society, as illustrated by Rick Springfield's song *Human Touch*. Growing numbers of workers across the country share singer his lament about alienation in the evolving high-tech work place.

They are being pulled into a world where the judgmental aspects of their jobs are being fragmented, put into a computer program, and sent directly to a machine," said Sidney Fine, an industrial psychologist in Washington, D.C.: "The result," he said, "is feeling of alienation and lack of responsibility. By alienation, I mean that workers don't feel significantly connected with the work they do," he explained. "They are working with machines and are separated from what is being produced. The machinist is working with numerically controlled computerized equipment that operates the machine that does the work."

Two prominent researchers on work redesign, Hackman and Oldham (1980), have observed that in the last several decades, organizations have increased the role of technology and automation in attaining organizational objectives, and have dramatically expanded the number of jobs that are specialized, simplified, standardized, and routinized. Moreover, organizations themselves have become larger and more bureaucratic in how they function.

The authors have reviewed much research on work and concluded that we may now have arrived at a state of organizational malaise. The way most organizations function is in severe conflict with the talents and aspirations of most of the people who work for them. Such conflict manifests itself in increased alienation from work and in decreased organizational effectiveness. Ways of structuring jobs and managing organizations must change (Hackman and Oldham, 1980).

Another researcher in the field of motivation and job design, Chung (1977), also makes reference to this phenomenon. Several recent studies on American workers and their work ethic generally agree that an increasing number of blue-collar as well as white-collar workers express unhappiness about their jobs. Contemporary workers, he observes, expect their jobs to be interesting and challenging, and desire more participation in shaping their work environment. There are, however, workers, both blue-collar and white-collar, who are bored and alienated from their jobs and search for every opportunity to decrease their work commitment. The problem, though, is that the new expectations and demands of workers, based on a humanistic and holistic view of the work world, are at odds with the work conditions and managerial practices of many organizations.

1.5 TECHNOLOGICAL CHANGE, WORK, AND ALIENATION: KARL MARX

1.5.1 The Role of Work

To understand the problem of alienation, it is necessary to understand the role which Marx attributes to labor. In Marx's opinion, labor constitutes man's most important activity—in fact he calls it "life activity." Through work, man creates his world, and as a consequence, he creates himself. A basic idea in Marx's concept of labor is that man "objectifies" himself, which means that through creative activity, man, by using his capacities in working on raw materials, transforms them into objects. Accordingly, these objects reflect his abilities. By means of his own work, man also experiences himself as an active, conscious being, as an active subject as opposed to a passive object. Since the objects of his work reflect his own "nature," he can evaluate himself through his activity (Israel, 1971). But this self-evaluation, may also becomes an object for himself, for his own perception. There is reciprocal interaction between the acting subject and the self-evaluating object.

Labor is the way in which man realizes his own nature, his *Gattaingswesen,* that is, the essence of the human species. Through work, man creates a world of objects. He molds and transforms nature and produces objects for his own needs and satisfaction. But the worlds of objects also includes social institutions through which the process of production is regulated and controlled. Technological change is a powerful object created by man for regulation and control of work. Marx notes the relationship between work and the essence of man. It is just in his work upon the objective world that man really proves himself as a species-being. This production is his active species-life. By means of it, nature appears as his work and his reality. The function of labor is, therefore, the objectification of man's species-life, for he no longer reproduces himself merely intellectually, as in his consciousness, but actively and in a real sense, and he sees his own reflection in a world he has constructed (Marx, 1975).

Man produces his own life. Through creative work, man achieves self-realization, that is, he realizes the potentialities of the species and at the same time gives expression to his basic social nature.

1.5.2 Human Nature and Alienation

Marx's theory of alienated labor presupposes a certain theory of human nature and a certain ideal of work, this ideal being part of the theory of human nature.

The first element in Marx's view of man is the notion that man is a natural being. This means that man is an active and acting being. As a natural being,

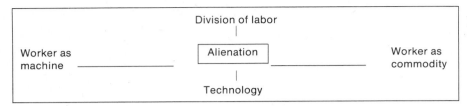

Figure 1.3. The elements of alienation.

and as a living natural being, he is, on the one hand, endowed with natural powers and faculties, which exist in him as tendencies and abilities, as drives. On the other hand, as a natural embodied sentient, objective being, he is a suffering, conditioned, and limited being (Marx, 1975).

The second element in Marx's view about human nature concerns man's basically social character. Man cannot exist without social relations. Therefore, society is a consequence of man's basic social character. The third element in Marx's view of man is human consciousness. Marx is here speaking of the role of labor in the context of material production as a means for self-creation of man, namely, the role of praxis. Finally, there is a fourth element in Marx's thinking about man's nature. Though man is a unique individual—and it is just this particularity which makes him an individual, a really individual communal being—he is equal to the totality, the ideal totality. He exists in reality as the representation and the real mind of social existence and as the sum of human manifestations of life. Total man is man using all his senses, his perception, thinking, and motivation in order to "appropriate the world" and to make it his own—a chance for the individual to realize all his capacities and talents. Such an ideal man is a consciously producing individual who rules the productive forces (technology and other artifacts) instead of being led by them. If the process of production is organized so that it allows man to realize himself, then external nature can be humanized, that is, by systematically transferring it into the sphere of human need satisfaction, which in turn facilitates human self-realization. Marx uses the term "humanism" to describe the self-realization of man through his productive activity. Marx maintains that the only way to create values is by human work. Conditions worthy for human nature are those conditions whereby man plans production rationally and keeps social and economic processes under his conscious control (Israel, 1971).

Comment

The critical question for us here is whether technological change impacts on these elements of human nature and if it enhances or subverts these elements. Technology can be designed to use a person's faculties, abilities, and tendencies and can also fragment and limit their utilization. Technology impairs the social character of man by organizing the work flow around the machine and not around factors that promote group cohesiveness or teamwork. In addition, if man is perceived as an appendage to the machine, this will detract from his consciousness or his essence. Finally, in terms of the totality of man, using all his senses, perception, thinking, and motivation to "appropriate the world," the individual is using a limited range of these attributes not necessarily because of technology but because of the division of labor, which is the genesis of the fragmentation of work. Technology may extend the division of labor and further contribute to alienation. But division of labor is not only a technical necessity for efficiency but also an organizational imperative for control, coordination, planning, and effectiveness. This is true not only of capitalist production but also socialist as well. If the genesis of alienation is the division of labor and its attendant specialization and fragmentation of tasks and roles, then alienation can also be expected in socialist production systems.

1.5.3 Alienation, Work, and Technology

For Marx, the concept of alienation is based on the distinction between existence and essence, and on the fact that man's existence is alienated from his essence; that is, in reality, he is not what he potentially is, or, to put it differently, that he is not what he ought to be and that he ought to be that which he could be.

For Marx, the process of alienation is expressed in work and in the division of labor. Work is for him the active relatedness of man to nature, the creation of a new world, including the creation of man himself. But as the division of labor develops, labor loses its character of being an expression of man's powers; labor and products assume an existence separate from man, his will, and his planning. The object produced by labor, its product, now stands opposed to it as an alien being, as a power independent of the producer. The production of labor is labor which has been embodied in an object and turned into a physical thing; this product is an "objectification of labour" (Fromm, 1975).

Marx stresses two points: (1) in the process of work and especially of work under the conditions of capitalism, man is estranged from his own creative powers; and (2) the objects of his own work become alien beings, and eventually rule over him and become powers independent of the producer. The laborer exists for the process of production, and not the process of production for the laborer. Marx's criticism of capitalist society is directed not at its method of distribution of income but its mode of production, its destruction of individuality, and its enslavement of man—not by the capitalist, but *of* worker *and* capitalist—by things and circumstances of his own making.

With reference to technology, Marx contends that owing to the extensive use of machinery as aids in the division of labor, work has lost all individual character and consequently all charm for the worker. He becomes an appendage of the machine, and it is only the most simple, monotonous, and easily acquired skills that are required of him. Machinery is adapted to the weakness of the human being in order to turn the weak human being into a machine (Fromm, 1975).

This theme if further developed in *Das Kapital* (1867) by Marx (1977).

> Within the capitalist system, all methods for raising the social productiveness of labor are brought about at the cost of the individual laborer; all means for the development of production transform themselves into means of domination over, and exploitation of, the producers; they mutilate the laborer into a fragment of a man, degrade him to the level of an appendage of a machine, destroy every remnant of charm of his work and turn it into a hated tool; they estrange from him the intellectual potentialities of the labor process in the same proportion as science is incorporated in it as an independent power
> (Marx, 1977).

The impact of technology on work is primarily reflected in its process of fragmenting work and its potential to produce monotony. According to Marx, work is creative (1) if man makes his life activity itself an object of his will and consciousness; (2) if man through work can express his capabilities in a comprehensive way; (3) if through this work he can express his social nature; and (4) if work is not simply a means for maintaining man's subsistence, that is, if it is not purely instrumental.

Marx's view is that the social combination of labor processes appears as an organized suppression of the worker's individual vitality, freedom, and autonomy. The machine does not free the worker from the work but rather deprives the work itself of all content. Another impact of technological change is that routinization of work leads to a partially developed individual who is merely the bearer of one specialized social function. This simplification of work causes endless drudgery. Work must give expression to the totally developed individual with a variety of modes of activity and social functions.

1.5.4 Technological Change

Modern industry never views or treats the existing form of a production process as the definitive one. Its technical basis is therefore revolutionary, whereas all earlier modes or production were essentially conservative. By means of machinery, chemical processes, and other methods, it is continually transforming not only the technical basis of production but also the functions of the worker and the social combinations of the labor process. At the same time, it thereby also revolutionizes the division of labor within society, and incessantly throws masses of capital and of workers from one branch of production to another. Thus, large-scale industry, by its very nature, necessitates variation of labor, fluidity of functions, and mobility of the worker in all directions (Marx, 1977).

1.6 ANATOMY OF ALIENATION

Walton (1972), a major researcher on job design, observes that alienation is expressed by passive withdrawal, tardiness, absenteeism and turnover, and inattention on the job. In other cases, it is expressed by active attacks, pilferage, sabotage, deliberate waste, assaults, bomb threats, and other disruptions of work routines. These trends all have been mentioned in the media, but one expression of alienation has been underreported: pilferage and violence against property and persons.

These acts of sabotage and other forms of protest are overt manifestations of a conflict between changing employee attitudes and organizational inertia, according to Walton. The expectations of workers conflict with the demands, conditions, and rewards of the organizations in at least six important ways (Walton, 1972):

1. Employees want challenge and personal growth, but work tends to be simplifed and specialties tend to be used repeatedly in work assignments. This pattern exploits the narrow skills of a worker, while limiting his opportunities to broaden or develop.
2. Employees want to be included in patterns of mutual influence; they want egalitarian treatment. But organizations are characterized by tall hierarchies, status differentials, and chains of command.
3. Employee commitment to an organization is increasingly influenced by the intrinsic interest of the work itself, the human dignity afforded by management, and the social responsibility reflected in the organization's products. Yet organization practices still emphasize material rewards and employment security and neglect other employee concerns.
4. What employees want from careers, they are apt to want right now. But when organizations design job hierarchies and career paths, they continue to assume that today's workers are as willing to postpone gratifications as were yesterday's workers.
5. Employees want more attention to the emotional aspects of organization life, such as individual self-esteem, openness between people, and expressions of warmth. Yet organizations emphasize rationality and seldom legitimize the emotional part of the organizational experience.
6. Employees are becoming less driven by competitive urges, less likely to identify competition as the "American way." Nevertheless, managers continue to plan career patterns, organize work, and design reward systems as if employees valued competition as highly as they used to.

The conflicts between the workers' expectations and the conditions of the traditional organizational forms produce an alienation syndrome. The bureaucratic organizational culture, with its emphasis on conformity, status differentials, tall hierarchies and chains of command, old-fashioned job hierarchies and career paths, and an inventory of rules, regulations, and procedures, helps produce alienation.

1.6.1 Modes of Alienation and Technology

One can evaluate the effects of technological change by focusing on the following modes of alienation: meaninglessness, powerlessness, isolation, and self-estrangement. If technological change increases any of these modes, there is a potential for alienation to occur. We must recognize at the beginning that alienation may be moderated by workers' growth needs for control, initiative, and meaning in work as well as by employees' educational levels. Furthermore, the modes of alienation vary with the type of technology adopted by the organization. We will base these modes of alienation on Blauner's (1964) classification.

1. *Powerlessness*. The split in man's existence and consciousness into subject and object underlies the idea of powerlessness. A person is powerless when he is an object controlled and manipulated by other persons or by an impersonal system (such as technology) and when he cannot assert himself as a subject to change or modify this domination. The nonalienated pole of the powerlessness dimension is the state of freedom and control.

2. *Meaninglessness*. Alienation reflects a split between the part and the whole. A person experiences alienation of this type when his individual acts seem to have no relation to a broader life program. Meaninglessness also occurs when the individual roles are not seen as fitting into the total system of goals of the organization but have become severed from any organic connection of the whole. The nonalienated state is understanding of a life plan or of an organization's total functioning and activity which is purposeful rather than meaningless.

3. *Isolation*. Results from fragmentation of the individual and social components of human behaviour and motivation. Isolation suggests the idea of general societal alienation, the feeling of being in, but not of, society, a sense of remoteness from the larger social order. The nonalienated opposite of isolation is a sense of belonging and membership in society or in specific communities which are integrated through the sharing of a normative system.

4. *Self-estrangement*. Refers to the fact that the worker may become alienated from his inner self in the activity of work. Particularly when an individual lacks control over the work process and a sense of purposeful connection to the work enterprise, he may experience a kind of depersonalized detachment rather than an immediate involvement in the job tasks. This lack of present-time involvement means that the work becomes primarily instrumental, a means toward future considerations rather than an end in itself. When work encourages self-estrangement, it does not express the unique abilities, potentialities, or personality of the worker. Further consequences of self-estranged work may be boredom and monotony, the absence of personal growth, and a threat to a self-approved occupational identity. Self-estrangement is absent in two main situations: when the work activity, satisfying such felt needs as those for control, meaning, and social connection, is inherently fulfilling in itself; or when the work activity is highly integrated into the totality of an individual's social commitments.

Comment
Technological change can produce a sense of powerlessness by placing control for one's work within the computer system; additionally, the pace of the work may be regulated by the system, performance feedback may be communicated through the system, and discretion and autonomy may be reduced by it as well. Meaningfulness can also be reduced through fragmented work, standardized and routinized task elements, and limited applications of skills, aptitudes, and abilities. Isolation can be produced through changes in interaction patterns and through remote supervision. Self-estrangement may be the result of powerlessness, meaninglessness, and isolation. It may be due also to the failure to develop one's potential, unique abilities, and personality on the job.

1.6.2 Robopathic Behavior and Alienation

Yablonsky (1972) uses the term "robopaths" to emphasize the degree of alienation of workers. Yablonsky argues that robots are machine-made simulations of people and uses the word robopath to describe people whose pathology entails robotlike behavior and existence that may be produced by technological change. He identifies eight interrelated characteristics that help define the robopath:

1. *Ritualism.* Robopaths enact ritualistic behaviour patterns in the context of precisely defined and accepted norms and rules. They have limited ability to be spontaneous, to be creative, to change direction or modify their behaviour in terms of new conditions.
2. *Past-Orientation.* Robopaths are oriented to the past, rather than to the here and now situation, or to the future. They often respond to situations and conditions that are no longer relevant or functional. If they are locked into a behavioral pattern, they will follow the same path even though it may ultimately be self-destructive.
3. *Conformity.* In a robopathic-producing social machine, conformity is a virtue. New or different behaviour is viewed as strange and threatening. Originality is suspect.
4. *Image-Involvement.* Robopaths are other-directed rather than inner-directed. They are forever trying to be superconformists. Their behaviour is dominated by image or status requirements set by the surrounding society.
5. *Compassion.* If compassion entails a concern for human interests of others at one's own personal expense, robopaths are acompassionate: they appear to be without any compassionate values or social conscience. Their role and its proper enactment becomes paramount over any concern for other people.
6. *Hostility.* Hostility both hidden and obvious is a significant trait of ro-

bopaths. People unable to act out their spontaneity and creativity develop repressed, venomous pockets of anger.
7. *Self-Righteousness.* Robopathic behavior is superconformist. It is never deviant or against the norms of the social machine. The robopath's behaviour is always right or is considered self-righteous.
8. *Alienation.* Despite the overall appearances of "togetherness," the typical robopath is in effect alienated from self, others, and the natural environment.

All of these traits contradict the humanistic ethic. Robopathic behavior patterns represent the extreme case of alienation in workers and managers.

1.6.3 Alienation in Retrospect

The complexity of the individual in terms of his needs, values, motivation, expectations, perception, frame of reference, self-concept, and orientations is indicated in two studies of alienation which we discussed earlier.

Argyris (1987), in studying the alienation of workers in typical industrial organizations, found cases of workers who were not alienated because their personal needs and predispositions made them comfortable in a highly authoritarian situation which demanded little of them, either because they did not seek challenge and autonomy or because they genuinely respected authority and status.

Blauner (1964) found evidence for different patterns of alienation depending on the nature of the technology which was involved in the work. Alienation is defined as being the result of four different psychological states which are, in principle, independent of each other: (1) a sense of powerlessness or inability to influence the work situation; (2) a loss of meaning in the work; (3) a sense of social isolation, lack of feeling of belonging to an organization, work group, or occupational group; and (4) self-estrangement or a sense that work is merely a means to an end, and a lack of any self-involvement with work. Technological change may not necessarily lead to alienation depending on the predispositions, needs, orientation, expectations, and motivation of the worker. Alienation is a complex contingent process and is also a multidimensional construct. Technological change may adversely impact one of the foregoing psychological states but may not be enough to produce a high degree of alienation. Moreover, different types of technologies involve different types of job design strategies, job scope and job depth, task specialization, work flow variability and complexity, and patterns of activities, interactions, and influence and therefore will produce varying degrees of job alienation and job satisfaction.

Furthermore, it has been difficult to determine, for example, whether an alienated worker was a person without achievement and self-actualization needs when he first joined an organization, or whether he became that way as a result of chronically frustrating work experiences. This point is critical, argues Schein (1970), because if motives are not capable of being elicited or stimulated, more emphasis should be placed on selecting those workers who initially display the patterns of motivation required by the organization; if, on the other hand, by changing organizational arrangements and managerial strategies it is possible to arouse the kinds of motives desired, more emphasis should be given to helping organizations change. We will consider some of these changes that are necessary to manage the new technologies and technological change in a more humanistic mode.

1.7 NEW TECHNOLOGIES: LUND AND HANSEN'S OPTIMISTIC PERSPECTIVE

Advanced technology presents us with a number of opportunities to develop new, more humane organizational forms and jobs leading to a high quality of working life. First, although it poses new problems, highly sophisticated technology possesses an unrecognized flexibility in relation to the social system. There is an extensive array of configurations of the technology that within limits can be designed to suit the social system's needs. Second, the new technology both increases the dependence of the organization on individuals and groups and requires more individual commitment and autonomous responsibility in the workplace. Such opportunities may now be at hand to overcome alienation and provide humanly meaningful work in sociotechnical systems providing for both organizational needs and the personal and social needs of those who work (Lund and Hansen, 1986).

1.7.1 Attitudes Toward Change

Attitudes play an important role in whether technological change proceeds smoothly or whether it stumbles along amid suspicion, fear, and controversy. Both the means to solve the various issues and the process of arriving at these solutions will influence attitudes.

Lund and Hansen (1986) further observe a number of possible negative feelings that may be encountered during the change to computer-based automation and advanced communications technology:

1. Personal doubts about ability to perform, because of skills or knowledge gaps.
2. Loss of sense of control over meaningful aspects of the job.
3. Distrust of the use that superiors might make of systems that carry far more information about employees performance.
4. Loss of opportunity for personal growth and advancement.
5. Antagonism toward new job parameters.
6. Fear of displacement.

These concerns are universal; they are not confined to the blue-collar work force. Each item is equally applicable to staff employees and middle managers.

On the other hand, these researchers observe that with conscious and sensitive planning, the adoption of these new technologies can be a source of very positive reactions because of the

1. Ability to use higher-level skills and intelligence on the job.
2. Increased sense of control and involvement in a broader job.
3. Higher individual visibility and recognition in smaller organizations.
4. Opportunity to learn, grow, and advance.
5. Relief in getting away from dangerous, harmful, and dehumanizing jobs.
6. Opportunity to see results of contributions to the economic health of the firm.

Lund and Hansen (1986) note that those who experience the negative impacts of technological change may come to regard technology as the enemy. They should not. Their problems are caused by human decisions, made in the context of corporate and public policies, that dictate how technology is to be applied and how people are to be treated. Their quarrel should be with those who, consciously or by default, fail to foresee the consequences of technological change. Their quarrel is with those who fail to redefine the nature of jobs, roles, and organizational relationships in the light of impending change. Instead of blaming technology, employees should expect that managers at every level will work with technologists to ensure that technology is used creatively and constructively to further human capabilities.

As regards motivation, Lund and Hansen pose the following question: What must be part of the job to encourage a person to work at the highest level of capability? We must exploit the extraordinary flexibility of computer/telecommunications technology in designing systems of people, machines, and information in which the capabilities of both humans and machines are maximized. To accomplish this requires management to set policies for design and selection of technology that recognize human needs and attributes. Corporate policy should require managers and engineers to pay attention to the principles of good job design before the system hardware design is irrevocably frozen.

Finally, there must be opportunity within the firm for individual participation, growth, and advancement concurrent with technological change. This principle relates partly to job design, but it looks beyond the design of individual production systems to the need for a corporate culture in which these values are considered essential to the well-being of the firm.

In addition to the purely technical assessment of the appropriateness of a technology to accomplish a given task, technology planning should include principles for designing and selecting processes consonant with the needs of people that will work with it. If organizational changes are implied, plans should provide for reorienting organizational relationships or reassigning people (Lund and Hansen, 1986). Adapting technology to human needs is a part of the planning process.

Engineers and machine designers hold the key to the nature of work in the new automated factories—they design jobs when they design production systems. Yet technical people are rarely trained to appreciate the value of designing machines in such a way that the jobs for workers associated with them contain meaningful, rewarding work (Lund and Hansen, 1986).

1.7.2 Relationships

The quality of human relationships in firms employing advanced manufacturing technology will depend on how well the total system of people, machines, and information is designed.

The results of technological change on this dimension are mixed. Lund and Hansen (1986) observe that there will be fewer people in the plant, so there will be the potential for greater isolation of individuals, each responsible for a segment of the system. Communications will frequently be remote rather than face-to-face. The nature of the work will shift from physical activity to monitoring and mental activity. Workers will have less collective power.

On the other hand, Lund and Hansen (1986) argue that there may be physical isolation, but there can also be substantial interdependence and interaction. What one portion of the system does is likely to influence the performance of other parts of the system directly. Workers will be monitoring machines, rather than physically controlling them; this will enable employees to become more mobile, help others in a crisis, consult with supervisors, trade off jobs, and act as participants in a team. A team spirit can be promoted by management through flexible or rotating work assignments, common training programs, group problem-solving sessions, and access to a communications network.

1.8 JOB STRESS AND TECHNOLOGY

In this section, we will examine a major contemporary topic in all fields of management, namely, stress and its relationship to technology. Work stress has become a central variable in organizational behavior and management. Individuals experiencing great work stress are likely to be less productive, have more work-related accidents, be absent or late for work more often, more frequently abuse drugs or alcohol, and also quit the organization. Thus, work stress is likely to be related to lowered individual and organizational performance (Burke, 1987). Organizations that search for excellence care about the health and well-being of their employees.

1.8.1 Technostress

Technostress is a condition resulting from the inability of an individual or an organization to adapt to the introduction and operation of new technology. Important variables affecting the probability of technostress are age of user, past experience with technology, perceived control over new tasks, and organizational climate (Brod, 1982).

Technostress has a negative impact on human performance by shifting attention from work-congruent stress to internal states of distress, reducing the ability to process information accurately, slowing the response time to computer-generated demands, and breaking up natural work–rest pauses that characterize normal work patterns.

Technostress, which begins as reduced performance, results in behavior that limits the usefulness of technology. Brod (1982) identifies some stress symptoms of workers. First, patterns of use are altered. For instance, if a worker who was initially excited about using a microprocessor then withdraws from it, the symptoms may manifest themselves in increased time on nonmachine tasks, manual calculation of data, and social activities away from the machine. Second, the normal flow of information is disrupted. Managers using new software packages push the information down the organizational hierarchy so that their subordinates are constantly reading and filing computer-generated reports. Decision making, on the other hand, will depend on information obtained from other executive sources. Third, error rates will increase. Employees will forget or violate new procedures. Some employees who learn quickly and are intellectually astute will try to alter work rules and procedures they perceive as too simple and repetitive. In order to avoid the stress caused by boredom, they will invent new rules and short-cut new procedures. Others will experience overload, which also create stress. Because of the visibility of input data errors within the information network also, performance feedback is instant and stressful. This may lead to sabotage of the system to alleviate accountability of performance.

Work stressors generally include such other components as job complexity, job overload or underload, rate and pace of change, role conflict and ambiguity, time pressures, job monotony or simplification, lack of participation in decision making, boredom, and lack of stimulation, autonomy, and social interaction. These in turn may produce degrees of alienation.

Another area of work stress deals with the concept of burnout. Burke (1987) proposed a set of components of burnout: reduced work goals, lowered personal responsibility for outcomes, less idealism, emotional detachment, work alienation, and greater self-interest.

The interest in organizational stress has grown remarkably in recent years, and performance, turnover, and absenteeism have often been included as organizational consequences in stress models. Many organizational variables have also been studied in relation to absenteeism. Significant correlations have been reported between job characteristics such as autonomy, responsibility, variety of tasks, role conflict, ambiguity and overload, skill underutilization, resource inadequacies, and absenteeism (Burke, 1987). Some of these organizational variables are associated with technostress as well.

1.8.2 Technological Change and Stress

With increasing technological change, participation, interaction, transaction, planning, and regulation become key issues, each with its own accompanying frustrations. Moreover, technological change can produce repetitive work and a less-than-congenial work environment that is conducive to stress. The way new incentive systems are designed can produce stress. The piece-rate payment system has its incumbent strain. Rapidly changing technology in the work setting creates the potential for stress.

The infrastructure and superstructure of the organizations that manage technological change may also produce stress. Such design elements as the hierarchy of authority, departmentalization, organizational goals, control and information systems, work flow, and degree of differentiation or integration, which technology generally changes, can be great sources of stress. The designs of the infra- and superstructures will determine the degree of role conflict and role ambiguity experienced by an individual working in an organization. Additionally, work overload and underload can be stressful as technology changes job scope and depth. What a new job requires in ability, skills, and knowledge may be beyond or below the person's capacity. In both cases, the extremes in the continuum of too much work–too little work, and too easy work–too difficult work, can induce stress. (Ket de Vries, 1980).

Further, since technological changes can produce changes in interpersonal relationships, group processes, and leadership style, these also have the potential to create stress. For example, signs of organizational stress originating from difficulties in interpersonal processes can be numerous. Sudden increases in vertical and lateral interpersonal communication, withdrawal from interaction, change in content and type of interaction, frequent man–machine interface, and limited interpersonal interaction are all potential stressors. Changes in group norms and cohesiveness, interaction patterns, work flow, and values as a result of technological change can similarly be stressors. Leadership style that is directive, bureaucratic, and authoritarian, a kind of close supervision, has the potential to be stressful as much as remote supervision by the alien tools of technology. Research findings also indicate that responsibility for people, as opposed to things, can be stress-inducing.

The individual, with his values, needs, frame of reference, expectations, predispositions, rhythms, attitudes, and perception, is capable of producing more conflict, tension, and strain on the supervisor, hence the attractiveness of remote supervision. But by the same token, the individual's uniqueness, vision, intuition, imagination, creative visualization, and powers to think, love, plan, and choose make him a more interesting co-worker, colleague, and friend than a machine.

The adaptability of the individual to change, his capacity to anticipate the need for change, to cope with the rate and magnitude of that change, and to initiate change in technologies is the common theme of human and societal evolution. Some level of stress is also necessary for achievement, creativity, innovation, and the entrepreneurial function. Such a degree of stress which puts one in a state of disequilibrium may cause one to reexamine one's values, goals, priorities, plans, expectations, frame of reference, and orientations and to explore new paths for a creative adjustment to one's environment (Sai Baba, 1985). In such a case, then, homeostasis is not the ideal state of being and stress may be the catalyst for a new beginning, a new search for meaning in one's creative, experiential, and instrumental values.

1.9 HUMANISM AND TECHNOLOGICAL CHANGE

In a task-oriented society, one can manage people by assigning a task, supervising the process, and auditing results. The ultimate realization of this management technique is an automated assembly line with computerized remote supervision. To many, this progression toward less immediate control over an environment and less direct personal involvement in a job is interpreted as a dehumanization of the workplace. Your boss is no longer standing over you; the tools of technology are. The use of these technological tools of remote supervision is accelerating, and some critics argue, so is the process of dehumanization. High-tech centers such as Silicon Valley, the Research Triangle, and Boston's Route 128 area are producing evidence that high-tech industries are more likely to reduce overall skill requirements than to increase or upgrade them. This new phenomenon and a work force whose composite educational achievement level is higher than ever combine to produce the newest and possibly most devastating dimension of technological alienation—mindlessness. Benson (1985) further notes that while some executives, managers, engineers, and programmers enjoy enriched and challenging jobs, the vast majority of high-tech employees (office workers, assembly-line workers, and low-level technicians) push buttons, watch gauges, and monitor equipment—jobs as boring and mundane as any work environment ever known to the world.

The result is jobs and work environments in which people do not have to use their minds, an environment where computers, robots, and other forms of high technology do the "thinking" and people are only appendages to the process. Such an environment produces drug and alcohol abuse and an epidemic of job stress. Mindlessness has become the newest dimension of technological alienation, according to Benson (1985).

While some time has been spent on reviewing the negative effects of technological change from a conventional and contemporary perspective, we now emerge with a perspective that can only enhance the positive dimensions of technological change and minimize its negative effects, namely, the humanistic perspective.

Humanism is a philosophy that asserts the dignity and worth of a person and his capacity for self-realization through reason. Technology allows the worker to complete his job and meet his quota without ever considering his dignity, worth, and capacity. Technology must facilitate communication, a key component of humanism, the psychological need for control over one's work environment, and the actualization of human values.

1.10 INDUSTRIAL HUMANISM: A CONTEMPORARY MANAGEMENT CREED

Industrial humanism is composed of a series of assumptions about the nature of man and the outlook for the "human condition." The industrial humanists, according to Scott et al. (1981),

1. Assume the dignity of man plus the need of protecting and cultivating personality on an equal rather than hierarchical basis.
2. Assume there is a steady trend in the human condition toward the perfectibility of man.
3. Assume that organizational gains are basically the gains of people in them and that the benefits (or satisfactions) flowing from these gains should be distributed as rapidly as possible to those responsible for new technologies.
4. Assume that those who are in organizations should be, in the last analysis, the source of consent for those who make policy and establish controls.
5. Assume that change in organizations should be the result of full awareness of alternatives and consensus by participants.

From the standpoint of the managerial posture toward organizations, this means that autocracy must be replaced by democracy, flexibility must be built into the organization, power equalization must be sought, and the climate of the organization must be designed to satisfy a wide range of human needs. Organizations must be designed to help everyone in them realize their desires for self-determination and self-realization. Industrial humanism's plans for organizational change rest upon assumptions about the nature of man, the direction of the movement of social forces, and the ideals of liberalism. Both job and organizational design must be organized around these humanistic imperatives (Scott et al. 1981).

Some of the major elements of humanism are reviewed in the following sections. These are the elements that can be utilized to humanize technology. The elements are based predominantly on works by Fromm, Frankl, Maslow, Elias and Merriam, and Herzberg, some of the leading scholars in the field of humanistic psychology.

By examining the critical elements of humanism, one can then plot strategies for effectively managing technological change. The strategies designed to humanize technology will then reflect these imperatives, which will

1.11 PERSPECTIVES ON HUMANISM

1.11.1 Erich Fromm

Compassion and Empathy. The essence of compassion is that one "suffers with," or in a broader sense, "feels with" another person. This means that one does not look at the person from outside—the person being the "object" of my interest or concern. This means I experience within myself what he experiences. This is a relatedness which is not from the "I" to the "thou," but one which is characterized by the phrase: "I am thou" (*"Tat Twam Asi"*). Compassion or empathy implies that I experience in myself that which is experienced by the other person and hence that in this experience he and I are one. All knowledge of another person is real knowledge only if it is based on my experiencing in myself that which he experiences. If this is not the case and the person remains an object, I may know a lot *about* him but I do not *know him*.

Knowing, in the sense of compassionate and empathetic knowledge, requires that we get rid of the narrowing ties of a given society, race, or culture and penetrate to the depth of that human reality in which we are all nothing but human. True compassion and knowledge of man have been largely underrated as revolutionary factors in the development of man, just as art has been, according to Fromm (1968).

True compassion arises out of unity. Albert Einstein refers to this unity:

A human being is part of the whole (called by us universe); a part limited in time and space. He experiences himself, his thoughts, and feelings as something separated from the rest, a kind of optical delusion of his consciousness. This delusion is a kind of prison for us, restricting us to our personal desires and to affection for a few persons nearest to us. Our task must be to free ourselves from this prison by widening our circle of compassion to embrace all living creatures and the whole of nature in its beauty.
(Fromm, 1968)

Comment
Compassion and empathy mean, therefore, that we cannot manipulate another human being as an object, we cannot view him as an appendage to a machine, and we cannot compromise his needs for self-esteem, love, belonging, and self-actualization when we introduce and manage change. The dynamics of caring is explicit in the principles of compassion and empathy. The dynamics of caring is now emerging as a critical factor in leadership excellence.

Identity. In industrial society, men are transformed into things, and things have no identity. When we speak of identity, we speak of a quality which does not pertain to things, but only to man. Identity is the experience which permits a person to say legitimately "I"—"I" as an organizing active centre of the structure of all my actual or potential activities. This experience of "I" exists only in a state of spontaneous activity, but it does not exist in the state of passiveness and half-awakeness, a state in which people are sufficiently awake enough to sense an "I" as an active centre within themselves. The concept of "I" is different from the concept of ego. The experience of my "ego" is the experience of myself as a thing, of the body I have, the memory I have, the money, the house, the social position, the power, the problems *I have*. I look at myself as a thing, and my social role is another attribute of thingness. Many people easily confuse the identity of ego with the identity of "I" or self. The difference is fundamental and unmistakable. The experience of ego and ego identity is based on the concept of having. I have "me" as I have all other things that this "me" owns. Identity of "I" or self refers to the category of being and not of having. I am "I" only to the extent to which I am alive, interested, related, active, and to which I have achieved an integration between my appearance—to others and/or to myself—and the spiritual core of my personality. (Fromm, 1968).

Comment
The identity crisis of our time is based essentially on the increasing alienation and reification of man, and it can be solved only to the extent to which man comes to life again, becomes active again. The organization man has an identity crisis. This label represents the increasing emphasis on ego versus self, on having versus being. The organizational man *has* power, status, prestige, influence, fringe benefits, money, and organizational values of rationality and efficiency, conformity, predictability, consistency, uniformity—the values emphasized in technological change.

Integrity. A concept which presupposes that of identity is that of integrity. Integrity simply means a willingness not to violate one's identity, in the many ways in which such violation is possible. Today, the main temptations for violation of one's identity are the opportunities for advancement in industrial society. Since the life within the society tends to make man experience himself as a thing anyway, a sense of identity is a rare phenomenon. But the problem is complicated by the fact that aside from identity as a conscious phenomenon, there is a kind of unconscious identity. That is, some people while consciously they have turned into things, carry unconsciously a sense of their identity precisely because the social process has not succeeded in transforming them completely into things. These people, when yielding to the temptation of violating their integrity, may have a sense of guilt which is unconscious and which gives them a feeling of uneasiness, although they are not aware of its cause. (Fromm, 1968).

The person who experiences himself as an ego and whose sense of identity is that of ego identity naturally wants to protect this thing—him, his body, memory, property, and so on—but also his opinions and emotional investments which have become part of his ego. He is constantly on the defensive against anyone or any experience that can disturb the permanence and solidity of his

ego boundaries. In contrast, the person who experiences himself not as having but as being permits himself to be vulnerable. Nothing belongs to him except what he has by being alive. But at every moment in which he loses his sense of identity, in which he is unconcentrated, he is in danger of neither having anything nor being anybody. This danger he can only meet by constant alertness, awakeness, and aliveness, and he is vulnerable compared with the ego-man, who is safe because he *has* without *being*.
(Fromm, 1968).

Comment

Our assumption about man's nature, identity, values, and being will influence what type of motivational strategies we adopt to enhance his growth. His needs for having, while important, will be subservient to his needs for being, for growth, for love, for self-actualization. Technological change must enhance the needs that will maximize one's authenticity while still providing for one's basic needs at the survival level. One's identity must not be compromised by a lack of integrity in the agents of change who manage the change process.

Survival and Transsurvival Needs. Because man has awareness and imagination, and because he has the potential of freedom, he has an inherent tendency not to be, as Einstein once put it, "dice thrown out of the cup." He wants not only to know what is necessary in order to survive, but also to understand what human life is all about. He is the only case of life being aware of itself. Man wants to make use of those faculties which he has developed in the process of history and which serve more than the process of mere biological survival. Hunger and sex, as purely physiological phenomena, belong to the sphere of survival. But man has passions that are specifically human and transcend the function of survival.

The dynamism of human nature, inasmuch as it is human, is primarily rooted in the end of man to express his faculties in relation to the world rather than in his need to use the world as means for the satisfaction of his physiological necessities.

Man's drives, inasmuch as they are transutilitarian, are an expression of a fundamental and specifically human need: the need to be related to man and to confirm himself in this relatedness.

The two forms of human existence, that of food gathering for the purpose of survival in a narrow or broader sense and that of free and spontaneous activity expressing man's faculties and seeking for meaning beyond utilitarian work, are inherent in man's existence. Each society and each man has its own particular rhythm in which these two forms of living make their appearance. What matters is the relative strength that each of these two have and which one dominates the other.

Both action and thought share in the double nature of this polarity. Activity on the level of survival is what one usually calls work. Activeness on the transsurvival level is what one calls play, or all those activities related to cult, ritual, and art. Thought also appears in two forms, one serving the function of survival and the other serving the function of knowledge in the sense of understanding and intuiting.

Comment
The categories of thought in the industrial age are that of quantification, abstraction, and comparison, of profit and loss, of efficiency and inefficiency. The cultures of most organizations create rituals, stories, cultural heroes, symbols, and artifacts that are geared toward the utilitarian ethos. The transsurvival needs are expressed in our search for meaning, authenticity, autonomy, and challenge, which most organizational cultures do not promote.

1.11.2 Viktor Frankl

The Will to Meaning. According to most of the current motivation theories, man is a being basically concerned with gratifying needs and satisfying drives and instincts; he does so, in the final analysis, only in order to alleviate the inner tension created by them, to the end of maintaining, or restoring, an inner equilibrium which is called homeostasis. Man is never primarily concerned with an inner condition, such as the inner equilibrium, but rather with something or someone out there in the world, be it a cause to serve or a partner to love—and, if he really loves the partner, he certainly does not just use him/her as a more or less apt means to the end of satisfying his/her own needs. If we saw in the will to meaning just another drive, or need, a man would again be seen as a being basically concerned with his inner equilibrium. Obviously, he would then fulfill his meaning, in order to satisfy a drive to meaning or need for meaning—that is to say, in order to restore his inner equilibrium. In any event, he would then fulfill meaning not for its own sake. It is one of the immediate data of life experience that man is pushed by drives but pulled by meaning, and this implies that it is always up to him to decide whether or not he wishes to fulfill it. Therefore, meaning fulfillment always implies decision making.

Thus human existence is always directed to something, or someone, other than itself (be it a meaning to fulfill or another human being to encounter lovingly). This constitutive characteristic of human existence is termed "self-transcendence." What is called "self-actualization" is ultimately an effect, the unintentional by-product, of self-transcendence. Self-actualization is not man's ultimate destination, not even his primary intention. Self-actualization, if made an end in itself, contradicts the self-transcending quality of human existence. Also, self-actualization is, and most remain, an effect, namely, the effect of meaning fulfillment. Only to the extent to which man fulfills a meaning out there in the world, does he fulfill himself. Conversely, if he sets out to actualize himself rather than fulfill a meaning, self-actualization would immediately lose its justification.

Just as self-actualization can be obtained only through a detour, through the fulfillment of meaning, so identity is available only through responsibility, through being responsible for the fulfillment of meaning.

One of the two aspects of self-transcendence, that is, reaching out for a meaning to fulfill, is identical with the will to meaning. This concept, which occupies a central place in the motivation theory of logotheraphy, denotes the fundamental fact that man is striving to find, and fulfill, meaning and purpose in life. This concept of a will to meaning has been empirically corroborated and validated by several authors.

The will to meaning is not only a time manifestation of man's humanness but also a reliable criterion of mental health. On the other hand, lack of meaning and purpose is indicative of emotional maladjustment. The will to power and the will to pleasure are substitutes for a frustrated will to meaning.

Primarily and normally, man does not seek pleasure. Instead, pleasure—or, for that matter, happiness—is the side effect of living out the self-transcendence of existence. Once one has served a cause or is involved in loving another human being, happiness occurs by itself. The will to pleasure, however, contradicts the self-transcendent quality of human reality. And it also defeats itself. Pleasure and happiness are by-products, happiness must ensue, it cannot be pursued. It is the very pursuit of happiness that thwarts happiness. The more one makes happiness an aim, the more he misses the aim.

The will to power or the status drive, and the pleasure principle or the will to pleasure, are mere derivatives of man's primary concern, that is, his will to meaning. It turns out that pleasure, rather than being an end of man's striving, actually is the effect of meaning fulfillment. And power, rather than being an end in itself, actually is a means to an end, inasmuch as, if man is to live out and exert his will to meaning, a certain amount of power (say, economic or financial power) by and large will be an indispensable prerequisite. Only if one's original concern with meaning fulfillment is frustrated is one either content with power or intent on pleasure (Frankl, 1969).

Today, the will to meaning is often frustrated. This is referred to as the existential vacuum, which derives from the following conditions. Unlike an animal, man is no longer told by drives and instincts what he must do. And in contrast to man in former times, he is no longer told by traditions and values what he should do. Now, knowing neither what he must nor should do, he sometimes does not even know what he basically wishes to do. However, being human means striving to fulfill meaning and realize values. It means living in the polar field of tension established between reality and ideals; man lives by ideals and values. Human existence is not authentic unless it is lived in terms of self-transcendence. Self-transcendence is the essence of existence.

Today, however, boredom and apathy are spreading. So are feelings of emptiness and meaninglessness. What I have called the existential vacuum—crisis in meaning—may be termed the mass neurosis of today. A sound philosophy of life is what is needed to overcome the existential vacuum (Frankl, 1975).

Comment
The will to meaning expresses itself in the value potentialities of the individual. One can find meaning from creative, experiential, and attitudinal values. The first involve interesting purposeful activity at work, intrinsically satisfying activity; autonomy, sense of achievement, self-actualization; learning potential of the job; and dignity of work, pride, and identification. Technological change must be monitored to see what changes in the elements of this value category are initiated by job design. Experiential values involve meaningful, interesting, and purposeful experiences at work; interactions providing a way of achieving recognition, friendship, feelings of happiness, positive role perception, high morale at work, affiliative needs, a sense of identity with work group norms; and positive sentiments and emotions that arise from such interaction (e.g., love, beauty, truth, trust, and harmony). Technological change can also affect these elements in this value category by changing job experiences, role perception, patterns of interaction, affiliative needs, work group norms. The third category, attitudinal values, is associated with stress, frustration, alienation, crisis, the Future Shock syndrome, adversity, and growth through crisis episodes. These moments of job stress and feelings of alienation must be capable of providing growth experiences and activating the will to meaning. Technological change, because of its rate, scope, and magnitude, can produce technostress and a sense of alienation. One's meaning as derived from work can be enhanced by each category of values.

1.11.3 Frederick Herzberg

Character Values. One must struggle to create meaning by developing one's character—the badge that designates one as a person (Herzberg, 1982). The need for status, power, and security must be integrated with the need for personal growth and character. Values are the foundations of one's character. Moral and ethical values produce integrity, truth, and love that create splendor in one's character. Character is the concretization of morality (Sai Baba, 1985). It is a source of power. Character is more powerful than information, expertise, and knowledge because it relates fundamentally to one's values, belief systems, and ideal self-concept. It is the spiritual core of one's being and the synthesis of all behavior and action. In the information age, what we need is not more information or disinformation, but transformation of our values, goals, and character. Information can then be used ethically to make decisions regarding life choices. Disinformation will be then perceived as a strategy of deceit that violates the basic principle of truth.

Herzberg (1982) suggests six levels of growth which can be used to assess one's character development:

1. *Knowing More (i.e., Learning).* Is learning conducive to growth, change, feedback to monitor's changes in one's values, attitudes and desires? Is learning only related to routine job procedures and work methods?

2. *Understanding More.* The assembling of knowledge must be accompanied by integration of this knowledge. Understanding provides feedback, clarification of values and priorities and the priorities assigned to one's goals.
3. *Creativity.* This is the leap of insight that makes it possible to fall in love with the product of one's learning. It is the use of a variety of cognitive styles to process information.
4. *Effectiveness in Ambiguity.* The ability to make good decisions in gray areas. Tolerance for ambiguity must be high for creative decision making.
5. *Individuation.* This brings happiness in the form of unique abilities. Exceptional values that set one apart.
6. *Real Growth.* The ability to look into two identity mirrors to differentiate personality adjustment and status from character accomplishments and self-respect. Do you behave ethically towards others or do you rely on illusory growth for satisfaction?

Comment

Character growth is not necessarily the growth of one's ego, which can compromise one's values and ideals. The power of the ego, in its play for recognition, status, power, and aggrandizement, can make one's values an ideals subordinate in the pursuit of these goals. The incentives associated with these goals and the desire for power can make one a prisoner of the ego. Character is the mechanism for liberation from one's ego and its demands and expectations. It provides the moral values to monitor the power play of the ego. Character as the concretization of morality is the most valid and reliable measure of one's spiritual growth. With integrity as the central value in one's character, one's goals, self-concept, and attitudes can be evaluated against this central value. Similarly, the absence in one's character of such blemishes as envy, lust, conceit, greed, and malice also indicates the growth of truth and integrity as central values in that character, according to Sai Baba (1985). His viewpoint is also reinforced by Fromm (1970), who states quite categorically that the subject matter of ethics is character, and only in reference to the character structure as a whole can value statements be made about the single traits or actions of people. The virtuous or vicious character, rather than single virtues or vices, is the true subject matter of ethical inquiry. Another humanistic imperative is therefore reflected in the character of the manager or change agent to manage the change process. Hidden agendas for change are no longer part of the change agent's repertoire. Power play and intimidation are de-emphasized. Ethical excellence in mapping the objectives of change and its process, effects, and benefits is stressed for managerial effectiveness.

1.11.4 Abraham Maslow

Self-Actualization. Growth, self-actualization, or self-transcendence are innate human characteristics according to humanistic philosophy and psychology. A person continuously strives to realize his unique potentials. Maslow's model of motivation, which arranges human needs in terms of a hierarchy, indicates that the higher-level needs of the individual are belongingness, love, and esteem (to feel that one is useful and one's life has worth), and finally, the need for self-actualization. Self-actualization manifests itself in a desire for self-fulfillment, for becoming what one has the potential to become.

Some characteristics of the self-actualized person are identified by Maslow and cited by Elias and Merriam (1980):

1. They are realistically oriented.
2. They accept themselves, other people, and the natural world for what they are.
3. They are spontaneous in thinking, emotions, and behavior.
4. They are problem-centered rather than self-centered in the sense of being able to devote their attention to a task, duty, or mission that seems cut out for them.
5. They have a need for privacy and even seek it out on occasion, needing it for periods of intense concentration on subjects of interest to them.
6. They are autonomous, independent, and able to remain true to themselves in the face of rejection or unpopularity.
7. They have a continuous freshness of appreciation and capacity to stand in awe again and again of the basic goods of life: a sunset, a flower, a baby, a melody, a person.
8. They have frequent "mystic" or "oceanic" experiences, although not necessarily religious in character.
9. They feel an identification with mankind as a whole in the sense of being concerned not only with the lot of their own immediate families, but with the welfare of the world as a whole.
10. Their intimate relationships with a few specifically loved people tend to be profound and deeply emotional rather than superficial.
11. They have democratic character structures in the sense of judging people and being friendly not on the basis of race, status, and religion, but rather on the basis of who other people are as individuals.
12. They have highly developed ethics.
13. They resist total conformity to culture.

The tasks to which self-actualized people are dedicated seem to be interpretable as embodiments or incarnation of intrinsic values rather than as means to ends outside the work itself, and rather than as functionally autonomous. The tasks are loved and introjected because they embody these values. That is, ultimately it is the values that are loved, rather than the job as such. People who are reasonably gratified in all their basic needs now become "metamotivated" by the B-values, or at least by "final" ultimate values in greater or lesser degree, and in one or another combination of these ultimate values. The gratifications of the B-values enhance human potential. These intrinsic and ultimate values may be taken as instinctoid needs. Some of the B-values that motivate behavior are truth, goodness, beauty, unity, uniqueness, perfection, necessity, completion, justice, order, simplicity, totality, meaningfulness, and so on. The metaneeds or B-values are not arranged in a hierarchy of prepotency, but seem, all of them, to be equally potent on the average. It is my (uncertain) impression that any B-value is fully and adequately defined by the total of the other B-values. That is, truth, to be fully and completely defined, must be beautiful, good, perfect, just, simple, orderly, lawful, alive, comprehensive, unitary, and amusing. It is as if all the B-values have some kind of unity, with each single value being something like a facet of this whole (Maslow, 1969).

Comment
Developing one's authentic needs, such as self-esteem, belonging, self-actualization, love, creativity, security, safety, and survival, is important for human excellence and growth. Technology must serve these authentic needs. By subverting these needs, technology produces alienation and job stress. By creating synthetic needs, technology dilutes the essence of the individual. The B-values must be actualized by managers for ethical excellence, not only self-actualization. In fact, self-actualization is based on ethical and moral excellence, and technological change must be compatible with this need to develop one's value potentialities.

1.11.5 J. Elias and S. Merriam

Freedom and Autonomy. Unlike behavioristic psychology, which argues that humans are controlled by forces in the environment and are in essence programmed to respond in predetermined ways, humanists believe that the person is truly a free creature. A person's behavior is not determined by external forces or internal urges. Behavior is the consequence of human choice which individuals can freely exercise.

This notion of freedom and autonomy means that human beings are capable of making significant personal choices within the constraints imposed by heredity, personal history, and environment. Not all is predetermined. The person can be proactive rather than reactive and, in so being, exert an influence on his situation.

Technological change that inpairs this freedom and autonomy is therefore problematic for the individual's growth.

Individuality and Potentiality. In humanistic philosophy, the individuality or uniqueness of each person is recognized and valued. Behavior then is not as predictable as some would desire. It is important to promote individuality by nurturing each person's special talents, facilities, and skills. The potential of each person is unlimited. The ideals, values, and aspirations of each person must be actualized. Reason and intelligence are the tools for the development of potentiality. Intuition and emotion are also crucial components for this growth.

> Critical intelligence, infused by a sense of human caring, is the best method humanity has for solving problems. Reason should be balanced with compassion and empathy and the whole person fulfilled. (Elias and Merriam, 1980)

Perception. Another important concept of humanism is that behavior is the result of selective perception. The world is known and stimuli (external and internal) are reacted to as a result of one's individual perception. Perception is a key concept in humanism, for it explains behavior. A person's overt behavior as well as his attitudes, beliefs, and values are all a product of personal perceptions. To understand another's behavior, one must enter that person's world. Empathy and trust become the critical variables in management.

Responsibility and Humanity. Elias and Merriam (1980) further note that humanism's emphasis upon the self, the individual, and the free autonomous person carries with it a strong sense of responsibility both to the self and to other people. Interaction with others is essential because the person is not only a social being, but needs others in order to satisfy drives for love, recognition, esteem, etc.

Comment
The growth of self does not occur in isolation from others. A culture of narcissism is to be devalued; egocentric behavior and impulses are dysfunctional. While each person is unique, the humanist vision is predicated upon the recognition of the common humanity of all people. Freedom, autonomy, individuality and potentiality, responsibility, and humanity are all criteria to be used in managing technological change. Perception of its effects on these elements must be visualized before embarking on it or designing jobs to manage it.

1.12 HUMANISM AND MANAGEMENT

We will summarize some of the imperatives that are derived from humanism and some of the imperatives of technological change in this section. Some of the implications of humanism for the management of technological change will also be examined.

1.12.1 Humanistic Versus Technological Imperatives

The imperatives of humanism conflict with the imperatives of technology and thus have the potential to produce stress. The humanistic perspective argues for actualization of the self and the development of value potentialities of the individual. The development is facilitated by freedom and autonomy in the work setting. The worker brings to his job situations certain hopes, expectations, and values. Emotion and perception are important elements of the individual complex. The "inner world" of the worker is viewed as just as important as external reality in determining productivity. The worker obtains his sense of identity through interpersonal relationships and is more responsive to the norms of the work group. He is responsive to management to the extent that the supervisor can meet his needs for esteem and self-actualization, his will to meaning, and his character development for ethical excellence.

The technological imperatives conflict with the humanistic imperatives because technology requires specialization for efficiency—the fragmenting of jobs into tasks that utilize limited aptitudes, skills, abilities, and talents. Technical rationality must be observed in programming means and ends of the work flow for optimal performance. Planning to reduce uncertainty is critical in this programming of the work flow. Also critical to this process are (1) the predictability of decision outcomes and performance expectations, (2) control of the work flow through operations manuals and programs, (3) stability of management systems to cope with complexity, and (4) order and conformity as major organizational values. Technology leads to the standardization of the work process through uniform application of job design principles. Next, formalization of the work process through rules, regulations, programs, and procedural specifications further increases standardization, which limits individual autonomy, discretion, and potential. The dysfunctional elements of these imperatives are noted by Scott et al. (1981), who contend that from the human standpoint, the individualizing characteristics of the doer must be separated as much as possible from the application of labor and intelligence to tasks:

> Individuality of the worker has no place in making automobile pistons. One reason why the machine is interposed between the worker and the work is to protect against deviation from specifications by aberrant individuals. The analogue of the machine in the factory is the rule in the office.

These writers further observe that rules, like machines, separate humans from work. Rules evoke standards and predictable performances so that the imprint of the individual will not disrupt the smooth and uniform flow of information through the organization. In the modern organization, the rules of bureaucracy have invaded the factory, and the office has become mechanized. Rules and machines, in nearly every function of the large organization, have in effect united to remove the opportunity for individual discretion in work performance.

Generally speaking, the logic of organization and management has long held that if personal work bears too much individual imprint, it disrupts

Humanistic Imperatives	Technological Imperatives
Freedom and autonomy Individuality and potentiality Character values Self-actualization Perception of totality Responsibility and humanity The will to meaning Dynamics of caring Identity and integrity	Specialization of task Technical rationality Predictability of outcomes Control over work flow Programmed sequence Stability of systems Order and conformity Standardization Formalization

Figure 1.4. Humanistic imperatives versus technological imperatives.

carefully calculated, interlocked efforts of others, thus impeding the efficiency of the entire organization. The interposition of rules and machines between people and work has had much to do with this concern. In addition, it has thwarted individual self-expression and the sense of individual worth (Scott et al., 1981).

1.12.2 Humanistic Managerial Strategies

These assumptions of humanism have major implications for management strategy:

1. Managers should not limit their attention only to the parameters of the task and job design principles but also to the needs, values, and expectations of workers.
2. The manager must design a psychological contract for the worker; it should focus on his sense of identity, self-esteem, self-actualization, sense of belonging, and cognitive and spiritual growth.
3. Managers must accept the work group as a reality and emphasize group incentives rather than individual incentives.
4. The manager's role shifts from planning, organizing, and controlling to a facilitator of individual and group goals. The dynamic of caring for workers' needs, values, expectations, and feelings emerges as an ethic for managerial excellence. The initiative for work rests with the worker.
5. Managers must encourage a participative style of management over hierarchical authority and the chain of command.
6. Managers can use moral incentives to enhance the self-esteem, worth, and ethical excellence, sources of individual power.
7. Managers should design work to be more intrinsically challenging and

meaningful through job enrichment strategies to develop a sense of pride and self-esteem among workers.

8. Managers should revise the reward system for work by emphasizing more intrinsic rewards and the theme that "the joy of work is its own reward." The psychological contract involves the exchange of opportunities to obtain intrinsic rewards (satisfaction from accomplishment and use of one's capacities) for high-quality performance and creativity. This, by definition, creates a moral rather than a calculative economic involvement, according to Schein (1970).

These strategies, if implemented, can reduce the incidence of alienation in managing technological change. A major theme of this chapter is that theories of alienation presuppose theories, or at least conceptions, about human nature, as well as human needs, values, and expectations. If human needs are perceived to include those of self-esteem, autonomy and independence, and self-actualization, then jobs must be designed to give expression to them. Additionally, participative management and the democratization of organizational culture can further reduce the incidence of alienation associated with technological change. A new psychological contract must be designed around humanistic assumptions and principles in the management of technological change. This is an imperative for management.

Organization and management theory, according to Schein (1970), has tended toward simplified and generalized conceptions of the individual. Contemporary research has emphasized the complexity of the individual, the organization, and management. The individual is more complex than assumed by the rational economic model, the social model, and the self-actualizing model. Not only is he more complex in terms of needs, values, expectations, and potentials, but he may also differ from his co-worker in the patterns of that complexity.

The major assumptions of the complex person are stated by Schein (1970):

1. Man is not only complex, but also highly variable; he has many motives which are arranged in some sort of hierarchy of importance to him, but this hierarchy is subject to change from time to time and situation to situation; furthermore, motives interact and combine into complex motive patterns (for example, since money can facilitate self-actualization, for some people economic strivings are equivalent to self-actualization).

2. Man is capable of learning new motives through his organizational experiences, hence ultimately his pattern of motivation and the psychological contract which he established with the organization is the result of a complex interaction between initial needs and organizational experiences.

3. Man's motives in different organizations or different subparts of the same organization may be different: the person who is alienated in the formal organization may find fulfillment of his social and self-actualization needs in the union or the informal organization; if the job itself is complex, such as that of a

manager, some parts of the job may engage some motives, while other parts engage other motives.

4. Man can become productively involved with organizations on the basis of many different kinds of motives; his ultimate satisfaction and the ultimate effectiveness of the organization depends only in part on the nature of his motivation. The nature of the task to be performed, the abilities and experience of the person on the job, and the nature of the other people in the organization all interact to produce a certain pattern of work and feelings. For example, a highly skilled but poorly motivated worker may be as effective and satisfied as a very unskilled but highly motivated worker.

5. Man can respond to many different kinds of managerial strategies, depending on his own motives and abilities and the nature of the task; in other words, there is no one correct managerial strategy that will work for all men at all times.

1.12.3 Implied Managerial Strategy

The foregoing assumptions by Schein (1970) have major implications for the management strategy. First, the successful manager must be a good diagnostician and must value a spirit of inquiry. If the abilities and motives of people are so variable, the manager must have the sensitivity and diagnostic ability to sense and appreciate the differences when managing technological change. Second, the manager must learn to live with individual differences in aptitudes, learning styles, perception, values, needs, orientations, and expectations, and design a repertoire of strategies and styles to meet this imperative for variety produced by change. Third, the manager must be able to change his style of leadership for flexibility and the range of skills necessary to vary his own behavior. He may use pure engineering criteria in the design of some jobs but let a work group completely design another set of jobs. A variety of interpersonal relationships, patterns of authority, value systems, frames of reference, and psychological contracts must be used to cater to the degrees of complexity he encounters in the management of technological change.

Walton (1972), after reviewing a number of job design programs and work innovations in the United States and examining work restructuring programs with which he was associated, has identified a number of principles for implementing work innovations:

1. In designing work structures, it is imperative to be absolutely committed to the results one chooses. One should become pragmatic in the choice of techniques to achieve these ends.

2. Recognize that no universally applicable set of human preferences and priorities regarding quality of work life exists. Hypotheses about what would enhance human experience at work may be useful, provided that they are tested with the people in question and are revised or discarded and replaced on the basis of that experience. The same points apply to the determination of the business results that the work culture should promote.

3. Accept that most techniques affect business and human results indirectly, altering first the culture of the organization. Even if in their designs, planners ignore cultural considerations, the latter will nevertheless surface as the most important elements of the operation. Participants and visiting observers are quick to appreciate the motivation, cooperation, problem solving, openness, and candor that often mark a successful effort in practice.

4. Imagine the attitudes, relationships, and capabilities that would promote both business achievement and quality of work life in a particular setting, and then use these cultural attributes as proximate criteria for guiding the design of the work structure. In many cases, duality of goals is absent, or the step of idealizing a work culture is omitted, or both. An elaborate methodology is not required, but a certain type of thinking is advantageous.

5. Be sure that at the technique level the many different elements of design and management practice—reward scheme, division of labour, performance reporting scheme, status symbols, leadership style—are consistent with each other, each reinforcing or complementing the other. When these elements of the work structure serve common or compatible signals, the culture will be internally consistent; if they send "mixed signals" people will feel ambivalent; the more the design elements are aligned with each other, the more powerful the structure will be in shaping a distinctive work culture.

1.12.4 The Ethics of Technological Change and Human Needs

One of the functions of the ethics of change is to provide principles for managing change. These principles of change must be based on the humanistic imperatives that were developed in earlier sections. The humanistic imperatives that have relevance for ethics are identity and integrity, self-actualization, the will to meaning, character values, the dynamics of caring, individuality and potentiality, responsibility and humanity, freedom and autonomy, survival and transsurvival needs, and compassion and empathy. These imperatives are exemplified ethically in Kant's categorical imperative, which states that the individual must always be treated as an end and not as a means to an end, and Herzberg's dynamics of caring for leadership excellence. It is not the information domain, nor the dominant norm of efficiency, that is at the center of the change strategy, but the human element. The individual is not a constraint on systems efficiency, but the central factor in the equation or in the formula for change. To conceptualize him as a constraint is to reduce him to a marginal value that must be negotiated in the change process. Any change program that compromises any of the imperatives above is problematic for the individual. The change must have as one of its objectives the actualization of the needs and values of individuals who interact with the new technologies. The principles of change can be derived from the questions posed as one initiates technological change and innovation. For example, what change in creative values will this technological change program create for workers and managers? What changes in experiential values will be necessary to facilitate technological change? What attitudinal values must be articulated in this program of change which may have the potential to produce stress and alienation? Finally, what ethical values, truth, integrity, justice, equity, and so on must I as a manager articulate in my principles for managing change?

The greatest problem for technological ethics is to provide conditions that enable human beings to actualize their potentialities and to reinforce those values that enhance our search for meaning.

The right of an employee to be treated like a human being is a moral right, according to De George (1982). It is an extremely broad and in many ways a vague right. But it is a central right. Its foundation is straightforwardly the fact that each person is a human being and a moral agent deserving of respect. Kant's categorical imperative states that to treat a human being as a means only is immoral. Thus an employer who treats his workers only as a means to his profit, or a way of getting done what he wants done, treats them immorally. They are not machines or objects.

Technological change involves rational actions of change agents, technocrats, managers, engineers, scientists, and so on. They act rationally according to a rational decision-making procedure. Their rational actions affect people and the ecosystem. Hence, these actions can be evaluated from a moral point of view. The consequences and outcomes of technological change can be evaluated from a moral perspective. Some of these consequences are good, bad, or both, and therefore can be subject to an ethical assessment.

The concept of satisfaction of human need has been eroded by technological change, according to Schon (1967):

> With the important exception of the abiding poor, we have become so well satisfied, to the extent that products satisfy, that it is progressively harder for industry to discover new needs for products to fill. The imperative of industrial growth forces companies to ever more frenetic proliferation of products, to finer and finer product differences, and to the exercise of increasing pressures on consumers to consume.

Marcuse (1977) states that technology must serve human needs. We may thus distinguish between true and false needs. False are those which are superimposed upon the individual by particular social interests in his repression: the needs which perpetuate toil, aggressiveness, misery, and injustice. Their satisfaction might be most gratifying for the individuals, but this happiness is not a condition which has to be maintained and protected if it serves to arrest the development of the ability (his own and others) to recognise the malaise of the whole and grasp the chances of curing the malaise. The result then is euphoria in unhappiness. Most of the prevailing needs to relax, to have fun, to behave and consume in accordance with advertisements, to love and hate what others love and hate, belong to this category of false needs.

The only needs that have an unqualified claim for satisfaction are the vital ones—nourishment, clothing, lodging at the attainable level of culture. The satisfaction of these needs is the prerequisite for the realization of all needs, of the unsublimated as well as the sublimated ones.

As Marcuse (1977) says,

> in the last analysis, the question of what are true and false needs must be answered by the individuals themselves, but only in the last analysis; that is, if and when they are free to give their own answer. As long as they are kept incapable of being autonomous, as long as they are indoctrinated and manipulated (down to their very instincts), their answer to this question cannot be taken as their own.

The ethics of change means that technology cannot be used to superimpose needs on the individual and hence reduces his autonomy:

> The people recognize themselves in their commodities; they find their soul in their automobile, hi-fi set, split-level home, kitchen equipment. The very mechanism which ties the individual to his society has changed, and social control is anchored in the new needs which it has produced. The need for brands and gadgets is sometimes more crucial than the need for love, belongingness, identity, self esteem or self actualization.
>
> —Fromm (1968)

1.13 CONCLUSION

It is our position that technology must and can be humanized. The elements of humanism can be reflected in the job design strategies used by managers and industrial engineers. New technologies have the potential to promote a more humanistic perspective in managerial thinking. Technology must serve the authentic needs of the individual and thereby promote harmony and excellence. Corporate cultures must reflect the value potentialities of the person by incorporating the elements of humanism to reduce the incidence of job stress and alienation. We are on the threshold of a new frontier with potential for growth, growth in managerial and corporate excellence. This is our theme, our mission. We now must design our flight plan for takeoff. Managers, scientists, engineers, entrepreneurs, philosophers, and theologians must together embark on the mission and the design of the flight plan.

REFERENCES

Argyris, C., "The Individual and the Organization: Some Problems of Mutual Adjustment," in L. Boone and D. Bowen (Eds.), *The Great Writings in Management and Organizational Behavior*, Random House, New York, 1987.

REFERENCES

Benson, G., "Mindlessness as Next to Mechanicalness," *Training and Development Journal* (1985).

Blauner, R., *Alienation and Freedom*. Chicago: University of Chicago Press, 1964.

Brod, C., "Managing Technostress: Optimizing the Use of Computer Technology," *Personnel Journal* (1982).

Burke, R., "Stress and Burnout in Organizations: Implications for Personnel and Human Resource Management," in S. Dolan and R. Schuler (Eds.), *Canadian Readings in Personnel and Human Resource Management*, West Publishing, St. Paul, Minn., 1987.

Chung, Kae H., *Motivational Theories and Practices*, Grid Inc., Columbus, Ohio, 1977.

De George, R., *Business Ethics*, Macmillan, New York, 1982.

Drucker, P., *People and Performance: The Best of Peter Drucker on Management*, Harper & Row, New York, 1977.

Drucker, P., *Innovation and Entrepreneurship: Practice and Principles*. Harper & Row, New York, 1986.

Elias, John, and S. Merriam, *Philosophical Foundations of Adult Education*, Krieger, Malabar, Fl., 1980.

Frankl, V., "Self-Transcendence as a Human Phenomenon," in A. Sutich and M. Vich (Eds.), *Readings in Humanistic Psychology*, Free Press, New York, 1969.

Frankl, Viktor E., *The Unconscious God*, Washington Square Press, 1975.

Fromm, Erich, *The Revolution of Hope: Toward a Humanized Technology*, Harper & Row, New York, 1968.

Fromm, Erich, *Man for Himself: An Inquiry into the Psychology of Ethics*, Fawcett World Library, New York, 1970.

Fromm, Erich, *Marx's Concept of Man*, Frederick Ungar, New York, 1975.

Hackman, R., and G. Oldham, *Work Redesign*, Addison-Wesley, Reading, Mass., 1980.

Herzberg, F., "The Lonely Struggle to Develop Character," *Industry Week*, 1982.

Israel, J., *Alienation from Marx to Modern Sociology*, Allyn & Bacon, Boston, 1971.

Ket de Vries, M. "Organizational Stress: A Call for Management Action" in F. Luthans and K. Thompson eds. Contemporary Readings in Organizational Behavior, McGraw-Hill, New York, 1980.

Leeman, M., "On the Meaning of Alienation," *American Sociological Review*, 24(6), December (1959).

Lund, R. T., and J. A. Hansen, *Keeping America at Work: Strategies for Employing the New Technologies*, Wiley, New York, 1986.

Mansfield, F., and Ket de Vries, "Organization Stress: A Call for Management Action," in F. Luthans and K. R. Thompson, *Contemporary Readings in Organizational Behavior*. McGraw-Hill, New York, 1981.

Marcuse, H., "The New Forms of Control," in A. Teich (Ed.), *Technology and Man's Future*, St. Martin's Press, New York, 1977.

Marx, Karl, *Economic and Philosophical Manuscripts*, T. B. Battomore (Ed.), Frederick Ungar, New York, 1975.

Marx, Karl, *Capital: A Critique of Political Economy* [*Das Kapital*, 1867], Vol. 1, Vantage Books, New York, 1977.

Maslow, A., *Motivation and Personality*, Harper & Row, New York, 1954.

Maslow, A., "A Theory of Meta-Motivation: The Biological Rooting of the Value-Life," in *Readings in Humanistic Psychology*, A. Sutich and M. Vich (Eds.), Free Press, New York, 1969.

Mezaros, I., *Marx's Theory of Alienation*, Merlin Press, London, 1970.

Misiak, H., *Phenomenological, Existential and Humanistic Psychologies: A Historical Survey*, Grune and Stratton, New York, 1973.

Sai Baba, Sri Sathya, "Education and Transformation" Sanathana Sarathi, Prasanthi Nilyam, Puttarparthi, India, 1985.

Schein, E. H., *Organizational Psychology*, 2nd ed. Prentice-Hall, Englewood Cliffs, N.J., 1970.

Schon, D., *Technological Change*, Harper & Row, New York, 1967.

Scott, W., Mitchell, T. and Birnbaum, P. *Organizational Theory: A Structural and Behavioral Analysis*, Irwin, Homewood, Ill., 1981.

Walton, R., "How to Counter Alienation in the Plant," *Harvard Business Review*, November–December (1972).

Yablonsky, L. Robopaths, Bobbs-Merrill, Indianapolis, 1972.

2
JOB DESIGN AND TECHNOLOGICAL CHANGE

2.1 ADVANCE ORGANIZER

The greatest challenge in the area of job design remains for manufacturing managers. Not only must they reverse the precepts about the nature of the work held by generation of managers, but they will also have to revise their concepts about the relationships between workers and managers. When individual responsibility becomes a key skill requirement for employees, managerial behavior and attitude will have to reflect a realization of this change through organizational realignment, improved communication and above all trust (Lund and Hansen, 1986).

Classical job design based on scientific management and industrial engineering methods involves a set of mechanisms for improving organizational efficiency. While major contributions are made to the goal of efficiency, the design of jobs based on principles of simplification, standardization, and formalization produce a high incidence of job stress and alienation and a dilution of the meaning of work.

The socio-technical systems (STS) design, an alternative job design strategy, provides a more integrated and humanistic perspective to job design. It enhances the meaning of work. This approach, which integrates social, human, and technical elements, can also produce some positive effects, as noted by Davis (1971). We examine the effects of a new technology, robotics, on workers and suggest some sociotechnical design strategies for managing technological change.

Next, we focus on the characteristics of technology, including task uncertainty, work flow uncertainty, interdependence, discretion, and technical uncertainty in terms of input, conversion, and output uncertainty. The impact of these characteristics on job design is reviewed. Some effects of new technologies on job design from the perspective of Lund and Hansen (1986) are also summarized.

The imperative in job design based on the STS design is to link the design to human needs, such as for variety, challenge, learning, autonomy, discretion, recognition, meaning, belonging, and identity. Two models of job design by Chung (1977), a job enlargement model and a job enrichment model, are presented. Some job design action steps for new technologies are suggested. The future scenarios for job design from the viewpoint of Hackman and Oldham (1980) are more positive than negative in terms of the humanistic imperatives.

The major observations on job design are summarized below; also refer to Figure 2.1 for a flowchart of the issues involved.

> Machines alone do not give us mass production. Mass production is achieved by both machines and men. And while we have gone a long way toward perfecting our mechanical operations, we have not successfully written into our equations whatever complex factors represent Man, the human element.
> —Henry Ford

> The computer revolution is changing the way people at all levels of the organization perform their work. In factories workers are using (or being replaced by) robots and computer-controlled tools. Engineers are making greater use of computer-aided design systems, while secretaries are using word processors or electronic memory typewriters. Finally, the advent of the personal desktop computer has caused a fundamental revolution in the way managerial work is performed.
> —Daft, 1989

Companies adopting new computer/telecommunications technology can choose one of two job design strategies. The first is founded on the classical job design strategy, which has been proven dysfunctional because it limits individual skills, ignores higher-level needs of individuals, primarily bases its psychological contract on economic incentives, and emphasizes efficient task performance as the predominant goal. This is the classical industrial engineering approach based on scientific management.

The second approach, a more humane, integrated, and comprehensive avenue to job design, involves designing the human jobs at the same time as the technology. The principles of job design attempt to satisfy human needs for security, self-esteem, belonging, meaning, and self-actualization. A variety of skills, such as creative visualization, analysis, problem solving, communications, and planning, are utilized in this approach to complement the computer/telecommuncations technology. This strategy provides for growth, meaning, dignity, identity and integrity, individuality, and poten-

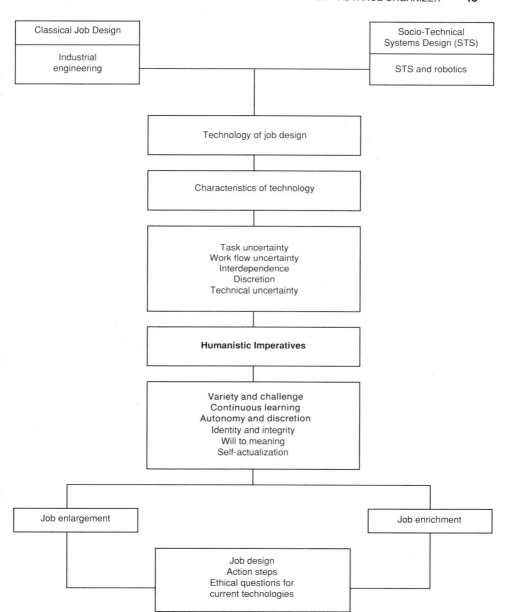

Figure 2.1. Job design and technological change.

tiality to the individual. It is referred to as the SocioTechnical Systems (STS) strategy of job design. The psychological contract is based on trust, integrity, and responsibility.

2.2 THE TECHNOLOGY OF JOB DESIGN

2.2.1 Classical Job Design: Industrial Engineering

The classical job design philosophy was articulated by an engineer, Frederick Taylor (1911), and became known as scientific management. Classical design used the full division, rigid hierarchy, and standardization of labor to reach its objectives. The idea was to segment the tasks through job simplification procedures into small parts of the job which required repetitive labor. Job performance was tightly controlled by a large hierarchy that strictly enforced the one best way of work. Predictability, precision, speed, control, programmed sequences, and consistency—essentially the attributes of a machine—were the attributes that defined the job.

Classical job design based on scientific management involves applying the scientific method of study, analysis, and problem solving to organizational problems. It is also viewed as a set of mechanisms or techniques for improving organizational efficiency. An assumption of scientific management is that by using the tools of definition, analysis, measurement, experiment, and proof, organizational efficiency can be improved. Another assumption is that the unit, or object, of analysis is the job itself, not the worker. The primary focus is on the work—its module, work flow, and procedures—through time and motion studies. A good employee is viewed as one who accepts orders but does not initiate actions. The worker is told how to do his job based on the scientific analysis of it by job analysts. Each worker is assumed to be the classical economic man, interested in maximizing his monetary income. The organization is seen as a rational instrument of production.

Taylor's codified principles of job design are summarized by Szilagyi and Wallace (1983)

1. The work to be done should be studied scientifically to determine in quantitative terms if possible (a) how the work should be partitioned among various workers for maximum simplicity and efficiency and (b) how each segment of the work should be done most efficiently. The tools of job definition, analysis, measurement, experiment and proof—the elements of the scientific method—will ensure maximum efficiency if applied to job analysis.
2. The worker must be perfectly matched to the demands of the task through systematic procedures.
3. Workers should be trained carefully by managers to ensure that they perform the work exactly as specified by the prior scientific analysis of the work. Also many planners and supervisors must ensure through close

supervision that workers attend to the productive work itself. The work of the supervisors is subdivided into functional specialties just as is done for rank-and-file workers.
4. To ensure that workers follow the detailed procedures and work practices that are specified and enforced by supervisors a substantial monetary bonus is provided to motivate workers.

From these assumptions and principles, a number of managerial strategies, as well as the type of psychological contract between the organization and its employees, are implied.

Job design is one of the most important components of scientific management. In its basic format, its assumes that job should be simplified, standardized, and specialized for each component of the required work. In general, organizations operationalized this basic job design format by breaking each job into very small but workable units, standardizing the necessary procedures for performing the work units, and teaching workers to perform their jobs under conditions of high efficiency (Szilagyi and Wallace, 1983).

Primary emphasis is on efficient task performance. Management's responsibility for the feelings and morale of people is secondary. The managerial strategy that emerges encompasses four principal functions that the manager must perform: (1) plan, (2) organize, (3) motivate, and (4) control.

Thus, an industrial organization operating by these principles will seek to improve its overall effectiveness according to Schein (1970), by worrying first about the organization itself—who reports to whom, who does what job, if the jobs are designed properly in terms of efficiency and economy, and so on. Second, it will reexamine its incentive plans, the system by which it tries to motivate and reward performance. Third, it will reexamine its control structure. Are supervisors putting enough pressure on the workers to produce? Are there adequate information-gathering mechanisms to enable management to identify which part of the organization is failing to carry its proper share of the load?

Scientific Management and Technological Change. Perhaps the most extensive adaptation of Taylor's approach to job design has been in manufacturing firms, particularly those with assembly lines. The following characteristics of the jobs of assembly-line workers are noted by Szilagyi and Wallace (1983):

1. *Mechanical Pacing.* The speed at which the employees work is determined by the speed of the conveyor line rather than by their natural rhythm or inclination.
2. *Repetitiveness.* Individual employees perform the same short cycle operations over and over again during the workday.
3. *Low Skill Requirements.* The jobs are designed to be easily learned to minimize training costs and provide maximum flexibility in assigning individuals to positions.

4. *Concentration on Only a Fraction of the Product.* Each job consists of only a few of the hundreds or thousands of operations necessary to complete the product.
5. *Limited Social Interaction.* The workplace, noise levels, and physical separation of workers spaced along a moving line make it difficult for workers to develop meaningful relationships with other employees.
6. *Predetermination of Tools and Techniques.* The manner in which an employee performs his job is determined by staff specialists. The worker may never influence these individuals.

According to Hackman and Oldham (1980), the procedures recommended by Taylor have been developed and perfected by industrial engineers over the last six decades. There now are highly sophisticated procedures for analyzing jobs (including systematic observation techniques using closed-circuit television) to determine the most efficient movements to be used in carrying out the work. Yet, the goal of industrial engineering remains the same: improved efficiency through standardized operations and simplified work.

The scientific management approach to industrial problems still prevails as an integral part of modern management techniques, according to Chung (1977). The need for high efficiency and reliable work measurements and the persistence of incentive pay necessitate the use of industrial engineering techniques such as motion and time study, work method analysis, work sampling, and ergonomics. However, to make industrial jobs more efficient, some industrial engineers go beyond the realm of ergonomics or work physiology and incorporate knowledge of work psychology in task design and work measurement. Industrial engineering techniques can be valuable tools for designing jobs with optimal levels of motivational elements if the point of optimization of the elements of the job and the needs of the individual is considered in managing technological change.

On the other hand, Davis (1971) argues that classical job design did result in substantial improvements, contrary to our views. There were remarkable increases in economic productivity; however, they were achieved at considerable human cost. There was excessive division of labor and overdependence on rules, procedures, and hierarchy. The worker became isolated from fellow workers. The result was higher turnover and absenteeism. Conflict arose as employees tried to improve the job situation. Management's response was to tighten the controls, increase supervision, and organize work rigidly. Management made a common error in treating the symptoms rather than the cause of the problem: The job itself was simply not meaningful or satisfying.

2.2.2 The Sociotechnical Systems (STS) Job Design Strategy

The "meaning of work" constitutes one of the most relevant issues in the further development of the socioeconomic structures of countries in the postindustrial era and is equally germane to the information age and the implementation of new technologies.

Steers and Porter (1975) incorporate some of the following dimensions of work. From psychological perspective, they identify four main functions of work that new technologies must incorporate in any job design strategy.

1. *Exchange:* Each person receives same form of compensation in exchange for the services that he/she gives.
2. *Social Function:* Interaction with other people.
3. *Status:* Work provides status and rank in society for workers.
4. *Personal Meaning:* Source of identity, self-esteem, self-actualization, and fulfillment; on the other hand work can also provide frustration, boredom, and a sense of alienation.

2.3 SOME POSITIVE EFFECTS OF TECHNOLOGY ON JOB DESIGN

Davis (1971) gives a summary of empirical studies of the United States and England regarding technological change and its effects on job design and organizational design.

The studies sought to find conditions in organization structure and job contents leading to cooperation, commitment, learning and growth, ability to change, and improved performance. The findings are summarized under four categories of requirements by Davis (1971): responsible autonomy, adaptability, variety, and participation. When these factors were present, they led to learning and behaviors that seemed to provide the sought-for organization- and job-response qualities. These studies lend support to the general model of responsible autonomous job and group behavior as a key aspect in sociotechnological relationships in production organizations.

Autonomy signifies that the content, structure, and organization of jobs are such that individuals or groups performing those jobs can plan, regulate, and control their own worlds (Davis, 1971).

The results obtained indicated that when the attributes and characteristic of jobs were such that the individual or group became largely autonomous in the working situation, then meaningfulness, satisfaction, and learning increased significantly, as did the knowledge of process, identification with product, commitment to desired action, and responsibility for outcomes. These supported the development of a job structure that permitted social interaction among employees and communication with peers and supervisors, particularly when the maintenance of continuity of operation was required. Simultaneously, high performance in quantity and quality of product or service outcomes was achieved.

The second requirements category, which has mainly been the province of psychologists, is concerned with "adaptation." The contents of the job have to be such that the individual can learn from what is going on around him and can grow, develop, and adjust. Slighted, but not overlooked, is the psychological concept of self-actualization or personal growth, which appears to be central to the development of motivation and commitment through satisfaction of higher-order intrinsic needs of individuals. The most potent way of satisfying intrinsic needs may well be through job design. Too often, jobs in conventional industrial organizations have simply required people to adapt to restricted, fractionated activities: their enormous capacity to learn and adapt to complexity is overlooked.

The third category is concerned with variety. Man, surely, has always known it, but only lately has it been demonstrated that part of what a living organism requires to function effectively is a variety of experiences. If people are to be alert and responsive to their working environments, they need variety in the work situation. Routine, repetitious tasks tend to extinguish the individual. Davis (1971)

There is another aspect of the need for variety that is less well-recognized in the industrial setting today, but that will become increasingly important in the emergent technological environment. The cyberneticist Ashby (1968) has described this aspect of variety as a general criterion for intelligent behavior of any kind. To Ashby, adequate adaptation is only possible if an organism already has a stored set of responses of the requisite variety. This implies that in the work situation, where unexpected things happen, the task content of a job and the training for that job should match this potential variability.

The last category concerns participation of the individual in the decisions affecting his work. Participation in development of job content and organization relations, as well as in planning of changes, was fundamental to positive work outcomes. Participation plays a role in learning and growth and permits those affected by changes in their roles and environments to develop assessments of the effects. Extensive literature on the process and dynamics of change supports the findings of the field studies. Davis (1971)

Technological change also affects social interaction patterns at work. Whyte found that increased automation decreased the opportunities workers had to interact with their co-workers. Additionally, more coordination may be required to manage the work flow. A decreased opportunity for interaction is generally associated with increases in worker alienation, stress, and absenteeism. New technologies also change work activities. If the change decreases variety, autonomy, and challenge in the jobs, or if it introduces activities incompatible with the workers' abilities and preferences, employees' attitudes will probably become more negative and their motivation is likely to fall. Alternatively, technological change can also increase variety, autonomy, and challenge in the jobs.

2.4 ROBOTICS: SOME EFFECTS OF A NEW TECHNOLOGY

Argote et al. (1983) studied the effects of workers' reactions to the introduction of a robot in the factory. Since robots are being increasingly used in offices and factories worldwide and represent a type of the new technologies being developed, it is important to consider some of the effects it produces for sociotechnical design. The following description of the effects of robotics is based on Argote et al. (1983) study.

2.4.1 Work Activities and Stress

The major change in activities in the new manufacturing cell came in the material handling activities. In the old system, the operator would pick up the stock, place it on machine one, clamp it in, start the machine, and then the milling machine would perform its work. The operator would then stop the machine, unclamp the stock, place it on the second machine, and the cycle would repeat. Each new stock would follow this cycle. When we asked the operators about the differences between their jobs before and after the robot introduction, they said:

- Now it's mainly watching . . . walking around the machines to be sure everything is running.
- We do more activities. Now you have to set up all three machines.
- There are also more functions . . . you need to program the robot.

The operators also reported that they were doing more activities and that the total work cycle had increased. They attributed the increased work cycle to more setups and delays in getting the new robot operational.

The change in activities was related to a change in skill requirements. The operators said:

- The job now requires more skills. . . . You have to learn how to program the robot and run it. . . . With more skills of course, comes more responsibility.
- Operating the robot requires more skills . . . the job is more sophisticated.

If we combine the ideas in these quotations with those in the preceding ones, it is clear that the new skills are in observing and detecting problems in the interface between the three machines and in programming and operating the robot.

The operators in our study experienced more stress or pressure. Two of the operators said:

- There is more stress now. . . . We have more responsibility. . . . They want the robot to run and we have to keep it going. . . . That's hard because it is still relatively new.
- It's nerve-racking . . . there are lots of details . . . it's an expensive piece of equipment.

This stress stems partly from the new tasks and responsibilities of the operators and partly from operating a new costly piece of equipment. There was also a more subtle source of stress when workers compared themselves to the robot. There was much speculation during our first visit to the plant about whether an operator who was particularly quick would be able to beat the robot. By our second visit, workers seemed resigned to the fact that the robot would always be able to outproduce a human worker.

One of the operators in our sample said that although he found observing and monitoring more boring than manual activities, he was currently satisfied with his job because it still included many manual setup activities. He commented, however, that without these setup activities, the current job would be more boring than his previous job. We suspect that this incompatibility between activities required by the job and preferences of the worker was another source of potential stress.

2.4.2 Work Interactions

The formal and informal interactions that develop around job activities are important in the workplace. The introduction of a robot can change these interaction patterns, which in turn may have psychological and behavioral consequences. For example, if the new technology breaks up existing social interactions and isolates the worker, we would expect increases in feelings of alienation and more resistance to the new technology.

The operators reported at Time 2 that they had less opportunity to talk with people on the job than they had before the robot was introduced. Two of the operators said:

- I haven't been able to talk as much. . . . I'm too involved with the robot. . . . You really have to concentrate.
- I don't have time to talk with anyone . . . I don't want them breaking my concentration. . . . I'm isolated now.

The decreased opportunity to interact with others seemed to come mainly from the increased mental demands of the job. Because workers had to concentrate more, they did not have time to talk with their coworkers.

The introduction of the robot did not change the work flow in the department. All the workers, including the robot operator, were located in the same area and participated in the same part of the work flow. Thus, while our operators reported less opportunity to interact with others in the department, the set of people they interacted with in the department remained roughly the same. This might have provided built-in support mechanisms to buffer the workers from some of the effects of the change.

2.5 SOME RECOMMENDATIONS FOR MANAGERS INTRODUCING NEW TECHNOLOGIES: SOCIOTECHNICAL DESIGN

These recommendations were developed by Argote et al. (1983):

1. It is critical to analyze the organization before introducing change. What effects will the technological change have on the activities, interactions, and beliefs of workers? Managers must anticipate potential problems that the change may bring—both obvious problems (e.g., job loss) and subtle ones (e.g., new job activities).

2. Management must develop a strategy for worker involvement in introducing the new technology. There is a wide variety of possible strategies. In our

study, management provided virtually no opportunities for worker involvement in the robot introduction. The workers wanted some level of involvement in certain decisions but not in others. Some involvement is likely to increase understanding about the robot and may perhaps lead to greater commitment to the change process.

3. Certain communication techniques seem more effective than others in introducing robots. Demonstrations that illustrate the operations of a robot seem to be a powerful technique.

4. Some feedback mechanism to monitor communication effectiveness in introducing this technology is necessary. Our study showed a discrepancy between what management was trying to communicate to workers and what the workers actually received.

5. It is vital that first-line supervisors be given information about the robot and support from upper management in dealing with workers' reactions to the robot. In times of change, workers are likely to go to their supervisors more frequently for information and advice. The attitudes and behaviors of supervisors are likely to have a big effect on the success of the robot introduction.

6. The robot will create job activities. It is very important to do a careful analysis of the new job and to maximize the fit between job characteristics and the personal characteristics of the worker. The literature indicates that if there is a poor fit between a job and particular worker, this may have a dysfunctional effect on both the individual and the organization. The question is not just whether the worker can do the new activities, but whether the worker *can do* and *prefers* these activities. If there is a poor fit between the job and the worker, management must consider alternatives in job redesign or in the selection procedures.

7. If the change in activities is from "doing" to "observing," workers may experience more boredom on the job. If this occurs, some mechanism to alleviate boredom, such as job rotation, may be helpful. Job rotation would increase task variety and build up a backlog of skills for future expansion of robotics.

8. It is important to train backup operators of the robot. In our study only one person per shift was initially trained to operate the robot. This led to disruptions in the work process when one of the operators was absent. Training backup operators would provide the organization with more flexibility and individual workers with more job variety.

2.6 TECHNOLOGY AND JOB DESIGN: SLOCUM AND SIMS MODEL

In this section, we focus on the characteristics of technology in terms of Slocum and Sims' model involving elements of task uncertainty, work flow uncertainty, and interdependence (1980); Hackman and Oldham's concept of discretion (1980); and Bass's model of technological uncertainty (1985). It is very important to define the dimensions of technology to monitor their impact on job design. At a general level, these dimensions of technology may affect job scope in terms of job range and depth. The range of a job refers to the

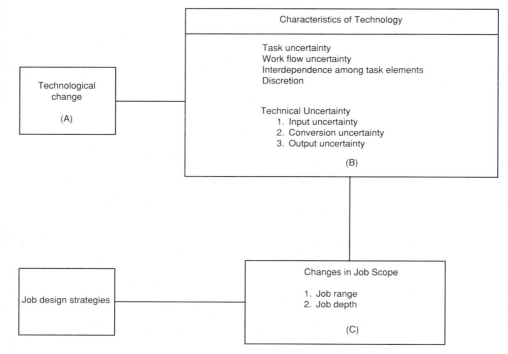

Figure 2.2. Dimensions of technology.

number of tasks to be performed, and job depth to the amount of discretion which an individual has to alter the job (Figure 2.2).

In a general sense, technology refers to the actions, knowledge, techniques, and physical elements (computers, tools, equipment) used to transform inputs into outputs. A number of models relate technology to job design. We will examine the model by Slocum and Sims (1980) first.

Slocum and Sims have argued that technology as it related to job design and organizational design can be analyzed in terms of three technology dimensions: work flow uncertainty, task uncertainty, and task interdependence.

2.6.1 Work Flow and Task Uncertainty

Uncertainty has two major components, work flow uncertainty and task uncertainty. Work flow uncertainty refers to knowledge about when inputs will arrive at an individual's station to be processed. Work flow uncertainty is likely to be high when the work system's external environment is complex and changing. The extent to which management can program responses to work flow uncertainty depends upon the type of environment, goals, and size of the firm. Work flow uncertainty is lower for employees producing standardized products because inventories (buffers) permit definite schedules for workers. When work flow uncertainty is low, a worker has little discretion to decide which, when, or where tasks will be performed. Task uncertainty refers to the degree to which the individual employee has a lack of knowledge about how to go about accomplishing the task. Task uncertainty is likely to be high where there is incomplete technical knowledge about how to produce the desired outcome. If there are no prespecified ways to deal with the transformation process, then experience, judgment, intuition, or problem-solving search activities are usually required by the individual. When work flow uncertainty is low, we would expect that employees could be programmed to execute tasks at specific times with little discretion. The employees follow rules and regulations established by the organization. The standardization of policies and procedures and the centralization of decision making are related to low boundary-transaction uncertainty. High uncertainty in the work flow create strategic contingencies for managers to the mix of tasks from one period to another. This degree of uncertainty may facilitate some degree of decentralization of decision making and participation.

Jobs having little task uncertainty permit extensive division of work and task specialization, increased skills through repetition of task-related activities, and ease of external programming by management. In contrast, jobs with high task uncertainty cannot be easily differentiated by management; there are no formulas or established procedures to aid in problem solving.

The interaction between these two types of technological uncertainty and their impact on job design are illustrated in Figure 2.3. Cell 1 represents conditions of both low task uncertainty and low work flow uncertainty for the individual. In such conditions, management has the option of programming activities and installing a control system that allows little individual discretion. Cell 2 represents conditions when management is able to reduce task uncertainty but not work flow uncertainty. Task activities can be easily programmed, but not their timing and/or sequence. As a result, the employees cannot predict what, when, and where inputs will arrive at his workplace, nor when outputs leave as finished products. Cell 3 represents conditions when management can reduce work flow uncertainty, but not task uncertainty. Input flow can be programmed by management, but task conversion activities cannot be specified in advance. Individual judgment, discretion, and experience are required for the individual to complete the transformation process. Managers may develop sophisticated scheduling algorithms to control work flow uncertainty. In cell 4, both tasks and work flow uncertainties are high. Neither the input flow nor conversion activities are programmable by managers external to the individual worker.

From a technology perspective, job enrichment generally involves increasing the degree of task uncertainty and/or degree of work flow uncertainty. This reduces the repetitiveness, specification, monotony, and predictability of task outcomes. The exercise of discretion and judgment in managing the work flow is now feasible. This framework also suggests how jobs could become too enriched. Some people who occupy cell-4 types of jobs could experience problems of stress, possibly caused by too much uncertainty.

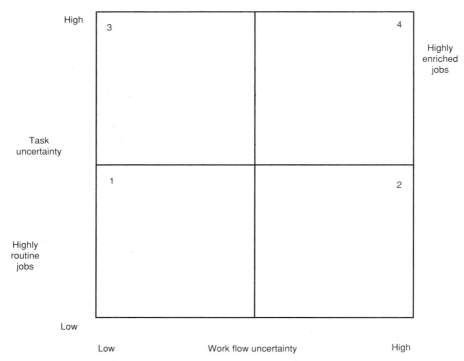

Figure 2.3. Technology and uncertainty. (Adapted from Slocum and Sims, 1980. Reprinted with permission from Plenum Publishing Corp.)

2.6.2 Task Interdependence

Another dimension of technology is interdependence. The tendency in many job redesign strategies is to group employees who perform interdependent jobs into a common work unit that is differentiated from other work units. Such a grouping appears appropriate when the technology requires interdependence among workers. Task interdependence is the degree to which decision making and cooperation between two or more employees (groups) is necessary for them to perform their jobs.

Interdependent task relations include the pooled, sequential, and reciprocal types. Pooled interdependence occurs when each employee is not required to interact with other individuals to complete the task or tasks. Sequential interdependence occurs when one employee must complete certain tasks before other employees can perform their tasks. The outputs of one employee become the inputs for other employees. The sequence of interdependencies can be a long chain in some mass-production technologies. The traditional auto assembly line is the best example of sequential interdependence. Reciprocal interdependence occurs when outputs from one individual (or group) become the input for others and vice versa. Reciprocal interdependence usually requires a high degree of collaboration, communication, group decision making, and integration among interdependent individuals.

In the design of new jobs or the redesign of existing jobs, it is often necessary to consider and make changes in one or more of the three technological dimensions: task uncertainty, work flow uncertainty, and task interdependence (pooled, sequential, or reciprocal). For example, *increases* in pooled interdependence *decrease* the amount of required job integration. The decrease in required job integration is achieved through a reduction in sequential and/or reciprocal interdependence. *Decreases* in the amount of job integration require *increases* in the amount of task uncertainty and/or work flow uncertainty for the individuals involved.

2.6.3 Technological Uncertainty and Job Design: Bass (1985)

Technological uncertainties act as constraints on the structuring of jobs. Thus, the structuring of jobs in relation to the uncertainty results in different interdependencies among tasks and different amounts of such job characteristics as autonomy, skill, variety, and task identity for individual tasks. Some positions are also so structured that employees are faced with technological uncertainty (Bass, 1985). Technology is viewed as consisting of three phases. *Input uncertainty* refers to the extent to which an employee can accurately predict what the inputs to his job will be, and when and where those inputs will arrive. *Output uncertainty* is defined as the degree to which an employee can accurately predict when and where he will be able to distribute the outputs of his task. *Conversion uncertainty* refers to the predictability of the transformation process from input acquisition and output distribution. These three phases may be highly interdependent. Bass (1985) used these dimensions of uncertainty in an empirical study and related them to characteristics of the job design model developed by Hackman and Oldham (1980) and to the integrative framework of technology and job design by Slocum and Sims (1980). This is shown in Figure 2.4.

The following reasons regarding the effects of technological uncertainty on job design characteristics are developed by Bass (1985).

When technological uncertainty is high, it will be difficult for the organization to provide prespecified programs or routines for the employees to follow. Uncertainty requires flexibility on the part of the employee, whether in acquiring and distributing the inputs and outputs or in the actual conversion process. Solving unpredictable technological problems may require a trial-and-error strategy when no specified routines are available. Therefore, it is expected that task autonomy, the extent to which an employee has freedom or discretion in carrying out the work, will be higher under conditions of high uncertainty than under conditions of low uncertainty. In this case, high technological uncertainty constrains the structuring of jobs. It prevents the organization from standardizing the work process and thereby limiting the discretion of the employee (Bass 1985).

In the case of low technological uncertainty, the organization may choose to standardize and formalize work procedures. In such a situation, the only technically rational choice is to increase technical efficiency by eliminating

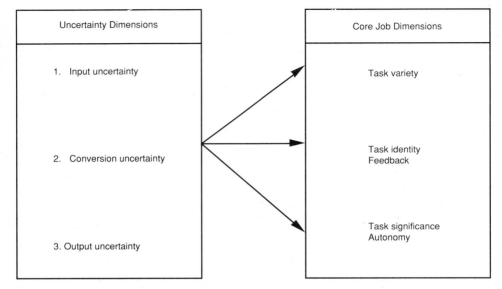

Figure 2.4. Technological uncertainty and job design.

individual discretion whenever possible. The assumption is made that organizations will attempt to reduce internal uncertainties so that the overall organizational task is as predictable as possible. Therefore, task autonomy will be limited to the minimum amount required by the technological uncertainty.

One might also expect that high technological uncertainty would require a greater variety of skills and talents on the part of the employee than would conditions of low uncertainty. The worker faced with unpredictable problems would need to develop a variety of skills for handling these situations. Likewise, an employee facing high technological uncertainty may perceive his or her job as highly significant. Thus, skill variety and task significance will be positively associated with uncertainty.

These predicted associations among task autonomy, skill variety, task significance, and technological uncertainty would seem equally plausible whether one is considering input, output, or conversion uncertainty. However, task identity, the extent to which an employee performs a complete, whole task, rather than just a small part of the task, will be primarily affected by the degree of conversion uncertainty (Slocum and Sims, 1980). Jobs with little conversion uncertainty would permit extensive division of labor. However, when the conversion process is not well understood, it is very difficult for the organization to break up the task into small component parts. Again, management would choose to divide the task whenever possible in order to achieve the advantages of division of labor (short training periods, ease of replacement, increased efficiency through repetition of the same small task, and ease of supervision). Therefore, a positive relationship between conversion uncertainty and task identity is predicted. Input or output uncertainty would not be expected to be related to task identity (Bass 1985).

While it is assumed that the successful completion of a highly uncertain task would be enhanced by feedback derived directly from the task activities, this does not guarantee that such feedback will be inherent in the task, or that workers will not acquire feedback from other sources, such as supervisors or co-workers. For example, when the arrival of inputs is uncertain, it is likely that a worker would seek feedback from other sources—most likely the persons providing the inputs. Thus, no systematic relationship is predicted concerning task feedback.

The results of Bass's study (1985) provide general empirical support for the proposition that technological uncertainty acts as a constraint on the structuring of jobs. As predicted, conversion uncertainty related positively to all the job characteristics except task feedback. Input uncertainty related to all the job characteristics, although the positive correlations with task identity and task feedback were unexpected. The task identity/input uncertainty relationship was not significant when controlling for conversion uncertainty, indicating that the relationship may be due to the strong overlap between the two uncertainty dimensions. The input uncertainty/task feedback relationship may be due to considering feedback from persons as feedback from the task itself in cases of high input uncertainty. Output uncertainty showed no significant positive correlations with any of the job characteristics. Also as predicted, technological uncertainty related negatively to satisfaction, whereas the job characteristics related positively to satisfaction. The conclusion from this study reinforces our observations that not only must organizations remain flexible in order to adapt to uncertain environments, but individuals must maintain autonomy and variety in order to successfully adapt to uncertain technologies.

2.7 JOB DESIGN FOR NEW TECHNOLOGIES

Lund and Hansen (1986) observe that the greatest challenge in job design remains for manufacturing managers. Not only must they reverse the precepts about the nature of work held by generations of managers, but they must revise their concepts about the relationships between workers and managers. When individual responsibility becomes a key skill requirement for employers, managerial behavior and attitude will have to reflect a realization of this change through organizational realignment, improved communication, and, above all, trust.

The concept of utilizing the maximum capabilities of both the person and the technology implies that there must be some sort of interplay between human aspects and technological factors at the very start of the process design sequence. It implies that at the time the process is being designed or selected there may have to be deliberate trade-offs between the positive human attributes (e.g., ability to sense problems) and those of a machine (consistency, speed).

To obtain optimal performance from the total person/machine system, Lund and Hansen provide the following job design principles:

1. *Absence of High Levels of Risks.* Computer-integrated manufacturing systems offer unprecedented opportunities to reduce the need for human beings to be in workplaces where there are high levels of risk. The change possible in these jobs is not only important from the standpoint of health and safety, but it also makes possible more equal job opportunities between men and women and for physically impaired peoples as well.

2. *Worker Participation.* The worker should have a say in how the job is structured. Workers' perspectives are highly valuable in spotting weaknesses in a given design and in improving efficiency.

3. *Information to Orient Actions.* Designing the system so it provides feedback on how a given stage in the process is affecting subsequent stages not only provides information for making timely adjustments to the process, but also stimulates the worker's interest in seeing that the overall system works well.

4. *Comprehensiveness.* Each job should, as far as possible, include all types of activity inherent in a productive work. Davis defines these as "auxiliary or service (supply, tooling), preparatory (setup), processing or transformation, control, and improvement." Given the rapid changes occurring in both product and process technology, the worker must be an integral part of the learning curve for that portion of the operation in which he is involved.

5. *Social Relatedness.* It should be possible for workers to establish relationships among co-workers, technical and support staff, and supervisors. They should be able to interact with them at work and to solve problems through group processes. One approach has been the autonomous work group, in which a large measure of the responsibility for assignments, scheduling, and conduct of production of operations is vested in the work group itself. Companies that have employed this process involve Volvo's Kalmar automobile assembly plant, Polaroid Corp., General Electric Co., Digital Equipment Corp., and Ingersol-Rand Co.

6. *Opportunity for Individual Growth.* Lund and Hansen further argue that there should be a chance for personal involvement on the job. Jobs that have opportunities for growth are likely to have long learning periods, a circumstance that traditional job designers seek to avoid. With virtually all of the pacing of a computer-based system under the computer's control, early output problems will tend to center on machine downtime and product quality. If there is appropriate technical assistance available during start-up, the production operator's learning time may not be a critical factor. Each job should also be designed so that a person can see opportunity for progression from job to job. Bridges or career paths should exist for virtually every position in the firm (Lund and Hansen, 1986).

If a firm is to succeed in establishing new job designs to match its new technologies, it may have to do more than simply adopt a set of principles. Every firm builds up a complicated structure of policies and practices that determines what can and cannot be done. Wherever these existing policies are at odds with new job design principles, changes will have to be made to bring the policies into harmony with the plans for changes in jobs.

Automated computer-aided production systems are bringing about crucial changes in the skill requirements of the job occupants. There is a reduction in requirements for the traditional high-coordination motor skills or craft precision skills, in favor of mental skills, perceptual skills in monitoring dials and gauges, as well as decision-making skills regarding machine adjustment or repair. Other studies have found it useful to consider increased responsibility, increased need for attention, and decision making discretion, conclude Lund and Hansen (1986).

2.8 JOB DESIGN, HUMAN NEEDS, AND TECHNOLOGY: HUMANISTIC IMPERATIVES

According to Trist (1978), alienation, a sort of nonwork ethic, has been increasing in the postwar period, especially among the younger generation whose expectations and experiences are different from those of the generation that lived under the conditions of scarcity that characterized the Depression years. Attitude surveys in several countries indicate that only the older worker continues to be willing to trade off dehumanizing work simply for good wages and employment security.

Trist (1978) makes the following observations on technology, job design, and human needs:

> The problem is not simply that of either "adjusting" people to technology or technology to people; it consists of organizing the man–machine interface so that the best match can be obtained between both. Rather, it is essential that the whole sociotechnical system—of which both technologies and humans are parts—be effectively "optimized." This is necessary at all levels and in all types of technology; the "mediating" technologies with which the service industries are concerned, the intensive technologies with which professional and R&D organizations are concerned, as well as the "long linked" technologies which form the basis of manufacturing. The task, in essence, is one of *joint optimization*.
>
> The human individual has work-related needs other than those specified in a contract of employment (e.g., wages, hours, safety, security of tenure, etc.) These *extrinsic* requirements form the legacy of the old work ethic. In addition, a variety of psychological requirements, or *intrinsic* factors, must also be met if the new work ethic is to develop. These intrinsic factors include
>
>   1. The need for the job to be reasonably demanding in terms other than sheer endurance and to provide a minimum of variety (not necessarily novelty, which is too much for some people though the spice of life for others). This is to recognize enfranchisement in problem solving as a human right.

62 JOB DESIGN AND TECHNOLOGICAL CHANGE

2. The need to be able to learn on the job on a continuing basis. Again, this is a question of neither too much nor too little, but of matching solutions to personal requirements. This is to acknowledge personal growth as a human right.
3. The need for some area of decisionmaking that the individual can call his own. This recognizes the opportunity to use one's own judgment as a human right.
4. The need for some degree of social support and recognition in the workplace, from both fellow workers and bosses. This acknowledges "group belongingness" as a human right.

These psychological needs and associated human rights are not confined to any one level of employment. Managers also need a high quality of work life (QWL). However, it must be realized that it is not only always possible to meet these needs to the same extent in all work settings; nor, indeed, do all kinds of people need them to the same degree—individual differences are considerable. Nevertheless, a high QWL may still be built in terms of these factors—so long as the factors on the extrinsic list are also satisfied. Together, they constitute the necessary and sufficient conditions for a high QWL at the job or task level, concludes Trist (1978).

Job design according to Hellrigel et al. (1986) is the formal and informal specification of tasks that are performed by employees, including expected interpersonal relationships and task interdependencies with others. Ideally, the needs and goals of employees and the organization are taken into account in the design or redesign of jobs. Any job design must give opportunity for employees to realize their needs for self-actualization, esteem, status, ego enhancement, belongingness, love, security, and so on. These human needs are the humanistic imperatives that technological change must address.

Herzberg (1982), a prominent researcher on motivation and job design, questioned the trend in modern industry toward work simplification introduced by scientific management and industrial engineers. This has led to fragmentation of jobs into tasks consisting of highly repetitive cycles of simple operations. Herzberg argued that jobs must be enriched to give the worker an opportunity to experience a sense of achievement. The job must therefore incorporate complexity, autonomy, and challenge, which technological change has reduced considerably.

Ket de Vries (1980) argues that organizations have historically far too often shown greater concern about the technology of work than social inputs and human assets. The emphasis has usually been on production, technology, and markets within a framework of an acceptable physical work environment. Restructuring an organization to meet the psychological needs of its members frequently has been a relatively low-priority item. Top management has been more interested in fitting the man to the job than fitting the job to the man.

Technological change must therefore ensure that jobs are designed as to allow for the expression of individual needs. It is the frustration of these needs through job simplification, job specialization, job fragmentation, and limited job depth and job scope that produces alienation, boredom, and stress. While individual needs vary, an effort to diagnose the level in the needs hierarchy at which the employee is functioning must be instituted so that job design strategies can be developed to cater to identified needs. The needs express the imperatives that a manager must consider in designing jobs such as: growth, achievement, advancement, recognition, status, self-esteem, self-respect, companionship, affection, self-actualization, and meaning.

2.9 TWO MODELS FOR DESIGNING JOBS: CHUNG

The following two models, for job enlargement and job enrichment, were developed by Chung (1977). The description of the job dimensions is based on Chung's work.

2.9.1 Model One: Job Enlargement

The problems created by the scientific management approach to job design developed to such an extent that "assembly-line blues" became a common malaise of workers. In response, managers and behavioral scientists then began to emphasize more humanistic imperatives in the design of workers' job. Job enlargement represented the first attempt by managers to redesign jobs (Figure 2.5).

Job enlargement is often called *horizontal job loading* because it expands task elements horizontally. Instead of performing a fragmented job, the worker produces a whole unit or a major portion of a job. Though often viewed as interchangeable concepts, job enlargement and job enrichment can be separately implemented as distinct managerial strategies. The difference is that the former requires changes in the technical aspects of a job, while the latter involves changes in the behavioral aspects of an organization. Although enriching a job is, technically, said to have a stronger motivational impact on employees than enlarging a job, it is extremely complex and time-consuming to implement because it requires some changes in attitudes and values of organizational as well as societal members.

Task Variety. Because of the influence of scientific management, many industrial jobs are so simplified that they can be easily standardized and mechanized. But the problem with such mechanized jobs is that as workers become satisfied with lower-order needs, they demand satisfaction of higher-order needs. Industrial jobs designed according to the principles of scientific management are at odds with the human needs for being creative and imaginative.

Performance decrement is expected when performing repetitive tasks because such tasks will force repetition of a limited number of responses to a

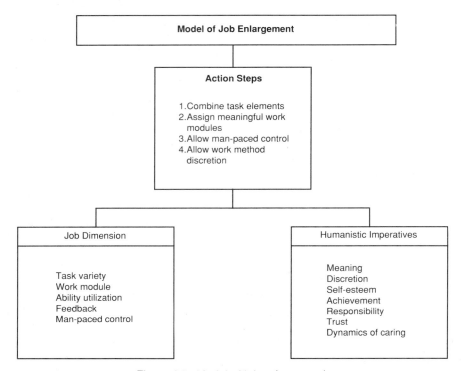

Figure 2.5. Model of job enlargement.

limited number of unchanging stimuli. Repetitive jobs increase boredom and daydreaming, which, in turn, increase errors and accidents. Job enlargement is supported by the activation theory because the number of task variations is related to the level of activation; an intermediate number of task variations should be able to sustain an optimal level of activation as well as performance.

Meaningful Work Module. By combining related task components, the existing job becomes larger and closer to the whole work unit. As the worker performs the whole work unit or at least a major portion of a product or project, he can see his contribution to its completion. When the employee completes the given work module within a time unit that is psychologically meaningful and technically sound, he may find the job interesting and worthwhile.

Establishing work modules also makes it possible to introduce a job rotation program. An employee can perform several work modules, each having different sets of stimuli. Job rotation not only reduces boredom and monotony, but also may increase the level of mental and physical activities by introducing new sets of stimuli requiring different sets of responses. The enlarged job will have an especially high motivational value when the completion of these different work modules leads to the accomplishment of a final result, such as completing a major project.

Performance Feedback. When a worker performs a fractionated job with short performance cycles, he repeats the same set of motions endlessly without a meaningful finishing point. It is difficult to count the number of finished performance cycles, but even if counted, the information is meaningless. On the other hand, knowledge of results (KR) on enlarged jobs is psychologically meaningful because the worker's level of accomplishment can be measured and evaluated for organizational rewards.

Ability Utilization. People derive satisfaction from jobs that permit the utilization of skills and abilities. Enlarged jobs usually require more mental and physical abilities than nonenlarged jobs.

Man-Paced Control. Mechanization of production methods has reduced the amount of control exercised by workers over their production speeds and work methods. The work pace is controlled mechanically, and workers are expected to exert their energies at regular and unchangeable rates of speed set by machines. Furthermore, work methods are standardized and uniformly applied to all workers regardless of individual differences in skills, abilities, and work habits.

Job enlargement makes it difficult to place workers on a machine-paced production line. Since work modules are completed by workers with different dispositions, work habits, and skill and ability levels, production speeds and work methods cannot be completely standardized. The man-paced production line has several advantages over the machine-paced one. First, it is motivational because it matches the worker's desire to control his work environment. Not only can the employee develop his own work methods and habits, which are suitable to his personality, but he can also choose the work pace that best reflects his own work rhythm. Second, enlarged jobs organized around the man-paced production line may help to reduce employee turnover and absenteeism. If the worker achieves greater satisfaction when performing enlarged jobs, he will be willing to show up more consistently and will exert more effort.

The man-paced system, however, has several disadvantages. The rate of production can be lower than the rate which results from the machine-paced production line. Workers may have to exert more energy to maintain about the same rate of production as before the job enlargement. It may be difficult to produce large amounts of goods with the man-paced system. If the size or volume of a business increases substantially, it often pays to shift the production system toward the machine-paced operation. Nevertheless, the advantages of using the man-paced production system offset its disadvantages. The crucial question is under what circumstances management should consider the use of man-controlled production systems. Managerial criterion for adopting this system can be whether it pays off in terms of high productivity and job satisfaction.

In summary, a properly enlarged job possesses such motivational characteristics as task variety, meaningful work module, performance feedback, ability utilization, and man-paced control. These motivational properties can arouse job satisfaction in rank-and-file employees not particularly interested in performing overly demanding jobs. Furthermore, job enlargement is prerequisite for job enrichment. Unless the job is first enlarged to make it interesting, job enrichment, which gives managerial authority to workers, is meaningless. When the job is dull and repetitive, employees may want to work less rather than more. Although job enlargement incurs some costs, it can be recommended to employers on the basis of humanistic considerations and cost savings attributable to reduced absenteeism, lower turnover, and decreased product rejects.

2.9.2 Model Two: Job Enrichment

Job enrichment is another form of motivational work system. Often called *vertical job loading,* it permits workers to participate in managerial processes previously restricted to managerial and supervisory personnel. It allows workers not only to perform more task components, but also to have more responsibility, autonomy, and control over the tasks they perform. Added to the motivational properties of enlarged jobs are such motivational characteristics of job enrichment as participation, goal setting, group management, and autonomy and control (Figure 2.6).

Employee Participation. Both the traditional and job enrichment approaches to management involve essentially the same managerial functions: planning, organizing, performing, and controlling organizational activities. Planning includes setting goals and specifying action plans and programs. Organizing includes division of managerial activities and assignment of responsibilities to accomplish the tasks. The performing function is the actual phase of implementing plans and carrying out the tasks, utilizing workers' skills and abilities. Controlling is the feedback process for assessing actual performance against plans. The feedback information becomes the basis for recycling managerial functions for future activities. However, the difference between the two approaches is that, in the traditional one, management alone plans, organizes, directs, and controls organizational activities, leaving only the performance function to employees. Job enrichment, on the other hand, allows employees to perform all these managerial functions regarding their own jobs. Its goal is to make employees managers of their own jobs.

Goal-Setting Responsibility. Motivation is goal-oriented behavior; it does not exist without goals or objectives. If a job enrichment program is to be successful, workers should be involved in the goal-setting process for their

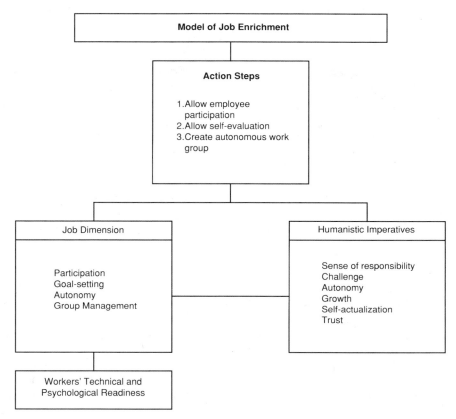

Figure 2.6. Model of job enrichment.

work group. High performance goals, as well as a supportive supervisory climate, must be present if an organization is to achieve high productivity. According to Chung (1977) several studies showing a positive relationship between participation and high productivity have involved the actual establishment of high performance goals by the participants. This finding implied that goal setting could be the key feature of employee participation.

In addition, performance goals set by employees affect the level of motivation by influencing the expectancy, instrumentality, and valence of behavioral outcomes. Challenging but attainable performance goals will positively affect the expectancy that workers' efforts are needed and will result in task accomplishment. Task accomplishment produces such intrinsic incentive rewards as a feeling of success, sense of achievement, and growth. These intrinsic rewards are highly instrumental in satisfying high-order needs such a self-respect and self-fulfillment. Employees with high achievement needs will especially be prone to become more ego-involved in achieving challenging goals because they may place higher value on such goal attainment.

Autonomy. To be most effective, job enrichment programs should go beyond the initial stage of allowing employees to participate in operational decisions. Workers should be given the authority to manage their own jobs; they should be allowed to have control over the means of achieving their task goals and should be permitted to take risks, learn from their mistakes, review their progress, and assess their own accomplishments. When workers are given such managerial authority, it should be unnecessary for managers to exercise close supervision, and they can then be available to employees for consultation, advice, guidance, and training. When autonomy is in fact working, managers can spend their time planning, troubleshooting, and helping their supervisors. In some cases, the number of supervisory personnel can be reduced.

Group Management. Managerial autonomy can be granted to employees collectively or individually. Most proponents and users of job enrichment programs prefer group action over the individualized approach. The work group defines its task goals, undertakes its tasks jointly, appraises its accomplishments and individual members' contributions to the group effort, and shares the outcomes of its accomplishments among its members. For example, self-managed work teams at General Foods in Topeka, Kansas, are given collective responsibility for managing day-to-day production problems. Assignments of individual tasks are subject to team consensus, and tasks can be redefined by the team to accommodate individual differences in skills, capacities, and interests. In addition, the work group is responsible for (1) coping with manufacturing problems that occur within or between work groups, (2) temporarily redistributing tasks to cover absentees, (3) selecting members for plant-wide task forces, (4) screening and selecting employees, and (5) counseling workers with performance problems.

There are several reasons for the group management approach. First, most jobs in industry are interdependent and require a high degree of interaction among work group members. Managers as well as workers have a responsibility for coordinating their efforts in achieving organizational or group goals. However, since these jobs are mutually interdependent, it is difficult to identify individualized goals, duties, and responsibilities. Second, a major strength of individualized management-by-objectives (MBO) programs is the strong personal responsibility an employee feels for his own goals. But such an individualized program can lead an individual employee to pursue personal goals for which he alone is responsible to the point where organizational joint responsibilities are neglected. It can also create an unhealthy competition among workers when organizational rewards such as pay increase and promotion are based on an individualized MBO program. Third, the group approach helps workers satisfy socialization needs on their jobs. Work groups control the means of satisfying employees' needs for socialization and affiliation. Groups provide a sense of belonging and identification, reinforce feelings of self-worth, and give emotional support and comfort in times of emotional stress. Interaction by itself does not necessarily lead to satisfaction: Interaction can yield either favorable or unfavorable sentiments among members. Since self-managing work group members are able to select their own peers as well as supervisors, their interactions will more likely be favorable. Furthermore, the group framework allows the members to accept and reinforce each other's feelings, thus ensuring positive relationships.

Job Enrichment Effects. The concept of job enrichment assumes that participation in decision making leads to greater acceptance of decisions by workers and thus increases employee motivation. However, there are a number of studies indicating that participation along does not necessarily lead to high motivation and productivity unless it results in high performance goals set by the participants for themselves. There are a number of individual and organizational constraints that may prevent the effective utilization of goal-setting systems. As indicated elsewhere, these constraints include workers' technical and psychological readiness to perform demanding jobs, pay, job security, and organizational climate. Further, it can be argued that workers will not set high performance goals unless their jobs have been horizontally enlarged to make them psychologically meaningful. Consequently, it is doubtful whether job enrichment alone can have a strong motivational impact on employee behavior. However, when these two types of work system are jointly applied under favorable circumstances, job enrichment can exert more influence on employee motivation than can job enlargement because the former provides workers with more opportunities for utilizing their abilities and for exerting control over their work environment.

2.10 SOME JOB DESIGN ACTION STEPS FOR NEW TECHNOLOGIES

If the job lacks motivational potential but the workers are ready to undertake demanding tasks, the following steps are suggested by Chung (1977) to increase workers' work motivation, job satisfaction, and performance. These action steps are proposed to enhance the motivational potential of various job dimensions. Selection of any step should be matched with workers' technical and psychological readiness (Chung, 1977):

1. *Combine Related Task Elements.* A number of related task elements are combined to form a meaningful work module so the workers can perform an identifiable work unit. Task combination elicits such motivational properties as task variety, meaningful work module, and ability utilization.
2. *Assign Work Modules to Workers.* Workers are assigned the responsibility of performing identifiable units of work. The completion of these work units will provide them with performance feedback, sense of accomplishment, and sense of doing something worthwhile.

3. *Allow Self-control of Work Pace.* A man-paced production system allows the workers to speed up or slow down as long as they get the job done. This will provide them with a sense of control over the production system and satisfy their needs for independence and autonomy.
4. *Allow Discretion for Work Method.* After basic training, workers are given the opportunity to express their differences in work methods. Discretion is especially necessary for performing nonprogrammable jobs. This will provide them with a sense of control over the production system. If the workers have the ability and desire to perform demanding tasks with additional challenge and responsibility, job enrichment elements can be introduced into the work system.
5. *Allow Workers to Manage Their Own Jobs.* Workers are allowed to participate in the managerial process. Since they are the ones who will carry out the tasks, they should have some influence in determining what, how much, how, and when they perform. This action step may help employees to internalize their task goals and satisfy their needs for autonomy, control, and power.
6. *Allow Workers to Evaluate Their Performance.* Workers are given the responsibility to assess their own performance. Self-generated performance feedback is immediate and real and thus psychologically more meaningful than the performance feedback given to them externally. This will provide workers with a sense of pride in their work and a feeling of responsibility.
7. *Create Autonomous Work Groups.* Self-managed work groups are given collective responsibility for managing their group activities. They perform such managerial functions as planning, organizing, performing, and controlling, which are usually reserved for managerial personnel. This will increase the motivational potential of the job by giving them more autonomy and responsibility.

The organizational culture must be congenial to such change in job design strategies. The bureaucratic organizational culture emerged from the scientific management job design framework. The movement to a participative organizational culture is not only imperative for the effective management of new technologies, but also for job satisfaction of workers and managers. The culture of the organization can act as a constraint on the implementation of technological change, as noted earlier. The attitudes, values, belief systems, and assumptions of managers must change from their bureaucratic orientations to permit a change in culture. Assumptions about human nature, human needs, values, and motivation are basic elements in an organization's culture and may facilitate and/or inhibit strategic change.

2.11 A SCENARIO IN JOB DESIGN

The following observations were made by Hackman and Oldham (1980) regarding the future of job design:

1. Technological and engineering considerations will dominate decision making about how jobs are designed. Technology is becoming increasingly central to many work activities, and that trend will accelerate. Also, major advances will be achieved in techniques for engineering work systems to make them ever more efficient. Together, these developments will greatly boost the productivity of individual workers and, in many cases, result in jobs that are nearly "people proof" (that is, work that is arranged to virtually eliminate the possibility of error due to faulty judgment, lapses of attention, or misdirected motivation). Large numbers of relatively mindless tasks, including many kinds of inspection operations, will be automated out of existence.

Simultaneous with these technological advances will be a further increase in the capability of industrial psychologists to analyze and specify in advance the knowledge and skills required for a person to perform satisfactorily almost any task that can be designed. Sophisticated employee assessment and placement procedures will be used to select people and assign them to tasks, and only rarely will individuals be placed on jobs for which they are not fully qualified.

The result of all of these developments will be a quantum improvement int he efficiency of most work systems, especially those that process physical materials or paper. And while employees will receive more pay for less work than they presently do, they will also experience substantially less discretion and challenge in their work activities.

2. Work performance and organizational productivity will be closely monitored and controlled by managers using highly sophisticated information systems. Integrated circuit microprocessors will provide the hardware needed to gather and summarize performance data for work processes that presently defy cost-efficient measurement. Software will be developed to provide managers with data about work performance and costs that are far more reliable, more valid, and more current than is possible with existing information systems. Managers increasingly will come to depend on these data for decision making and will use them to control production processes vigorously and continuously.

Because managerial control of work will increase substantially, responsibility for work outcomes will lie squarely in the laps of managers, and the gap between those to do the work and those who control it will grow. There will be accelerated movement toward a two-class society of people who work in organizations, with the challenge and intrinsic interest of key managerial and professional jobs increasing even as the work of rank-and-file employees becomes more controlled and less involving.

3. Questions of employee motivation and satisfaction will be considered explicitly when new technologies and work practices are invented and engineered (just as intellectual and motor capabilities are presently considered). No longer will work systems be designed solely to optimize technological or engineering efficiency, with motivational problems left for managers to deal with after the systems are installed. Moreover, there will be no single "right answer" about how best to design work and work systems. In many cases work will be "individualized" to improve the fit between the characteristics of an employee and the tasks that he performs. Standard managerial practices that apply equally

well to all individuals in a work unit will no longer be appropriate. Instead, managers will have to become as adept at adjusting jobs to people as they now are at adjusting people to fit the demands and requirements of fixed jobs.

2.12 ETHICAL QUESTIONS FOR CURRENT TECHNOLOGIES

The two ethical questions posed in this section are based on Ferré's (1988) observations. The major nonmoral good promised by industrial technologies is material security, in the sense of vastly increased productivity per unit of human labor.

What are the forms that should concern us? One frequent complaint voiced by those who work with industrial technology is that they thereby lose the good of their own (intrinsically valued) personal autonomy. Ruled by the production line, their sense of responsible agency is eroded so much that basic human dignities are lost. The workers' autonomy is replaced by a sense of dependency, an only partial exercise of abilities, and personal replaceability.

Can automation in the workplace avoid or minimize these harms while reforming the goods it provides? This is a key issue for an ethical assessment of industrial technology, but the answer is not clear, according to Ferré. Job enlargement, job rotation, job enrichment, and QWL (quality of work life) programs, as noted in earlier sections of this text, are some attempts to redefine work to provide variety, challenge, feedback, task identity, and self-actualization.

To discuss workers and managers raises, in one of its forms, the problem of justice. Are there relevant and justifying reasons for the distinctions of fulfillment, dignity, physical, and psychological well-being, and material reward between the different persons involved in the production process? While technology changes job depth and job range, it must attempt to provide some equity in intrinsic and extrinsic rewards. In any event, the ethical assessment of technology in the workplace will require that dislocations be compensated and that retraining be provided for those who lose their accustomed job. And the ethics of the situation would require an answer in case employment itself is reduced. Industrial engineering ethics dictate that the values and needs of workers be considered in any job design strategy.

The possibilities of robotics and increased automation in the workplace are largely dependent on the technologies of computerization. The nonmoral goods produced by the computer are many in terms of health diagnosis and hospital care, inventories planning, increases in productivity, and so on. Ferré notes that they also serve the creative values, like the quest for understanding in science, the pursuit of beauty in the fine arts, and some degree of individuality.

What are some of the harms? Expressed ethically, a serious change is that computers deprive persons of individuality by invading privacy and facilitating social centralization. Another harm that Ferré notes is the sense of helplessness and indignity that comes from attempting to argue with a computer. Such lost sense of responsible agency is even more a problem when computers are required to make decisions of great consequence before human decision makers would have time to work out conclusions from complex data. Issues of justice raised by computers range on the one hand from tort law over who is rightly to be held liable for errors made in medical diagnosis due to a software defect, to major social questions of the distribution of expertise and power on the other. Also, if workers are not using their minds, but simply monitoring gauges and controls, then the potential for stress and alienation is great, as noted in Chapter 1. Technostress and alienation are realities that must be ethically examined as a psychological harm. Strategies to enhance creative, experiential, and instrumental values at work are ethical imperatives in industrial engineering.

REFERENCES

Argote, L., P. Goodman, and D. Schkade, "The Human Side of Robotics: How Workers React to a Robot," *Sloan Management Review*, Spring (1983).

Ashby, R. Variety, Constraint, and the Law of Requisite Variety in W. Buckley, Modern Systems Research for Behavioral Scientists, Aldine Chicago, 1968.

Bass, D. J., "Technology and The Structuring of Jobs: Employee Satisfaction, Performance, and Influence," *Organizational Behavior and Human Decision Processes*, No. 35 (1985).

Chung, Kae H., *Motivational Theories and Practices*, Grid Inc., Columbus, Ohio, 1977.

Cummings, T. G., "Self-Regulating Work Groups: Socio-Technical Synthesis," *Academy of Management Review* (1978).

Cummings, T. G., "Designing Work for Productivity and Quality of Work Life," *Outlook*, No. 6 (1982).

Daft, R., *Organization Theory and Design*, West Publishing, St. Paul, Minn., 1989.

Davis, L., "The Coming Crisis for Production Management: Technology and Organization," *Journal of Production Research*, 9 (1971).

Davis, L. E., and J. C. Taylor, "Technology and Job Design," in L. Davis and J. Taylor (Eds.), *Design of Jobs*, Goodyear Publishing, Palo Alto, Calif., 1979.

Ferré, F. Philosophy of Technology Prentice Hall, Englewood Cliffs, NJ, 1988.

Gyllenhammar, P., "Technology as a Tool for Organizational Renewal," in K. Davis and L. Newstrom (Eds.), *Organizational Behavior: Readings and Exercises*, McGraw-Hill, New York, 1985.

Hackman, J. R., and G. K. Oldham, *Work Redesign*, Addison-Wesley, Reading, Mass., 1980.

Hellrigel, D., J., Slocum, and R. Woodman, *Organizational Behavior*, West Publishing, New York, 1986.

Herzberg, C. (1982) Industry Week Feature on Herzberg.

Ket de Vries, M, "Organizational Stress: A Call for Management Action," in F. Lutans and K. Thompson (Eds.), *Contemporary Readings in Organizational Behavior,* McGraw-Hill, New York, 1980.

Lund, R. T., and J. A., Hansen, *Keeping America at Work: Strategies for Employing the New Technologies,* Wiley, New York, 1986.

Schein, E. H., *Organizational Psychology,* 2nd ed., Prentice-Hall, Englewood Cliffs, N.J., 1970.

Slocum, J., Jr., and H. Sims, Jr., "A Typology for Integrating Technology, Organization and Job Design," in *Human Relations, 13*(3) (1980).

Steers, R., and L. Porter, *Motivation and Work Behavior,* McGraw-Hill, New York, 1975.

Szilagyi, A., Jr., and M. Wallace, Jr., *Organizational Behavior and Performance,* Scott, Foresman, Glenview, Ill., 1983.

Taylor, F, Principles of Scientific Management, Harper & Row, New York, 1911.

Trist, E., "Adapting to a Changing World," *Labour Gazette,* January (1978).

Walton, R., "Work Innovation in the U.S.," *Harvard Business Review,* July–August (1979).

3
CORPORATE CULTURE AND TECHNOLOGICAL CHANGE

3.1 ADVANCE ORGANIZER

Occupations and organizations typically build their practices, values, and basic self-image around their underlying technology. If the technology changes substantially, the organization or occupation must not only learn new practices, but must redefine itself in more profound ways that involve deep cultural assumptions (Schein, 1985). Companies must realize that culture change is only a starting point; it must be followed by the use of all other management tools, such as new leadership styles, structural changes, and changes in reward systems and value premises to promote an innovative organization.

When implementing innovations, ignoring culture invites disaster. Managing around culture may be realistic in the short run; over the long run, a change in culture is imperative (Stonich, 1982). Implementing a new technologies in any organization will impact on all elements of culture, such as behavioral patterns, norms, and values, the organizational climate, rules of the game, organizatonal ideology, and so on.

Three types of corporate cultures are identified: entrepreneurial, bureaucratic, and participative. All three are needed for different stages of technological change; the initiation stage requires an entrepreneurial culture, the implementation stage, a participative culture, and the routinization stage, a bureaucratic culture.

Evaluating an organization's culture is critical when managing an organization through a period of strategic change produced by the adoption and diffusion of new technologies. Various levels of an organization's culture, such as artifacts, creations, values, and basic assumptions, are audited for an assessment of its impact on technological change. From this culture audit comes the conclusion regarding high or low culture/value adjustment for new technologies.

The kind of change that is possible depends also on the developmental stage of the organization, for exmple, organizational mid-life or organizational maturity. The culture change mechanisms for each stage are identified by Schein (1985). A change in the culture paradigm at each stage may be necessary for the adoption and diffusion of technological innovations.

Strategies for changing culture must be examined when new technologies are introduced. Since the most dynamic element of culture is values and since ethics is the science of human values, the ethics of culture change is an important consideration. Ethics is not only the heart of organizational culture, but is also the fulcrum for producing change. The effect of culture change must be monitored in the implementation of new technologies.

The basic themes of the chapter are organized around the following framework (Figure 3.1).

Most corporations today are facing major problems due to the acceleration of technological change. Today's manager must assume the task of creating, developing, and complementing innovation and change. To execute this successfully, the manager must use all the tools of management, the most important of which are culture, strategy, structure, rewards, and style.

In this chapter we will examine the role of corporate culture in the dynamics of technological change, namely, how corporate culture may inhibit and/or facilitate change and how technological change impacts on culture.

3.2 NEW TECHNOLOGIES: SOME PERSPECTIVES

New technologies abruptly destroy the present while creating the future, according to Toffler (1985). It is impossible to be unaware that something extraordinary is happening to our way of life. The swift spread of microprocessors and biotechnology, the convergence of computer and telecommunications, the creation of startling new materials, the emergence of artificial intelligence, and other technological advances are accompanied by equally important social, demographic, and political changes. When so great a wave of change impacts on society, beliefs, values, cultures, the traditional management system, and organizational procedures and forms become obsolete (Toffler, 1985). Lund and Hansen (1986) identify some of the new technologies that are riding the crest of this wave of change: robotics, flexible manufacturing systems, telecommunications, laser, electronic beam, and plasma technology, acoustics, pattern recognition, materials innovation, and light holography.

A survey of information technology by *Business Week* (1983) compared its effects to those of the industrial revolution in terms of changing hierarchies,

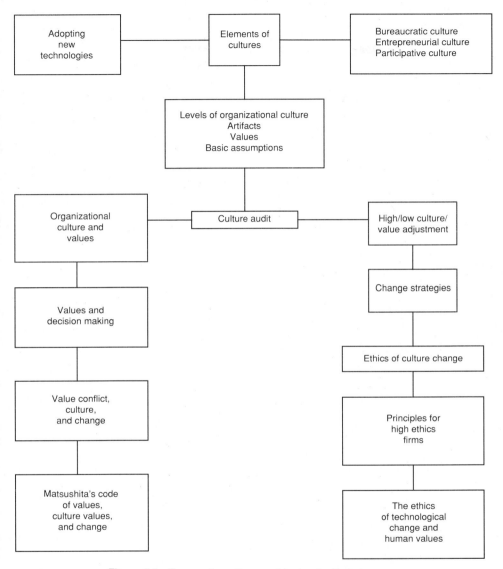

Figure 3.1. Corporate culture and technological change.

radicalizing labor, realigning political forces, and disrupting social and psychological patterns. The impact of new technologies on organizational culture is quite visible, and it may become critical to change culture to facilitate the adoption of new technologies. Organizations that are not receptive to new technologies and that do not modify themselves to absorb them will become corporate dinosaurs. Technological innovation is revolutionizing manufacturing. New technologies such as computer-aided design and manufacturing (CAD, CAM) robotics and computer-integrated manufacturing (CIM) are making the factory of the future possible. The problem for industry is twofold: first, they must become more receptive to the adoption of advanced technologies; second, managers must ensure that the performance of these technologies is not hindered by the culture and systems of the adopting organization. The strategic management of organizational culture is crucial because of its dynamic interaction with basic organizational processes such as communications, decision making, change, and power. Its potential to facilitate and/or inhibit the adoption of new technologies is therefore quite significant.

The impact of these technologies on the organization and the management of work, the design of organizations, and basic organizational processes further reinforces this significance. The integration and multiple use of these technologies also point to a dramatic realignment of these organizational variables, power centers, information systems, job design strategies, decision centers, leadership style, control systems, chain of command, and so on. For example, manufacturing processes are integrated with information-processing technology through use of low-cost computers and dedicated microprocessors linked by high-speed, high-capacity communications networks. An example by Lund and Hansen (1986) illustrates the unique advantages of integration. Yamazaki Machinery Works Ltd. has just completed a flexible manufacturing system outside of Nagoya, Japan. This new plant uses 65 computer-controlled machine tools and 34 robots linked by fiber-optic cable with a computer-aided design (CAD) center. By controlling the CAD data base, the automated factory can be directed to manufacture parts and make tools and fixtures to produce these parts. The plant has 215 employees producing what in a traditional manufacturing plant would take 2500 workers. At maximum capacity, the plant can produce $250 million worth of machine. Sales can be reduced to $85 million per year to maintain profitability (Lund and Hansen, 1986). The results of this integration of new technologies are, for example, that time horizons and product life cycles are shortened, inventories are reduced, great economies of scale are achieved, decision making is more decentralized, a participative organizational culture emerges, greater interdependence exists among elements of the work flow, and interaction patterns are changed, etc. Advanced technologies alter the very foundation of the firm using them. Hence, a change in culture is an organizational imperative for adoption of new technologies.

The cultural perspective in the adoption of new technologies is emphasized by Wilkins (1983). The elements of the philosophy of the organization in which new technologies are adopted must be consistent with the belief system of the organizational culture regarding the quality of work life, managerial values for managing change, ethical premises in decision making, flexibility in job mobility, normative control, job rotation, flexibility in job deployment patterns, and consensual decision making—some elements of Japanese management that promote the adoption of new technologies.

According to Child (1984), new technologies have become a significant vehicle for changes in the organization in England and Europe. The changes that management advocates reflect its strategic intentions in introducing new technologies; some of these changes include reduced operating costs, increased flexibility, improved quality of products or services, and enhanced control and integration. Possibilities for management control are extended by new technologies; in particular, such control facilitates unification of hitherto fragmented control systems. It can also assist integration through its enhancement of communication. These possibilities present opportunities for change within management in regard to hierarchical location of decision making, the complexity of coordinative arrangement, the size of middle management, the number of management levels, and changes in organizational culture.

3.3 ORGANIZATIONAL CULTURE: SOME PERSPECTIVES

Culture is a set of key values, beliefs, and understandings that are shared by members of an organization (Smircich, 1983). Culture defines basic organizational values and communicates to new members the correct way to think and act, and how things ought to be done (Schein, 1985). The purpose of culture is to provide members with a sense of identity and to generate a commitment to beliefs and values that are larger than themselves. Culture also enhances the stability of an organization and provides members with an understanding that can help them make sense of organizational events and activities. Culture can be a positive force when used to reinforce goals and strategy of the organization (Sathe, 1985). Chief executives can influence internal culture to be consistent with corporate strategy. Culture indicates the values employees should adopt to behave in a way consistent with organizational goals. Top executives deal in symbols, ceremonies, and images (Peters, 1978). Managers signal values, beliefs and goals to employees. Techniques top managers use to convey the appropriate values and beliefs are rites and ceremonies, stories, symbols, and slogans; these can be used to manage organizational culture that is hard to shape by conventional means (Daft, 1989).

Generally, the behavior of individuals and groups within an organization is strongly shaped by norms stemming from shared beliefs, expectations, and actions. Organizational culture also includes and is further defined by (Schein, 1985):

1. *Observed behavioral regularities* when people interact, such as the language used and the rituals around difference and demeanour.
2. The *norms* that evolve in working groups such as a fair day's work for a fair day's pay.
3. The *dominant values* espoused by an organization, such as product quality or price leadership.

4. The *philosophy* that guides an organization's policy toward employees and customers.
5. The *rules* of the games for getting along in the organization, or the "ropes" a newcomer must learn in order to become an accepted member.
6. The *feeling or climate* that is conveyed in an organization by the physical layout and the way in which members of the organization interact with customers or other outsiders.

None of these by itself represents the culture of the organization. Taken together, however, they reflect the culture of an organization and lend meaning to the concept of organizational culture. The common theme of a number of recent popular books is that strong, well-developed cultures are an important characteristic of companies with a record of high performance and innovative corporate strategies. The role of corporate culture in the change process is cogently expressed by Stonich (1982).

When implementing a new strategy, ignoring culture is an approach that invites disastrous consequences. Managing around it may be the most realistic course for an organization in the short term, but over the long term, a change in culture may be necessary.

It is clear that most organizations develop unique cultures and that a specific culture has a direct impact on how well new technologies are implemented. Recognizing the existence of culture and understanding its implications are essential steps when contemplating the implementation of strategic change.

Implementing new technologies in any organization will impact on all elements of corporate culture, such as (1) observed behavioral patterns, (2) norms, (3) dominant values, (4) organizational ideology, (5) rules of the game, and (6) the climate of the organization. These elements also have the potential to inhibit the effective implementation of new technologies. The adoption of new technologies is a strategic change, hence its impact on basic elements of organizational culture. The organization may have to redefine its philosophy, assumptions, values, and beliefs for effective implementation of advanced technologies that may revolutionize its work flow, job design, mode of supervision, information systems, communications networks, and decision making centers.

Culture is central to change. Managers must consider culture when they decide to adopt and implement new technologies. If the culture of the organization is flexible, change will take place much more easily. If, however, the change process conflicts with values, beliefs, attitudes, and other elements of the organizational culture, it will be difficult to implement.

There are three prerequisites for changing culture to accommodate new technologies:

1. The strategy for implementing new technologies and all its elements must be explicitly stated and easily understandable.
2. A culture audit of both objective and subjective elements is essential.

3. A culture change strategy must be articulated.

For effective implementation of new technologies, organizations should make changes operational and concise, and attempt to monitor the effects of change on culture. Strategic change will affect critical elements of corporate culture, including power, status, and prestige; therefore, resistance must be anticipated.

Culture can also inhibit change in two ways, according to Lorsch (1986). First, it can produce a strategic myopia. Because managers hold a set of beliefs and values, they see events through this prism. They frequently miss the significance of changing external conditions because they are blinded by strongly held beliefs. Second, even when managers can overcome such myopia, they respond to changing events in terms of their culture. Because the beliefs have been effective guides in the past, the natural response is to stick with them.

The area of strategic organizational change is one of the most important to analyze, understand, and manage. Management of an organization's culture should be one of the basic elements of a corporate strategy for an effective business enterprise. Peters and Waterman (1982) refer to excellent companies with strong corporate cultures. These companies are directed more toward their marketplace and make limited use of policy manuals, charts, and rules and regulations. Such cultures are more adept at adapting to environmental changes.

Without exception, the dominance and coherence of culture proved to be an essential quality of excellent companies. Moreover, the stronger the culture and the more it is directed to the marketplace, the less need there is for policy manuals, organizational charts, or detailed procedures and rules.

3.4 TYPES OF CULTURES: THREE PROFILES

There are critics, however, notably Pastin (1986), of corporate culture who believe a strong culture can be detrimental to the organization. Their main criticism is that there are elements of culture, such as belief systems or ideologies, rituals, value systems, assumptions, and climate, that are more resistant to change, and that change in the environment can lead to cultural obsolescence.

3.4.1 Bureaucratic Culture

Some cultures, because of the elements discussed above, can become bureaucratic, hierarchical, and compartmentalized; they are characterized by clear lines of authority and responsibility, zero-sum power assumptions, and an ethos promoting attitudinal and behavioral conformity and a value system rewarding stability, standardization, and order as organizational imperatives.

Such a bureaucratic culture facilitates incremental or consolidative change. However, because the bureaucratic culture comprises rules, operations manuals, control systems, a chain of command, and information networks, it can make the implementation of new technologies easier. It has the management infrastructure to implement change but not the imperatives for *initiation* of change because the culture creates attitudinal and behavioral conformity, rituals, standard behavioral repertoire, conservative value systems and assumptions about the organization's potential for growth.

3.4.2 Entrepreneurial Culture

This type of culture promotes innovative change. The ethos of the organization facilitates risk-taking behavior, exercise of discretion and autonomy in decision making, and the prevalence of "organizational slack" for absorbing the consequences of errors in risk-taking behaviors. The value system of the entrepreneurial culture rewards the exercise of initiative, judgment, creativity, and ethics in strategic decision making. This is the type of culture that organizations must design and nurture to enable the emergence of new technologies. The institutionalization of an entrepreneurial function within the complex organization is one of the critical priorities of contemporary management. The type of organizational culture that will reward and promote the creation of new technologies via the entrepreneurial function is still a problematic issue in the management of change (Sankar, 1988b).

The entrepreneurial culture predisposes an organization toward a high culture/value adjustment capacity for *initiation* of new technologies.

3.4.3 Participative Culture

This third type of organizational culture is based on the following philosophy articulated by R. Power (1986), chief executive officer of Reichold Chemicals, Inc. One tenet is that people act and are treated as professionals. A second is that creativity and reasonable risk taking are encouraged. A third is that decision making should be made at the lowest possible level. The fourth is that attaining goals requires the action of a team.

This philosophy supports the principles of participative management: encouraging the free flow of information and ideas essential to the team effort. Participative management fosters trust and support; improves communication; helps people develop; generates better decisions; leads to enriched jobs and increases motivation; generates a low-risk, high ROI; and attracts and retains better people. In participative management, a limited number of decisions still come from the top, but everyone involved has a real role in arriving at meaningful decisions. The participative culture promotes the commitment that is necessary to manage new technologies effectively. The consensus regarding the goals of technological change, the implementation of change, the monitoring of the effects of change, and the design of criteria for evaluating the effectiveness of change can emerge from a participative culture.

However, all three types of culture are necessary for different stages of technological change. At the initiation stage of the change process, the entrepreneurial culture is appropriate, since it involves the decision to adopt an innovation. At the implementation stage of the technological change process, the participative culture is more suitable, as conflict, stress, and disequilibrium of the organization will be at their maximum. The strategies for managing change, the process of change, and the integration of the new technology with the management system are some problem areas at this stage. A participative culture can be used to achieve consensus in the group's assumptions, beliefs, and values in the management of the change process. At the routinization stage of the change process, the bureaucratic culture is effective. Rules, regulations, procedures, standardization, formalization, and integration are some of the priorities at this point. Routinization of the change is necessary for predictability, control, and consistency in the flow of work, resources, and information. Stability then becomes a major organizational value which a bureaucratic culture promotes.

3.5 STRATEGIES FOR THE MANAGEMENT OF CORPORATE CULTURE

To match corporate culture with new technologies, three strategies could be used: (1) ignore the culture, (2) manage around the culture, and (3) change the culture to fit the strategy for adopting new technologies. The process of shaping culture is seen as a primary management role. The leader not only creates the rational and tangible aspects of organizations, such as technology and structure, but also symbols, ideologies, language, beliefs, rituals, and myths (Schein, 1985).

Many academics and practitioners believe culture can be positively influenced by consistent, thoughtful managerial action. Others are more skeptical as to whether culture is indeed manageable, and see corporate culture as possibly being another disappointing managerial tool.

To determine what method should be used, the cultural risk must be determined. Cultural risk is the extent that culture will put the institution at risk in terms of its ability to implement its strategy.

In determining the cultural risk, a manager must first know the organization's culture. Diagnosing the culture is the first step in managing corporate culture. This is done through a cultural audit.

3.6 THE CULTURAL AUDIT FOR NEW TECHNOLOGIES

Evaluating an organization's culture, or performing a cultural audit (Figure 3.2), is critical when managing an organization through a period of strategic change produced by the adoption and diffusion of new technologies.

84 CORPORATE CULTURE AND TECHNOLOGICAL CHANGE

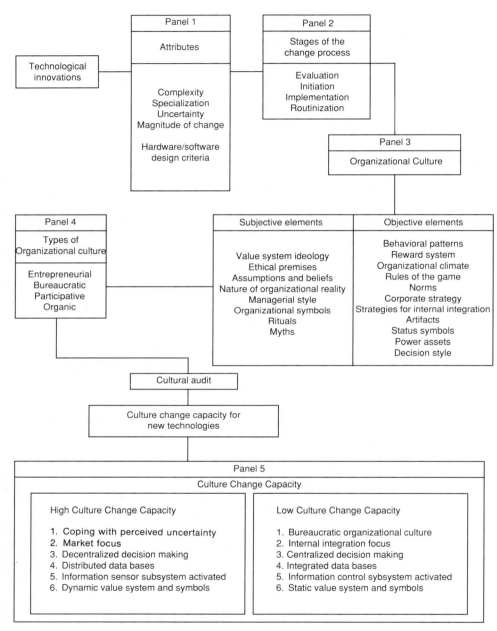

Figure 3.2. The cultural audit for managing technological change.

A culture audit involves the top management group developing a consensus about their shared beliefs. This process starts with each member of the group answering questions such as those identified by Sankar (1991) in Table 3.1. (In answering these questions, it is critical that the emphasis be on what top managers believe as evidenced by their practices, not on some idealized view of their company.) Then, individual answers to these questions are compared and provide a basis for discussion among managers. Through this process, the beliefs that are shared can be identified and codified.

Wilkins (1983) suggests that in performing a cultural audit, the manager should start with some basic notions about what kinds of assumptions are most important to the functioning of the organization, and then focus on those places in the organization its culture is most likely to assert itself. These focus areas could include (1) when employees change roles, (2) when subcultures conflict, and (3) when top management implements strategic decisions about company direction and style.

An organization's culture is best altered by gradually reducing perceived differences between current norms and the new behavior, increasing the value placed on adaptability, and enhancing managers' ability to effect the strategic management of change. Managers cannot be expected to change the manner in which they operate (i.e., in their tasks and relationships) until they know the behavior required in the new culture and what will enhance their personal development in the firm. Changing culture requires coordination with other planned internal changes in management processes and organizational structure. The company's reward system is a valuable tool in effecting culture change (Wilkins, 1983).

An effective culture audit cannot be delegated. It must involve the company's major decision makers, including the chief executive officer. Top managers must be able to identify the consistent pattern of their culture and how beliefs are related to each other. If the audit is successful, top managers will have made the once invisible barrier of the culture visible. Making culture explicit is one way of facilitating more rapid strategic change. A culture audit that emphasizes beliefs about (1) goals, (2) distinctive competencies, (3) product market guidelines, and (4) management assumptions will make the value premises of the culture more explicit and visible for effective management and effective introduction of new technologies (Lorsch, 1986).

3.7 THE LEVELS OF CULTURE AND TECHNOLOGICAL CHANGE

Some elements of culture are depicted in Figure 3.3 (Schein, 1985). Included in these are the physical layout of an organizaiton's offices, rules of interaction, basic values as reflected in the organization's ideology or philosophy, and the underlying conceptual categories and assumptions that enable people to communicate and interpret everyday occurrences. Basic assumptions are

TABLE 3.1 Cultural Audit Checklist for Managing Technological Change

Panel 1. Attributes of Technical Innovations

How complex are the changes in the work flow produced by new technologies?
How specialized are the information, task, and job design requirements?
Where is the greatest degree of technical uncertainty, input uncertainty, conversion uncertainty, or output uncertainty?
What is the magnitude of the change in the work flow produced by new technologies?
How can the software design be incorporated into coordination modes for the work flow?

Panel 2. Stages of the Change Process

What diagnostic data are important to determine the need for technical change?
How is the decision to adopt an innovation justified? (Sathe, 1985)
What are the dynamics and process of the implementation stages of the new technology?
What cultural constraints impact on the implementation stage? Sankar (1988b)
How can the new technology be integrated into the management systems of the organization?
What are the management principles for managing culture change?
How can the ethics of planned change be incorporated into these principles? (Pastin, 1986)

Panel 3. Organizational Culture

SUBJECTIVE ELEMENTS

What beliefs and assumptions are embedded in the culture regarding new product design, marketing strategies, and change?
What organizational symbols, artifacts, etc., are more conducive to change?
What value systems are more congenial to culture change?
What managerial style is predominant in the organization: theory X, theory Y, or theory Z?
What effects will technological change have on the beliefs of workers, their interaction patterns, and group culture?
What are the beliefs, values, and assumptions of culture creators or carriers?
What symbols must be revised for culture cohesion?
What humanistic imperatives must be emphasized (trust, caring, empathy, etc.) in cultural change?

OBJECTIVE ELEMENTS

What extrinsic rewards are available for motivating change agents?
What are the intrinsic rewards that must be used for influencing change?
What norms of the work group facilitate/inhibit change?
How does corporate culture relate to corporate strategy as an instrument of change?
What mechanisms of culture change are available to the leader?
Is the organization culture geared toward the problem of external adaptation?
What cultural risks are involved in changing the objective elements of culture?
What value premises are supportive of change?

TABLE 3.1 (Continued)

Panel 4. Types of Organizational Culture

What is the ideal culture for change?
Does the culture have an external market focus and an ethos for risk-taking behavior?
Does the culture focus on rules, procedures, regulations, and operations manuals and decision-making strategies?
Does the culture emphasize consensual decision making?

Panel 5. Culture Change Capacity

HIGH CULTURE CHANGE CAPACITY (HIGH CCC)
Does the environment of the organization face high uncertainty in its major strategic sectors?
Do the leaders of the organization or the dominant coalition have an external market focus?
Is decision-making authority delegated to lower levels of the organization?
Is the information domain segmented into different configurations?
Does the sensor subsystem scan the environment for changes in markets, technology, etc.?
Are the value systems in the subcultures of the organization oriented toward change and growth?

LOW CULTURE CHANGE CAPACITY (LOW CCC)
Are the subcultures programmed toward consistency, uniformity, and predictability of outcomes?
Is the culture oriented more toward internal integration than external adaptation?
Is the decision-making process centralized in the strategic apex?
Is the information domain patterned into a vertical configuration?
Is the control subsystem constantly and closely monitoring the management systems?
Are the value systems of the subcultures oriented toward stability of the system?

Source: Sankar (1991). Reprinted with permission from Harcourt Brace Jovanovich, Inc.

treated as the essence of culture, and values and behaviors are the observed manifestations of the cultural essence. These levels of culture range from visible to invisible. Technological change can impact all levels of an organization's culture as well as the physical and social environment, work flow, interaction patterns, group networks and configurations, technological output, and the overt behavior of group members. Every facet of group's life produces artifacts, creating the problem of classification. Another set of artifacts is comprised of the status symbols of the groups within organizations. Technological change effected by redistributing power and patterns of influence via information, knowledge, expertise, and uncertainty coping, etc., can also change the subjective and objective criteria for allocating

88 CORPORATE CULTURE AND TECHNOLOGICAL CHANGE

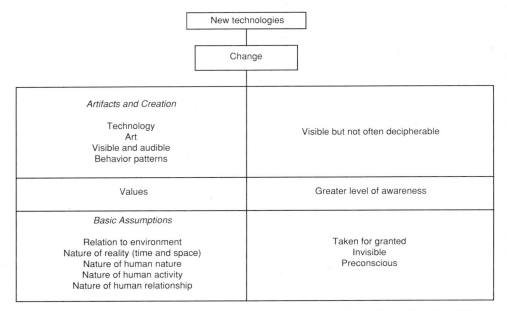

Figure 3.3. The levels of culture and technological change. (Adapted from Schein, 1985. Reprinted with permission from Jossey-Bass, Inc., Publishers.)

status. The meanings attached to visible behavior is more complex to decipher, but technological change does influence the behavior patterns of organizational members by creating changes in incentive systems, supervisory style (e.g., remote supervision), job design strategies (e.g., job modules as opposed to specialized fragmented job units), and so forth.

If one wants to achieve a greater understanding about the effect of technological change, one can attempt to analyze the central values that provide the day-to-day operating principles by which the members of the culture guide their behavior. In a sense, all cultural learning reflects someone's original values, their sense of what ought to be, as distinct from what is. When a group faces a new task, change, and uncertainty, the first solution proposed to deal with it can only have the status of a value because there is not as yet a shared basis for determining what is factual and real (Schein, 1985). Consensus through social validation may reinforce such values—values about how people should relate to each other, exercise power in the context of new technologies, manage uncertainty produced by such technologies, distribute the rewards generated by new technologies, and set standards of performance and norms for cooperative behavior in managing change.

To get to the deeper level of understanding, to decipher the pattern, to predict future behavior correctly, we must understand more fully the category of basic assumptions. Basic assumptions are different from dominant value orientations in that the latter reflects the preferred solution among several basic alternatives, but all the alternatives are still visible in the culture. At a somewhat deeper level lie the hidden assumptions—the fundamental beliefs behind all decisions and actions—that underlie culture (Kilmann 1985). These assumptions pertain to the nature of the environment and to what various stakeholders want and need, how they make decisions, and what action they are likely to take both now and in the future.

In deciding to adopt a new technology or engage in technological change, a new set of assumptions generally emerges: that productivity can be further enhanced by this change, producing a competitive advantage and so increasing profits, that it is a means of facilitating adaptability to a high-tech environment or reinforcing values and norms of excellence, that it will diversify product portfolios, that it will project a "good" corporate image for an avant-garde company or lead to more effective marketing strategies of a product, that will enhance professional pride and esteem of workers, and so on. If we examine an organization's artifacts and values, we can try to infer the underlying assumption that tie things together (Kilmann et al., 1985).

According to Kilmann et al (1985), when culture change involves changing surface-level behavioral norms, it can occur with relative ease because members can articulate what behaviors are required for success today in contrast to those required yesterday. In addition, closing the gap between actual and desired norms is easier if the desired norms are essentially the same throughout the organization, that is, if the environment is homogeneous. Even when multiple cultures exist, requiring different changes in each work group, change is still easier to effect when the focus of culture change is on behavioral norms rather than hidden assumptions or human nature. Technological change because of its magnitude is likely to influence behavioral norms, assumptions, and humanistic imperatives. The idea of managing corporate culture is still quite new to most practitioners. At best, most managers have a vague sense of what the term "corporate culture" means: something having to do with the people and the unique quality or character of organizations. But ask how culture can be identified and, if found dysfunctional, changed, and most managers are at a loss. This lack of knowledge about corporate culture is in sharp contrast to managers' intricate knowledge of goals, strategies, organization charts, policy statements and budgets—the more tangible aspects of their organizations (Kilmann et al., 1985).

3.8 A MODEL OF ORGANIZATIONAL CULTURE AND CHANGE

Figure 3.2 indicates that a technical innovation may possess a number of attributes, such as degree of complexity, specialization, and uncertainty, that may produce some magnitude of change for the organization. These attributes, coupled with the hardware and software design criteria, may increase the magnitude and rate of change for the organization. The stages of the change process for the organization and new technology involve evaluation, initiation, implementation, and routinization. Each stage has a different dynamic, task, function, and information-processing agenda for action. The subjective and objective elements of organizational culture will be affected differently by each stage of the change process.

At the implementation stage, for example, the disequilibrium of the organization is at its greatest because of major changes in power structure, decision centers, integration patterns, managerial styles, information systems, organizational artifacts and symbols, and managerial assumptions. The type of organizational culture must be diagnosed through a cultural audit to see what changes in subjective and/or objective elements must be changed to accommodate the new technologies.

High culture/value adjustment and change is likely if the organization faces high normative uncertainty, an external market focus, decentralized decision making, distributed data-processing configurations, an active information sensor subsystem, dynamic value systems, and organizational symbols. Low culture/value adjustment and change is likely if there is homogeneity of the elements of organizational culture, an internal integration focus, centralized decision making, integrated data bases, an activation of the control subsystem, and the prevalence of a conservative value system in the culture of the organization.

The effective *initiation* of new technologies depends on a high culture change, which is characteristic of the entrepreneurial culture. The bureaucratic culture is more predisposed toward a low culture change. This type of culture has the management infrastructure for *implementation* of the new technologies because of program repertoire, standard operating procedures, planning and control mechanisms, integrated management systems, and chain of command, some of the imperatives of bureaucracy. The strategic management of change involves alternating between flexibility and stability in organizational forms and organizational cultures (Sankar, 1991).

Some specific changes in organizational culture are as follows. Relationships among workers change when the nature of their work is modified. Lund and Hansen (1986), in their study of new technologies in the United States, suggest that social distance between employees will tend to increase as a consequence of greater physical isolation from each other and increased dependence on electronic forms of communication. Relationships with supervisors will also change. Workers in the factory culture of the future will have more decision-making discretion, will not be supervised as closely, and will intervene more directly in the production process. The communications networks, the interdependence among the systems of communication, and bypass of the chain in command all produce a less bureaucratic organizational culture. The roles of lower and middle managers will also be altered. In many cases levels of management will be eliminated because the transmission and processing of information, their primary functions, will reside largely in hardware and software, especially with greater use of expert systems to aid decision making. The information pathologies in the bureaucratic culture, such as information filtration, distortion, and overload, will be neutralized by information technology. Today's new technologies will create an impact substantially different and more profound in organizational design and culture than any form of technological change witnessed so far.

If the negative side effects of technological change on organizational culture are to be averted, then managers, information executives, systems analysts, and change agents must be actively involved in the design of a cultural audit for the adoption and diffusion of new technologies. Advanced technologies required fresh thinking about the ideal match between the structure of the firm and the technology it is using. The type of communications networks for data bases, the strategies for decoding the sectors in the environment, change in software designs, the restructuring of organizations, and the strategic management of organizational culture are all imperative in the management of new technologies.

3.9 CHANGING CULTURE: LORSCH'S PERSPECTIVES

Changing a given culture greatly depends on how deep-seated the existing culture is. The stronger and older a culture, the more difficult the process of cultural change. In most organization, however, the impetus to initiate the change in culture is problems in achieving financial and other valued goals when the pattern or principles that have worked well no longer do. Therefore, even for firmly entrenched cultures, the top executives realize that a cultural adjustment is necessary and that a change must occur (Lorsch, 1986).

The change of culture is often achieved in two ways: one method is using incremental change, while the other is a direct fundamental change. In looking at the need for cultural adjustment, top executives may first entertain only minor adjustments. By varying a vew of the cultural principles, the company may be able to respond in a new way. This process seems to be a natural growth process of reacting to the technological environment in order to survive. This method certainly would be easy to adjust to, as Lorsch (1986) has found: "It was easier to bend one less central principle than those at the core of top management culture."

This method, which entails periodic change retains the core of the cultures. The main problem with incremental change is that it may not cause enough fundamental change to react to the increasingly complex and changing technological environment of the 1980's and beyond. If not, it will be necessary to make drastic fundamental changes to the culture. The top executives becoming cognizant of the need for a fundamental change is the first step in the change process. While it has become obvious to the top managers that new direction is needed, the exact direction and dimension of these new beliefs may not be obvious. The changing of the culture will take time (Lorsch, 1986). The generation of new ideas and beliefs may require new people at top levels who aren't inflicted with the old cultural norms. The meeting of the old and new minds will spawn many new beliefs that have potential direction for the company. Over a period of time, often through trial and error, these beliefs and values will get consolidated into a new working group of shared beliefs. This will be the new culture for the company and should contain the flexible and innovative qualities that were sought for the period of strategic change generated by advanced technologies.

Creating a new culture does in no way mean discarding all the old standard beliefs, assumptions, and values that the organization has grown with. The culture has new elements that essentially ensure flexibility and innovation, but these must be intertwined with the existing cultural elements to maintain stability. It is the best of both worlds and helps the older managers, that is, the career man, to adhere to their new culture by allowing adherence to the old beliefs (Lorsch, 1986).

Obviously, the more technology and the environment change, the greater the degree of adoption of new standards. Even under times of dramatic change, the old beliefs that described the company's character and competence will still be vital and cannot be abandoned. The culture, once changed, must be instilled throughout the organization to provide each person with the new vision and belief in order to truly affect the organization. The process of changing the culture throughout the organization can be achieved from two opposite directions. The top–down approach, whereby management dictates the changes and encourages compliance, may be effective in simplistic situations. This method does broadcast that changes in culture are required and should signal the company's new beliefs (Lorsch, 1986).

Managers who still have a strong feeling for the original culture work by the conventional norms and standards. New innovation is, by definition, creative and will fall outside these old norms, and the new ideas may become suppressed. This is demonstrated in Sathe's (1985) culture article through four different situations. Although management was publicly promoting innovation as part of its culture, innovative ideas were suppressed, as they offended old existing norms that really had no place in the new cultural system. This resistance at the top level to adopt a new culture fully, is further reinforced by executives' fear of change, uncertainty, and redistribution of power that technological change may create. This fear must be overcome by their own recognition that innovation is necessary and that middle managers and other employees cannot start the innovative process until top management fully accepts it and realizes that they are responsible for its successful implementation.

The corporate culture needs support to promote change properly and completely. The proper incentives and communication networks must be created. These incentives should encompass both financial and intrinsic rewards. The latter should include such things as recognition, advancement, challenge, autonomy, discretion, long- and short-term achievements, and recognition through proper feedback channels. These incentives will show the employee that the top executives are truly committed to the new culture, and should provide each employee with individual motivation to change elements of the culture to accommodate new technologies and managerial practices, Lorsch (1986) concludes.

3.10 HOW TO ASSESS RESISTANCE TO CULTURE CHANGE

According to Sathe (1985), culture change may involve more or less a change in (1) the content and/or (2) the strength of the existing organizational culture.

3.10.1 Change in Culture's Content

Culture's content depends on what the shared assumptions are and how they are ordered, and it determines in what direction culture influences behavior. A change in culture's content is produced in one or more of four ways (Sathe, 1985): (1) existing beliefs and values are reordered, (2) existing values and beliefs are modified, (3) existing beliefs and values are removed, and (4) nonexisting beliefs and values are added.

Two important features of a change in cultural content should be noted. First, we are not talking about changes in people's attitudes, which more readily accompany changes in behavior, but rather the changes in their more central beliefs and values. Second, changes in organizational behavior patterns may be accompanied by changes in some beliefs and values for a certain number of people in the organization. Such changes will not be viewed as culture changes unless the beliefs and values in question are widely enough shared and highly enough placed relative to other beliefs and values.

3.10.2 Incremental Versus Radical Change in Culture's Content

The more radical the proposed change in culture's content, as in the case of technological change, the greater the resistance to it. A greater change in culture's content is involved to the extent that it involves

1. A greater number of important shared assumptions.
2. More central (highly ordered) shared assumptions.
3. A movement toward more alien (less intrinsically appealing) shared assumptions.

A radical change in culture's content is more difficult to accomplish than incremental change. Further, for a given degree of change in culture's content, cultural resistance will be greater in a strong culture because it involves change for a larger number of organizational members (Sathe, 1985).

3.11 ORGANIZATIONAL CULTURES AND VALUES

Strong, highly effective cultures are characterized by three major factors: (1) performance of managerial roles—particularly decisional, informational, and interpersonal roles—in ways desired by the company; (2) an internal environment that sets the tone and provides clear cries for desired employee behavior and performance; and (3) human resources program to implement and maintain the desired culture (Sathe, 1985). Another factor we will consider is a dynamic value system and ethical standards for decision making which can make the culture of the organization more dynamic.

Although ethics and culture are closely related, the infatuation with culture does not extend to ethics. One reason is that those promoting organizational culture are now competing in a small but quite profitable industry, the organizational culture of consulting. Introducing ethics into the discussion is perceived as not helpful in making sales. And those who sell culture, like the managers who buy it, have allowed themselves to become illiterate in the language of ethics. So they ignore it (Pastin, 1986). But the truth is that ethical ground rules are the heart of organizational cultures. In fact, if ethics is the heart of organizational culture (Pastin, 1986), then the myths, symbols, rituals, ideologies, and customs, the elements of culture, are the fat around the heart, strangling it and destroying its vitality. Ethics is not only the heart of organizational culture, but also the fulcrum in culture for producing change. Since technological change means change in a given direction that is determined by certain goals and values, then the value premises within an organization's culture must be diagnosed in managing strategic change.

According to Schein (1985) a set of values that becomes embodied in an ideology or organizational philosophy can serve as a guide and a way of dealing with the uncertainty of intrinsically complex events. Such values will predict much of the behavior that can be observed at the artifactual level. But if those values are not based on prior cultural learning, they may also come to be seen as "espoused values," which are not consistently applied to behavior. If espoused values are reasonably congruent with the underlying assumptions, then the articulation of those values into a philosophy of operating can be helpful in bringing the group together and serving as a source of identity and core mission (Ouchi, 1981; Pascale and Athos, 1981; Peters and Waterman, 1982).

Even after we have listed and articulated the major ethical values of an organization, we may still feel that it lacks coherence. Often such lists of values are not patterned, and sometimes they are even contradictory or incongruent with observed behavior. To get at that deeper level of understanding, to decipher the pattern, to predict future behavior correctly, we have to understand more fully the category of "basic assumptions." It is through diagosis of the value system of an organization's culture that we glimpse these basic assumptions in the company's belief system for managing technological change. It is during the process of strategic change that one's values are tested (Schein, 1985).

3.11.1 Culture, Values, and Decision Making

In every type of organization, business, educational, religious, government, medical, and voluntary managers engage in decision making. Decision making is the essence of management. It is therefore in the decision-making process that one's values, as well as the values of the organization, are tested.

Values provide a decision maker with a set of guidelines to steer him through the entire decision-making process. Value judgments taken from his personal values and overlaid by the values of the organization occur in the formulation of objectives, the search for relevant alternatives, the ranking and evaluation of alternatives, the moment of choice, and the point of implementation.

> Values within the culture of the organization influence the perception of situations and problems, the entire process of choice in decision making as well as sets limits to ethical behavior in decision making.

Values may be thought of as the guidance system a manager uses when confronted about choices among alternatives. A value can also be viewed as an explicit or implicit conception of what an individual or group, selecting from among available alternatives, regards as desirable ends and means to these ends. The role of values and ethics in the decision-making process is rather explicit (Harrison, 1975).

Values and ethics pervade the entire process of choice. In addition, they are an integral part of the decision maker's belief system as they are reflected in his behavior in arriving at a choice and implementing it. Decisions are complex sets of value premises which involve choosing among these value premises.

Ground Rules for Ethical Decision Making. Pastin (1986) develops the following ground rules for ethical decision making.

> Once we have surveyed the alternatives in the decision making process we must visualize what each alternative offers. There are two parts to this strategy. One part is to see what we can say *factually* about each alternative. In order to reach a decision we need to form a view or visualize what is likely to happen if we adopt each alternative. We evaluate each of the projected outcomes. In assessing outcomes there are two major questions. How much value do I see in each outcome and what do I have to do to get the outcome I desire.

Ground Rules for Value

> Answers to the first question reveal one kind of ground rule, ground rules for value. In reaching a decision, we rate some outcomes as more desirable than others, as more valuable. Here the factual issues concerning alternatives are

important. But once we have considered the facts, what we count as desirable (valuable) and undesirable (not valuable) orders the options.

Ground Rules for Evaluation

The second question in the process is what do I have to do to get the desirable outcome? Answers to this question reveal the second type of ground rule, ground rules for evaluation. We often judge one option to be most desirable yet choose to pursue another because the first would have required us to do things we just do not do. The point is not that the option involves too much risks but that we prefer to avoid these issues related to assessing how desirable the option is. Rather there are kinds of actions that we will not take or do not want to take in order to obtain very desirable outcomes. We evaluate these actions as unacceptable using a set of principles, economic, social, and ethical.

At the heart of every decision and thus of every action resulting from a decision are ground rules expressing what we value and what we will do to get what we value. "To find your ground rules squeeze the factual assumptions out of your decisions, the ground rules are the residue that drives the decision." (Pastin, 1986) The ground rules for the individual are to be found in the value system and frame of reference of the individual, the code of ethics of the organization embedded in the culture of the organization.

3.11.2 Value Conflict, Culture, and Change

Conflicts of values occur at all levels in the decision-making process and can serve as a mechanism for change. The individual frequently must compromise his personal values as part of the continued membership in the group. Groups must in turn, in the event of disagreement, generally defer to the values of the organization. There are many situations in which a manager's personal values are at variance with the goals and objectives of the organization. Norm conformity is not the ideal strategy for innovation and risk-taking behavior espoused by the organization's culture, so some degree of variance may be entertained in the same way that organizational slack is promoted. The organizational culture does place constraints on individual values through its socialization techniques, but at the same time does not reward complete conformity. The necessary attainment of consensus at the group level and the maintenance of the unity of action and commonality of purpose at the organization level requires a pattern of accommodation within the ethics of the organization. The ultimate values of the individual become the final arbiter of uncompromising choices according to Harrison (1975).

Value conflicts are inevitable in the process of decision making in formal organizations. Such conflicts in organizational values may cause a decision maker to (1) rethink and reformulate aspirations and objectives, (2) review and reject marginal alternatives, (3) revise and rescale relevant alternatives, (4) reflect and possibly reconsider impending choices, and (5) resist implementation practices. At the macro level, value conflicts may cause a reexamination of the ultimate and instrumental values of the organization, a change in socialization techniques, the institutionalization of values clarification mechanisms, the linkage between values and ethical standards for decision making, a change in organization strategy, a change in leadership style, and a new mechanism for problems of internal integration. The very characteristics of culture that Schein (1985) noted—behavior patterns, norms, dominant values, philosophy, rules of the game, and climate in the event of strategic value conflicts in the organizational culture—may be changed. This can be functional for organizational renewal.

3.11.3 Hierarchy of Values

Although values exist at many levels, it is useful initially to focus on the personal value system of the decision making as it forms the basis of the value systems of groups, organizations, and the total society. The personal value systems of individuals who function as organizational decision makers have the following general qualities (Sankar, 1991):

1. They influence the perception of situations and problems.
2. They influence the entire process of choice.
3. They influence interpersonal relationships.
4. They influence the perception of individual and organizational achievement and definitions of success.
5. They influence the other elements of organizational culture, symbols, artifacts, meanings.
6. They set the limits to ethical behavior within the organizational culture.
7. They influence the priorities assigned to organizational goals.
8. They impact on attitudes, belief systems, and style of organizational leaders.
9. They facilitate the transmission of the symbolic facets of organizational culture.

All decisions imbued with personal and organizational values reflect the orientation of the culture of an organization. Value judgments arising out of the personal value system of the decision maker conditioned by the values implicit in organizational objectives are the rule rather than the exception in the process of choice. One finds more value judgments in nonroutine than in routine decisions; in nonrecurring than in recurring decisions; and in decisions where the outcome has a high degree of uncertainty. Still, the essential properties of values encompass decisions under all types of conditions and with all kinds of constraints. It is in the decision-making process that one finds the essentials of organizational culture.

In order to apply a structure to the general concept, it is useful to delineate a hierarchy of values. That is, values commence at the level of the individual; the values of several individuals comprise the values of a group; several sets of group values make up organizational values. Harrison (1975) cites the following authors.

Bernthal (1962) has set forth a hierarchy of values that place human values at the top. He states the case for ultimate human values as follows:

> Man's search for ultimate values is the story of life itself. Implied in every decision he makes is a judgement in terms of a priority of values, when goals come into conflict, and in terms of the ultimate decision criterion. In every business decision, the manager acts upon his own response to the basic philosophical question, "What is the nature of man?"

The ultimate goals of the individual are generalized as freedom, opportunity, self-realization, and human dignity. These values assume a respect for the rights of others, a concern over justice and order under law, and a responsibility for maintaining and perpetuating systems that make this freedom possible. Bernthal identifies the values at the level of the business organization as (1) profits, (2) survival, and (3) growth. He places such values at the bottom of his hierarchy, below the values of the individual, society, and the economic system. In essence, his hierarchy suggests that

> . . . responsible business management requires decisions and actions that contribute to the goals of the firm without violating higher goals or values . . . places human values above economic values . . . assumes that economic objectives can be achieved without denying the individual his basic rights as a human being.

Another way of viewing the values of individuals is within a basic framework of values. Leys (1962) cites six independent values that guide the individual as a decision maker:

1. Happiness, which is associated with desirable results and maximized satisfactions.
2. Lawfulness, which is related to precedents, customs, contracts, and authorizations.
3. Harmony, which is analogous with logical consistency, Platonic justice, order, plan, and the common good.
4. Survival, which is influenced by political power and friend–foe relations.
5. Integrity, which is attained through self-respect, rationality of the individual, and peace of mind.
6. Loyalty, which is related to institutional trends and social causes.

These independent values identified by Leys are analogous to the ultimate values espoused by Bernthal; that is, they are both guides to action. Values such as freedom, opportunity, dignity, self-realization, harmony, integrity, loyalty, survival, and lawfulness can be incorporated into the culture of the organization as its value system. Such a value system represents the pulse of the organization, guidelines that it can articulate in its code of values.

Matsushita's Code of Values. Matsushita was the first company in Japan to have a code of values. It foresaw that a lifetime organizational experience (culture) shapes one's character indelibly. Work should not be denied its powerful role in shaping one's character. Matsushita provides two distinct kinds of training. One is basic skills training, but the second and more fundamental one is training in the Matsushita values.

The basic principles, beliefs, and values of the firm are as follows:

The Seven Spiritual Values

1. National service through industry.
2. Fairness.
3. Harmony and cooperation.
4. Struggle for betterment.
5. Courtesy and humility.
6. Adjustment and assimilation.
7. Gratitude.

These values provide a spiritual fabric of great resilience (Pascale and Athos, 1981). They foster consistent expectations among employees in a work force that reaches from continent to continent. They permit a highly complex and decentralized firm to evoke an enormous continuity that sustains it even when more operational guidance breaks down. This continuity is one of the major functions of organizational culture.

Pascale and Athos (1981) make the following observations regarding Matsushita and American companies:

> And when we compare Matsushita with American firms of the same age—firms that were born in the 1920's such as General Motors, American Telegraph and Telephone, Westinghouse and RCA—it is difficult to find many that have sustained their original vitality. Any inquiry into how Matsushita has sustained itself while so many others have fallen behind must surely turn to its value system as a central ingredient in its success.
>
> Perhaps the ultimate triumph of Matsushita is the balancing of the rationalism of the West with the spiritualism of the East.

As a general proposition, Matsushita has proved incredibly adept at perpetuating its aggressive strategy and corporate values for a period of over 60 years. Part of this arises from the strong imprint of the Matsushita style. Part of it derives from the firm's spiritual fabric which provides a kind of secular "religious system" that keeps the organization's members believing in the same things and marching in the same direction. Values are meticulously imparted to each successive generation of management, conclude Pascale and Athos.

3.11.4 Culture, Values, and Change

The adaptive organization seeks strong ethics and weak culture. A strong culture puts basic beliefs, attitudes, and ways of doing things beyond question (Pastin, 1986). But unquestionables must be questioned for an organization to be quick on its feet, strategic, and just plain smart. Because cultures are rooted in tradition, they reflect what has worked, not what will work. A bureaucratic culture is more preoccupied with precedents, rules, regulations, and operating procedures rather than ethical standards in decision making. If culture is an information system, then such rules, precedents, and procedures may contribute to information distortion, omission, filtration in the decision-making process, and disinformation.

Cultures are, by consensus, hard to change. The stronger the culture, the harder it is to change. And most established organizations have strong, if not attractive or "appropriate" cultures. The complaint about organizational culture is that it changes too slowly. But the complaint about ethics is that it is changing too fast: "What happened to the work ethic of integrity, honesty, and hard work?" (Pastin, 1986). This is puzzling, since the ethics of an organization compromises the basic ground rules by which it operates. What is the relation of ethics to culture? Are there strong-ethics, weak-culture companies? There are, and these companies have the best prospects for sustained success, unimpeded by tragedy (Pastin, 1986).

Ethics is closer to the surface of culture than any other components (myths, ideologies, and aesthetics) according to Pastin (1986). When an organization enforces its ethics, its principles come to the surface. Ethics has another role. Ethics is the forum in which societies, groups, and organizations argue fundamental changes in their ground rules. Criticism is as much part of ethics as enforcement is. For a culture to persist and serve those who work in it, the culture must learn. It must allow challenges to its basic principles in a setting that tolerates some change without threatening to undo the culture. Thus, culture promotes discussion of issues in ethical terms and mandates that these discussions are serious. This is functional in that it provides an opportunity for a culture to change while keeping intact what is valuable in the culture (Pastin, 1986).

Many of the effects of organizational culture are summarized by four key ideas put forth by Hellrigel et al. (1986). First, knowing the culture of an organization allows employees to understand the firm's history and current approach, which, in turn, provide guidance about expected behaviors in the future. Second, organizational culture can serve to establish commitment to corporate philosophy and values. This provides organizational members with shared feelings of working toward goals they value and believe in. Third, organizational culture with its related norms serves as a control mechanism to channel employee behaviors toward desired and away from undesired behavior patterns. Finally, certain kinds of organizational cultures, such as participative cultures, may be related to greater effectiveness and productivity. With specific reference to values and ethics, organizational culture can become more dynamic by periodic reappraisal of its value system and the ethical premises of organizational actions. Excellent companies are not only those that achieve their goals but those that conduct their business within an ethical framework. The ethical imperative is as crucial as the profit motive, or the bottom line. Excellence indicates that the rational decision-making procedures of the organization can and should be subject to moral evaluation. The character of an organization, like that of an individual, has as its spiritual fabric its value system, as in the case of Matsushita. The most dynamic aspects of organizational and societal cultures are not the rituals, rites, myths, customs, symbols, and heroes, but values and ethics. The code of values of an effective organizational culture is reflected in its strategy, mission, and philosophy. With reference to internal integration, values are the soul of the organization that promotes harmony, excellence, and authenticity for a cohesive work group and a culture with élan.

3.11.5 Principles for High-Ethics Firms

Pastin (1986) advances a set of principles for high-ethics, high-profit organizations. To have ethics is to have a set of ground rules, to which we have already alluded. These ground rules constitute the set of ethical principles—based on research. The research included a three-year international study of 25 firms recognized for both their economic and ethical performance. The firms studied included Cadbury Schweppes, 3M, Acro, Motorola, Hilby Wilson, Northern Chemical, Apple Computer, Interwestern Management, and many other. The objective was to discover what connections, if any, there are between high ethical performance and high economic performance. The principles are as follows (Pastin, 1986):

Principle 1: High-ethics firms are at ease interacting with diverse internal and external stakeholder groups. The ground rule of these firms make the good of these stakeholder groups part of the firm's own good.

Principle 2: High-ethics firms are obsessed with fairness. Their ground rules emphasize that the other person's interests cound as much as their own.

Principle 3: In high-ethics firms, responsibility is individual rather than collective, with individuals assuming personal responsibility for actions of the firm. These firms' ground rules mandate that individuals are responsible to themselves.

Principle 4: The high-ethics firms sees its activities in terms of a purpose. This purpose is a way of operating that members of the firm value. And purpose ties the firm to its environment.

The ethical firm operates on ground rules that deal fairly with diverse constituencies, promote individual responsibility, and enact a purpose within the culture of the organization. The mark of the ethical firm is that individuals enact the ethical standards and value system in its decision-making process. In this firm, ethical conduct is natural and needs no support from codes, slogans, and phony ceremonies. The agenda for action calls for the thinking manager to look at his own organization to ask whether its ground rules enact these principles.

3.12 ETHICS OF CULTURAL CHANGE

An important ethical question that managers engaged in culture change may confront is: Do managers have any business trying to change people's beliefs and values?

The following points may be made in addressing this issue, according to Schein (1985). First, questions about changing people's beliefs and values are laden with emotion because they connote "brainwashing." Second, such questions are personally threatening because a lot is at stake for the individual. One's beliefs and values are not a random assembly; changes in one or more of them require changes in related others. Such a reorganization is stressful and frequently resisted because of the learning of new skills and behaviors implied. Finally, it is important to note that we are talking here about organization-related beliefs and values, not such private beliefs and values as religious and political ones. The problem is that the two sets are interrelated. Changes in one set most likely affect the other; more theoretical and empirical work is needed to better understand the interrelationship between people's organization-related beliefs and values and their private ones.

It may be argued that just as it is in the nature of the manager's job to influence organizational behavior in a responsible and professional manner, so it is his job to conscientiously shape organizational beliefs and values as appropriate. Since ethics may be defined as the quest for truth and as the science of human values, it is essential that managers, in attempting cultural change, be ethical in their decisions in reordering, modifying, and changing existing values and beliefs and adding nonexisting beliefs and values.

At another level, Schein (1985) points to some ethical problems in culture change. In making strategic choices based on assessment of an organization's culture, these kinds of risks must be assessed:

1. *The analysis of the culture could be incorrect*. The analysis of what the basic assumptions are and how they fit into a pattern and paradigm may be wrong, and, if so, may give the decision maker incorrect data on which to base decisions. If decisions are made on the basis of incorrect assumptions about the culture, serious harm could be done to the organization.

2. *The organization might not be ready to receive feedback about its culture*. The analysis may be correct, but the insiders may not be prepared to digest what

has been learned about them. If culture is like character and functions in part as a set of defense mechanisms to help avoid anxiety and to provide positive direction, self-esteem, and pride, then various conditions may make an organization reluctant to accept the cultural truth about itself. If culture is to the organization what character is to the individual, then insight into one's own culture may remove defenses that had been operating and on which the organization had been relying. To study a culture and reveal that culture to insiders, then, can be likened to an "invasion of privacy," which under many conditions is not welcome. The consequences of diagnosing an organization's culture in the management of technological change must therefore be articulated by the cultural analyst.

3. *The organization could be made more vulnerable through having its culture revealed to outsiders.* If a correct analysis of the culture of an organization becomes known to outsiders, because it either is published or discussed among interested parties, the organization may become vulnerable or put at a disadvantage, since data that would ordinarily remain private now may become public. If the information is inaccurate potential employees, customers, suppliers and any other categories of outsiders may be adversely influenced.

Bureaucratic, entrepreneurial, and participative cultures must place a high value on ethics. In ethical organizations, right is distinguished from wrong, and honesty and fair play are valued. Amoral organizational cultures value success, no matter how it is achieved (Linder, 1985).

Executives set the examples, whatever the organization. They set the tone, create the spirit, and decide the values for an organization and for the people in it. They lead, or mislead.

These concepts can be used in managing corporate cultures through a period of strategic change. Where values, assumptions, norms, and belief systems are affected by the magnitude of the technological change, then ethics, the science of human values and the quest for truth, emerges as a vehicle for monitoring the effects of such a change.

3.13 MERGING CORPORATE CULTURES

Bennigson (1985) reports that companies were acquired, wholly or in part, at the frantic rate of 11 per day in the United States. With record dollar values at stake, the future performance of these mergers is a widely debated topic. A large number of these mergers will fail, many of them not because of financial or operational difficulties, but cultural clash. Since culture has been seen to limit and determine strategy, a cultural mismatch can be as severe a problem as a financial, product, or market mismatch. Many merger failures indicate the lack of insight the company had into its own culture. After, when the management find themselves face-to-face, it is often discovered that their companies' value systems are in conflict.

Corporate cultures can be merged, but it takes a lot of time and effort.

Bennigson (1985) outlines six steps which are crucial during the difficult period following a merger or acquisition:

1. *Build Knowledge.* A knowledge base is required to judge the feasibility of a "cultural fit." The insight gained here will be valuable throughout the remaining steps.
2. *Develop a Shared Vision.* The business strategy and its reasoning should be made clear so that managers do not have to imply it.
3. *Determine the Desired Changes in Beliefs.* Beliefs which will create barriers to change must be identified so that alterations can be made. Communicate these new values throughout the organization.
4. *Translate Values into Concrete Behavior.* Change must begin at the top, the CEO must behaviorally display the desired changes.
5. *Reorient Power to Support New Values and Behavior.* Power should be shifted to those people whose skills, values, or functions are critical to the new cultural direction.
6. *Harness High-impact Management Systems.* Commitment to the merger can be demonstrated by utilizing high-impact management systems to direct employee behavior.

These steps are similar to those that would be taken for any major organizational change. Handling the cultural merger strategically will certainly improve the odds of success. Hellrigel et al. comment as follows.

If innovation, growth, and revitalization are mandatory in a world characterized by rapid change and worldwide competition, organizations must find a way to breathe new life into their procedures, management styles, and cultures. Otherwise, the large, established firms in the United States will continue to lose ground to the more vibrant smaller companies in this country and to the newer firms in other nations. At the same time, if the large, established companies attempt to foster innovation by acquiring smaller, more dynamic firms, ways must be found to mesh the two merging corporations so that innovation is realized. If not, the large firms will stifle the innovation of the acquired companies just as they have done repeatedly in themselves (Hellrigel et al., 1986).

For both innovation and corporate mergers to succeed, corporate culture —the missing link to moving forward to today's world—must be managed, and all other strategies for managing complex organizations must be implemented as well (Hellrigel et al., 1986).

Managing corporate cultures is now possible. If our understanding of this important area continues to improve, we will be able to develop new methods that increase our control of corporate cultures, instead of vice versa. We should refrain from latching on to the next panacea and be careful not to overlook what we have already learned because it is called by a new name. If the current interest in corporate culture is taken seriously and lasts, we will not be reading about the demise of efforts toward innovation and revitalization. Rather, we will see that an integrated approach to managing organizations is possible, with people—and culture—at center stage (Hellrigel et al., 1986).

3.14 THE ETHICS OF TECHNOLOGICAL CHANGE AND HUMAN VALUES: MESTHENE

Values play a critical role in shaping the course of society and the decisions of individuals. Questions of values become more pointed and insistent in a society that organizes itself to control technology and that engages in deliberate social planning, according to Mesthene (1977), who makes the following observations on values:

> Planning demands explicit recognition of value hierarchies and often brings into open value conflicts. In economic planning, for example, we have to make choices between the values of leisure and increased productivity, without a common measure ot help us choose. In planning education, we have the value dilemma of equality versus achievement. The new science-based decision making techniques also call for clarity: in the specification of goals, thus serving to make value preferences explicit. The effectiveness of systems analysis, for example, depends on having explicitly stated objectives and criteria for evaluation to begin with, and the criteria and objectives of specific actions invariably relate to society's system of values. The increased awareness of conflicts among our values contributes to a general questioning attitude toward traditional values that appears to be endemic to a high technology, knowledge based society—a society in which the store of knowledge concerning the consequences of action is large and is rapidly increasing—a society in which received norms and their justifying values will be increasingly subjected to questioning and reformulation. The increased questioning and reformulation of values, coupled with a growing awareness that our values are in fact changing under the impact of technological change, lead many people to believe that technology is by nature destructive of values. But this belief presupposes a conception of values as eternal and unchanging and therefore tends to confuse the valuable with the stable. The fact that values come into question as our knowledge increases and that some traditional values cease to function adequately when technology leads to changes in social conditions does not mean that values per se are being destroyed by knowledge and technology.
>
> Technology has a direct impact on values by virtue of its capacity for creating new opportunities. By making possible what was not possible before, it offers individuals and society new options to choose from. For example, space technology makes it possible for the first time to go to the moon or to communicate by satellite and thereby adds those two new options to the spectrum of choices available to society. By adding new options in this way, technology can lead to changes in values. Specifically, technology can lead to value changes either (1)

by bringing some previously unattainable goal within the realm of choice or (2) by making some values easier than heretofore, that is, by changing the costs associated with them.

When technology facilitates implementation of some social ideal and value and society fails to act upon this new possibility, the conflict between principle and practice is sharpened and creates new tensions and ethical dilemmas. For example, the economic affluence that technology has helped to bring to American society makes possible fuller implementation of the traditional values of social and economic equality. Until it is acted upon, that possibility gives rise to tensions we associate with the rising expectations of the underprivileged and provokes activist response of the radical left for change and social justice.

In the case where technological change alters the relative costs of implementing different values, it impinges on inherent contradictions in our value system. Modern technology can enhance the value we associate with democracy via computer technology and television voting on issues. But it can also enhance another value—that of secular rationality by facilitating the use of scientific and technical expertise in the process of political decision making. This can in turn further reduce citizen participation in the democratic process. Technology thus has the effect of facing us with contradictions in our value system and of calling for deliberate attention to their resolution.

3.15 THE TECHNOLOGICAL SOCIETY: ITS VALUE ORIENTATIONS: FROMM (1968)

The technological system is programmed by two principles that direct the thoughts and efforts of everyone working in it. The first principle is the maxim that something ought to be done because it is technically possible to do it. If it is possible to build nuclear weapons, they must be built even if they might destroy us all. It is possible to travel to the moon or to the planets, it must be done, even if it is at the expense of many unfulfilled needs here on earth. This principle means the negation of all values which the humanist tradition has developed. This tradition said that something should be done because it is needed for man, for his growth, joy, and reason, because it is beautiful, good, or true. Once the principle is accepted that something ought to be done because it is technically possible to do it, all other values are dethroned, and technological development becomes the foundation of ethics.

The second principle is that of maximal efficiency and output. The requirement of maximal efficiency leads as a consequence to the requirement of minimal individuality. The social machine works more efficiently, so it is believed, if individuals are cut down to purely quantifiable units whose personalities can be expressed on punched cards. These units can be administered more easily by bureaucratic rules because they do not make trouble or create friction. In order to reach this result, men must be de-individualized and taught to find their identity in the corporation rather than in themselves.

The question of economic efficiency requires careful thought. The issue of being economically efficient, that is to say, using the smallest possible amount of resources to obtain maximal effect, should be placed in a historical and evolutionary context. The question is obviously more important in a society where real material scarcity is the prime fact of life, and its importance diminishes as the productive powers of society advance.

A second line of investigation should be a full consideration of the fact that efficiency is only a known element in already existing activities. Since we do not know much about the efficiency or inefficiency of untried approaches, one must be careful in pleading for things as they are on the grounds of efficiency. Furthermore, one must be very careful to think through and specify the area and time period being examined. What may appear efficient by a narrow definition can be highly inefficient if the time and scope of the discussions are broadened. In economics there is increasing awareness of what are called "neighborhood effects"; that is, effects that go beyond the immediate activity and are often neglected in considering benefits and costs. One example would be evaluating the efficiency of a particular industrial project only in terms of the immediate effects on this enterprise—forgetting, for instance, that waste materials deposited in nearby streams and the air represents a costly and a serious inefficiency that take account of time and society's interest as a whole. Eventually, the human element needs to be taken into account as a basic factor in the system whose efficiency we try to examine. Dehumanization in the name of efficiency is an all-too-common occurrence.

Efficiency is desirable in any kind of purposeful activity. But it should be examined in terms of the larger systems, of which the system under study is only a part; it should take account of the human factor within the system. Eventually efficiency as such should not be a dominant norm in any kind of enterprise.

The tendency to install technical progress as the highest value is linked up not only with our overemphasis on intellect but, most importantly, with a deep emotional attraction to the mechanical, to all that is not alive, to all that is man-made. This attraction to the non-alive leads in its less drastic form to indifference toward life instead of "reverence for life." Those who are attracted to the non-alive are the people who prefer "law and order" to living structure, bureaucratic to spontaneous methods, gadgets to living beings, repetition to originality, neatness to exuberance, hoarding to spending.

3.16 INDUSTRIAL ENGINEERING ETHICS: A NEW MODEL

The scientific management principles, as noted earlier, focused on the job as the primary unit of analysis and resulted in (1) mechanical pacing, (2) repetitiveness, (3) low skill requirements, (4) concentration on only a fraction of the product, (5) limited social interaction, and (6) predetermination of tools and techniques. These factors led to boredom, stress, and alienation of work. The individual was deprived of opportunities to develop his/her self esteem, self actualization, human dignity, potential, and so on. The individual was a means to the major industrial value of efficiency.

The Socio Technical Systems Design calls for a joint optimization of both technical and human characteristics and attributes. This joint optimization in

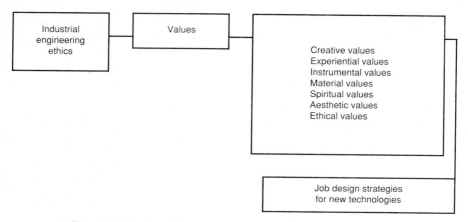

Figure 3.4. Industrial engineering ethics, values, and job design.

job design for new technologies is an ethical imperative and a human right, as noted by Trist. This human right finds expression in the advocacy of Quality of Work Life (QWL) programs.

The QWL programs invariably focus on elements of the psychological contract that organizations must initiate with their workers. The elements of the psychological contract are based on the right of individuals to expect that working conditions will facilitate their growth, continuous learning, social interaction, identity, integrity, self esteem, and human dignity.

Industrial engineering must, for the information age, design a value-based model for job design for new technologies. Such a model will view work as a medium through which workers affirm their values. There is meaning in our work and life to the extent that we actualize these values. Through the medium of work we realize the whole spectrum of human values. We realize our creative, experiential, instrumental, social, spiritual, material, ethical, and aesthetic values (Figure 3.4). It is, therefore, an ethical imperative for job designers to focus on these values in the implementation of new technologies.

Creative Values and Work

Creativity, according to Gardner (1965), is possible in most forms of human activity. In some activities the possibility is greatly limited by the nature of the task. The highest levels can be expected where performance is not severely constricted by the nature of the task to be accomplished and in those lines of endeavor that involves man's emotions, judgement, symbolizing powers, aesthetic perceptions and spiritual impulses.

Creative values at work encourage (1) dynamic learning, (2) creative visualization, (3) creative problem solving, (4) exercise of initiative, (5) variety in learning styles, (6) commitment to excellence, (7) perceptual restructuring, (8) new modes of thinking, (9) change, and (10) a sense of achievement.

By changing patterns of task variety, identity, feedback, autonomy, goal setting, modes of participation, group culture symbols, and communications rituals, it is possible to enhance creative values at work. Vertical job loading and horizontal job loading can be used as specific job design strategies for enhancing creative values.

Experiential Values and Work

The ways in which one experiences the act of working suggest that particular values color these perceptions at work. Experiential values involve meaningful, interesting and purposeful experiences at work; interactions that provide a way of achieving recognition, friendship, happiness, positive role perceptions, high morale and a sense of identity with workgroup norms and sentiments that promote trust and harmony.

Experiential values at work encourage (1) group cohesiveness (2) an open climate, (3) candid communications, (4) a helping relationship, (5) performance feedback, (6) the art of listening, (7) trust, (8) a positive attitude towards one's co-workers, (9) the expression of emotions and sentiments, and (10) the growth of self esteem.

The only meaningful way of life is activity in the world, not activity in general, but the activity of giving and caring as we experience the world of work. Fromm (1976) further notes that we as human beings have an inherent and deeply rooted desire to be: to express our faculties, to be active, to be related to others, to escape the prison cell of selfishness. The experience of sharing makes and keeps the relation between people alive. We find manifestation of the will to give in people who genuinely love their work.

Instrumental Values and Work

Instrumental values involve the attitudinal set one has towards one's job. Values are discernible in the attitudes workers have towards their job environment. In any work setting there develops a more or less stable set of attitudes toward what is possible. These attitudes invariably set ceilings for performance. In work settings where workers are considered as replaceable cogs in the machine, commitment to the organization and its goals will be low. On the other hand, organizations that have cultures that emphasize trust, empathy and a dynamic of caring will produce work attitudes that are instrumental in increasing meaning, motivation, and job satisfaction. Where growth and creativity are concerned, an attitude of optimism, renewal, and change is imperative.

To implement instrumental values at work, it is necessary to (1) conduct attitude surveys, (2) identify positive cluster of work attitudes, (3) target attitudinal set instrumental in enhancing commitment, (4) redesign task elements, (5) design attitude change strategies, (6) change work attitudes, (7) conduct culture audit re pattern of basic assumptions, communications rituals, supervisory style and organizational symbols, (8) target values that drive attitudes, and (9) use positive reinforcement schedules to change attitudes.

Material Values and Work

Pay for performance is the most obvious aspect of material values. The importance that people place on the artifacts of power, status, and prestige is also an indicator of material values. It may be carried to extreme where one's self esteem and one's worth are determined by these artifacts paraded in the organization's culture. The ego enhancing aspects of work is also associated with the exercise of power.

The individual who is motivated primarily by material values may perceive himself as a commodity in the market place guided by competing market values and financial incentives, which have the potential to neutralize the intrinsic values of work; the financial contract supplants the psychological contract. However, when employees are basically frustrated in their job and feel no sense of achievement or recognition, no amount of peripheral benefits or external palliatives will improve their attitudes at work.

To implement material values at work, it is necessary to (1) offer a variety of incentives, (2) link incentives to performance levels, (3) design an inventory of status symbols, (4) map the power assets in the organization, (5) target power and influence patterns to achieve job goals, (6) design a code of ethics for use of power and incentives, (7) use a variety of job titles, (8) use merit awards (9) use job rotation, (10) cultivate a sense of achievement and commitment, and (11) monitor equity in payment systems.

Spiritual Values and Work

Kahlil Gibran in his classic, *The Prophet,* dramatizes the spiritual values associated with work.

> Then a ploughman said, speak to Us of Work, And he answered, saying:
>
> You work that you may keep pace with the earth and the soul of the earth;
>
> For to be idle is to become a stranger unto the seasons, and to step out of life's procession that marches in majesty and proud submission towards the infinite.
>
> When you work you are a flute through whose heart the whispering of the hours turn to music.
>
> Which of you would be a reed, dumb and silent, when all else sings in unison?
>
> You have been told that life is darkness, and in your weariness you echo what was said by the weary.
>
> And I say that life is indeed darkness save when there is urge,
>
> And all urge is blind save when there is knowledge.
>
> And all knowledge is vain save when there is work.
>
> And all work is empty save when there is love;
>
> And when you work with love, you bind yourself to yourself, and to one another, and to God.
>
> Work is love made visible.

Work can provide the opportunity to test one's moral vision and imagination as well as one's ideals. Morality is designed to shape one's character. It is character, the concretization of morality, which determines one's moral vision and growth at work. The work ethic is generally expressed through one's character. Integrity is the measure of one's character. Hope, faith, devotion, equity, love, justice, truth, duty, and service are all associated with the work ethic and with ethical excellence. Pathological values such as greed, envy, conceit, arrogance, and malice are also expressed through the medium of work. These values lead to the disintegration of human personality at work and have the potential to create stress and alienation at work. On the other hand, a work culture infused with spiritual values promotes a dynamic of caring, trust, justice, love, peace and service, as well as an integrated personality.

To implement spiritual values at work, it is necessary to (1) promote trust in all interactions, (2) instill an ideal of service via work, (3) implement a code of ethics (4) re-orient the values of aggression, competition, and selfishness towards a helping relationship, (5) instill the use of moral incentives at work, (6) focus on the intrinsic values of work and the love of work (7) change work attitudes to enhance one's identity, (8) link work to the psychological contract, (9) promote work as a means to achieve growth, potential, and excellence, and (10) detach one's actions from their outcomes.

Aesthetic Values and Work

This set of values involves the pride of craftmanship, beauty, finesse, precision, logic, and harmony expressed through the medium of one's work. Aesthetic values are associated with intrinsic rewards, a performance that satisfies a need within the individual; it is the desire to please oneself, the satisfaction that comes from a job well done, namely, the pride in performance. Work can also satisfy a number of emotional needs. There are also many opportunities to find an aesthetic pleasure in our products, a sense of pride in design and execution.

To the extent that the worker uses powers of the mind such as insight, imagination, intuition, and intelligence, and aesthetics of the mind, the greater the identification with the work and pride in the performance. To the extent that work permits creative visualization, problem solving, and learning, the greater the challenge in the work and pride in performance. The spectrum of human values, creative, experiential, instrumental, spiritual, material, and ethical all enhance the perception of the aesthetics of work and the meaning of work. Meaning is positively correlated with the aesthetics of work. The corporate culture can also be used with its symbols, rituals, myths, and artifacts to enhance the aesthetic dimensions of work.

Ethical Values and Work

Organizations should have a code of ethics for their workers. Ethics is defined as the science of human values and as the quest for truth. It is in the decision-making process that one's moral vision and imagination are tested.

Ethics has been called the standards for decision making. What constitutes a "good" decision is a major concern of ethics. The system of values is important in evaluating the rationality of the process of choice. It is difficult to conceive of a rational approach to decision making in the absence of some system of values.

At work it is necessary to focus on: (1) the integrity of managerial action, (2) the honesty of intention, (3) justice and human rights, (4) fairness in allocating organizational rewards, (5) perceived equity in rewards, (6) the dynamics of caring, (7) the goodness of means and ends, (8) the duty of the manager and change agent, (9) the truth of management policies, (10) corporate social responsibility, and (11) harmony in the workplace.

The Value Audit and Job Design

The cluster of values can be used to produce a Value Audit of one's job. This is essentially a diagnosis of one's job in term of the seven value orientations. The diagnosis is based on three operational questions, e.g. (1) how much creative value, that is, opportunity for learning, problem solving, exercise of initiative and challenge *is there in your job?* (2) How much opportunity for these creative elements *should there be?* (3) *How important is this* opportunity for creative problem solving, etc. in your job? The discrepancy between the actual and the ideal can be adjusted if the value assigned to this variable is high. Some people will have a low value index for creative values, some high index for spiritual values, some high index for experiential values, some low for aesthetic values, others high for material values and so on. The value index is moderated by the level of education, growth need strength, locus of control, self esteem and so on. Not all workers will derive meaning from their work if all values were changed or enhanced; it may be a change in one value or combinations of value orientations.

Since the psychological contract at work is based on these value orientations, the job designer must consider it an ethical imperative to ensure that they are expressed in the job design. Strategies to enhance these values become critical in the implementation of new technologies. The socio technical systems design with its premise of joint optimization cannot be divorced from the value orientations of the workers and their search for meaning via these values at work.

REFERENCES

Bennigson, L., Managing Corporate Cultures Management Review, 74 (1985).

Bernthal, W., "Value Perspectives in Management Decisions," *Academy of Management Journal* (1962).

Business Week, Crisis in Middle Management (January, 1983).

Child, J., *Organizations,* Harper & Row, New York, 1984.

Daft, R., *Organization Theory and Design,* West Publishing, Boston, 1986.

Deal, T. and Kennedy A. Corporate Cultures: The Rites and Rituals of Corporate Life, Addison Wesley, Reading, Mass, 1982.

Fromm, E., *The Revolution of Hope: Toward a Humanized Technology,* Harper & Row, New York, 1968.

Fromm, E., *To Have or To Be,* Harper & Row, New York, 1976.

Harrison, F., *The Managerial Decision Making Process,* Houghton Mifflin, Boston, 1975.

Gardner, S., *Self Renewal,* Harper & Row, New York, 1976.

Hellrigel, D., J. Slocum, and R. Woodman, *Organizational Behavior,* West Publishing, New York, 1986.

Kilmann, R., "Corporate Culture," *Psychology Today,* April (1985).

Kilmann, R., R. Saxton, M. Serpa and Associates *Gaining Control of the Corporate Culture,* Jossey-Bass, San Francisco, 1985.

Leys, W., "The Value Framework of Decision Making," in S. Mailick and E. Van Ness (Eds.), *Concepts and Issues in Administrative Behavior,* Prentice Hall, Englewood Cliffs, N.J., 1962.

Linder, J., "Computers, Corporate Culture, and Change," *The Personnel Journal,* September (1985).

Lorsch, J., "Managing Culture: The Invisible Barrier to Strategic Change," *California Management Review,* (1986).

Lund, R. T., and J. A. Hansen, *Keeping America at Work: Strategies for Employing the New Technologies,* Wiley, New York, 1986.

Mesthene, E., "The Role of Technology in Society," in A. Teich (Ed.), *Technology and Man's Future,* St. Martin's Press, New York, 1977.

Ouchi, W. G., *Theory Z,* Avon Books, New York, 1981.

Pascale, R., and A. Athos, *The Art of Japanese Management,* Warner Books, New York, 1981.

Pastin, M., *The Hard Problems of Management: Gaining the Ethics Edge,* Jossey-Bass, San Francisco, 1986.

Peters, J., "Symbols, Patterns and Settings: An Optimistic Case for Getting Things Done," *Organizational Dynamics,* **7** (1978).

Peters, T., and R. Waterman, *In Search of Excellence,* Harper & Row, New York, 1982.

Power, R., *"Corporate Culture"* Conference Board Bulletin, Ottawa Canada, 1986.

Sankar, Y., "Management as the Symbolization of Values," *International Conference on Organizational Culture and Symbolism,* Milan, Italy, 1987.

Sankar, Y., "Corporate Culture, Values and Ethics," *International Journal of Value Based Management* (1988a).

Sankar, Y., "New Technologies and Corporate Culture," *Journal of Systems Management,* (1988b).

Sankar, Y., *Corporate Culture in Organizational Behavior,* Harcourt Brace Jovanovich, Orlando, Fla., 1991.

Sathe, V., *Culture and Related Corporate Realities,* Irwin, Homewood, Ill., 1985.

Schein, E., *Organizational Culture and Leadership*, Jossey-Bass, San Francisco, 1985.

Smircich, L., "Concepts of Culture and Organizational Analysis," *Administrative Science Quarterly*, **28**(3), 1983.

Stonich, P., *Implementing Strategy: Making Strategy Happen*, Bellinger, Cambridge, Mass., 1982.

Toffler, A., *The Adaptive Corporation*, McGraw-Hill, New York, 1985.

Wilkins, A., "The Cultural Audit: A Tool for Understanding Organizations," *Organizational Dynamics*, **12**(2) (1983).

4
A SYSTEMS APPROACH FOR MANAGING TECHNOLOGICAL CHANGE

4.1 ADVANCE ORGANIZER

To cope effectively with technological change, organizations must be designed as open systems and must behave as open systems. The systems paradigm provides a set of conceptual tools for the design of an adaptive corporation. The systems approach affords a method of planning and managing technological change (Figure 4.1).

One approach, which uses the elements of the systems paradigm, is the sociotechnical systems (STS) design. Because STS management focuses on the work processes, organizational culture, job design, environmental domain, subsystems interdependence, feedback loops, etc., rather than on the parts of the system, it is a systems approach. The STS approach in managing technological change is noted by many researchers, including Daft (1989) and Taylor et al. (1986), as quite an effective approach.

The general systems approach and the characteristics of an organizational system are described. These can be used to manage or design the system as an open system in a state of dynamic equilibrium with its environment.

From a systems perspective, the argument is made that an organization must adapt to its environments if it is to remain viable. The greater the degree of change, complexity, and uncertainty, the greater the degree of differentiation of the organization. This degree of differentiation creates a need for integration of the organization's resource, work, and information flow. It also inhibits the rate of diffusion of innovation in the system. With reference to major systems concepts, it is argued here that as organizational systems emphasize control, integration, and regulation of the operations, the amount of equifinality will be reduced because of the tendency to limit the search for

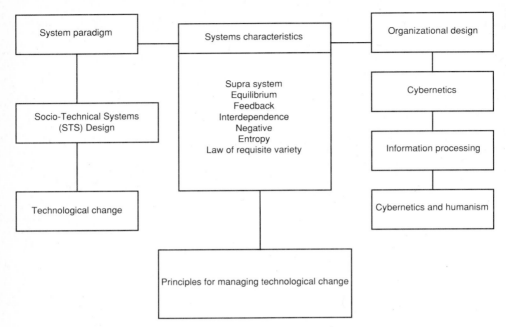

Figure 4.1. A systems approach for managing technological change.

options and innovative solutions and to codify the organization's information systems. Technological change sometimes decreases equifinality within the system through the process of job simplification and standardization of procedures, programs, and repertoire. With regard to entropy, the argument is made that organizations fostering creativity, innovation, and dynamism will be high in negative entropy. The ability of organizations to arrest the entropic process is one of the determinants of their viability and effectiveness. Another systems concept is requisite variety, which states that the complexity in one system is required to control complexity in another. Therefore, a change in the degree of complexity of the organization's environment will be reflected in the degree of vertical and horizontal differentiation of the organization and the degree of complexity of information systems, decision support systems, and data bases (both integrated and distributed data bases). This is another major principle for organizational change. The more complex the environment and the technology, the greater the degree of vertical and horizontal differentiation of the organization. Another design principle holds that equilibrium is not to be identified with a fixed structure, but must also focus on the capacity of the system to elaborate the structure through expansion and growth of its information-processing capabilities. Thus, an organization can adjust its functioning to achieve stability and a variable rate of change at the same time as it adjusts to technological change. Organizations maintain their equilibrium through feedback process. Feedback mechanisms permit various parts of the organization and work flow to adjust to other parts and segments of work flow. Feedback also facilitates learning and growth. Technological change may require new feedback loops within the work flow and the organizational system. A change in interdependence among the parts of a system or segments of the work flow will produce changes in the design of various coordination modes such as rules, procedures, plans, mutual adjustment, and so on. Technological change invariably alters the type of interdependence (pooled, sequential, reciprocal and team) and coordination modes.

Next, we discuss the systems approach as it relates to cybernetics and its use in designing the organization as an information-processing system. Some cybernetic principles for managing technological change are reviewed. We conclude with a section on cybernetics and humanism which calls for the need for a strategic focus on the individual as the center of the system and not merely as a systems constraint. To view the person as a systems constraint is equivalent to devaluing his centrality in information processing as the agent and beneficiary of change.

It seems established that when technological change is considerable, some effects on the social system must be recognized and planned for, but the question of coordination of change in these two systems is still unanswerable (Taylor, 1973). The systems approach provides one method of planning and managing technological change. Katz and Kahn (1978) view systemic change as the most powerful approach to changing organizations. Systemic change involves changed inputs from the environment, such as technology, which create internal strain and imbalance among the subunits of the system. Because of the interdependence among these subunits a change in one will produce changes in others. Managing interdependence among the subunits is one strategy in planning technological change. This strategy is generally called the sociotechnical systems (STS) approach.

The sociotechnical approach implicitly recognizes the role of the external environment of the enterprise in creating the need for internal change and in setting limits to the amount of choice available to the organization for adapting successfully.

In the words of a manager who has become famous for practicing the sociotechnical tradition (Miles, 1980):

> . . . systemic interdependence implies, among other things, that managerial policies and programs of limited scope—i.e., relevant to only one part of the organization or of its work environment—are no longer good enough. Policies and programs must possess the quality of a coherent whole and must refer to the whole of the organizational system and its related technical system as well as its environment. Fads and fancies, tricks and gimmicks, become more suspect than ever, however well clothed they may be in their own limited and often elegant logic. Management training programs have often been ineffective, not because they were "bad" training programs, but because they were not integrated with other aspects of the total organizational system. Computerized

operations control, programmed product innovations, "improved" cost accounting systems, to name some respectable examples, gain their merit and utility in large part from the coherence of the sociotechnical system of which they become a part. They fail, usually, when introduced with insufficient regard for the establishment of a wide network of new and modified system interdependencies of kinds necessary for the innovation to have a "place" in the whole.

4.2 CHARACTERISTICS OF THE SOCIOTECHNICAL SYSTEMS THEORY

Significantly, sociotechnical systems theory provides a basis for analysis and design overcoming the greatest inhibition to development of organization and job strategies in a growing turbulent environment. It breaks through the long-existing tight compartments between the worlds of those who plan, study, and manage social systems and those who do so for technological systems. At once it makes nonsensical the existing positions of psychologists and sociologists that in purposive organizations the technology is unalterable and must be accepted as a given requirement. Most frequently, therefore, only variables and relationships not influenced by technology are examined and altered. Without inclusion of technology, which considerably determines what work is about and what demands exist for the individual and organization, not only are peripheral relations examined, but they tend to become disproportionately magnified, making interpretation and use of findings difficult, if not impossible. Similarly, it makes nonsensical the "technological imperative" position of engineers, economists, and managers who consider psychological and social requirements as constraints and at best as boundary conditions of technical systems. That a substantial part of technological system design includes social system design is neither understood nor appreciated. Frightful assumptions, supported by societal values, are made about humans and groups and become built into machines and processes as requirements. (Davis, 1979).

Sociotechnical systems analysis provides a basis for determining the appropriate boundaries of systems containing people, machines, materials, and information. It considers the operation of such systems within the framework of an environment that is made an overt and specific object of the sociotechnical study. It concerns itself with spontaneous reorganization or adaptation, with control of system variance, with growth, self-regulation, etc. These are aspects of system study that will become increasingly important as organizations in the postindustrial era are required to develop strategies that focus on adaptability and commitment. For these reasons, sociotechnical systems analysis is felt to offer one of the best current approaches to meeting the postindustrial challenge (Davis, 1979).

Sociotechnical systems theory appears to be at the same time very general and very specific. It is general in the sense that the principles of the theory are framed and discussed at a rather general level of analysis and few explicit conceptual links are forged between the tenets of the theory and the particular actions taken or outcomes observed in various applications. It is specific

4.2 CHARACTERISTICS OF THE SOCIOTECHNICAL SYSTEMS THEORY

in the sense that characteristics of the organizations where experiments in work design are conducted are usually described in rich and complete detail (Hackman and Oldham, 1980).

Taylor et al. (1986) report an STS application at Zilog Inc., a California affiliate of Exxon Enterprises which built a new semiconductor (S/C) circuit plant in Nampa, Idaho. The manufacturing processes were designed using a sociotechnical systems approach that resulted in product-related work groups rather than the technology-based or functional work groups for the industry. Groups were arranged to produce an identifiable part of the completed S/C circuit chip (and to control variances associated with product quality), in contrast to merely being grouped around a type of machine or technical function. The S/C product-design department, known as Component Design Engineering (CDE), had successfully provided the company with two generations of microprocessor devices (8-bit and 16-bit) without paying much attention to how it was organized. But times were changing in the S/C industry. What had begun as a strictly high-tech industry was quickly changing from being "engineering driven" to being "market driven." This paradox of engineering—no longer being the sole arbiter of design but being expected to design even more complicated S/C products—was apparent in late 1979 and early 1980.

Taylor et al. (1986) argue that this case was the first application of STS analysis and design technique to the computer-assisted design (CAD) technology of circuit layout drafting. This case also emphasizes the role of professional engineers in the organization and how they are organized. In addition, the STS analysis places CAD technology within the total structure of the CDE department as an engineering system and places CDE within the context of the other departments in the company and of its industry environment. The upper limit in capability of the CAD system or any technology is not achieved because of an exclusive focus on the technology itself, but is placed as an extension of the organization that applies it.

Because STS management focuses on the work process as a whole than on its parts, it is a systems approach. It starts with the particular mission or purpose of the organization and develops a design for and system of management tailored to that purpose. This purpose orientation, according to Taylor et al. (1986), is much more powerful than the more common problem-oriented approaches (e.g., office automation and quality circles), which address only part of the system and which often ignore strategic management issues and overall mission.

Demands from the external environment of an organization often compel changes in the technical subsystem. Likewise, changes in the social environment and the embedding culture may induce changes in an organization's social subsystems. It is therefore necessary for organizational managers and designers to consider both the independent and joint effects of technical and social subsystems in the design, creation, maintenance, and change of complex organizations.

To understand the sociotechnical system better, we must focus on the basic elements of the systems approach. Further, to appreciate the dynamics of systemic change and the magnitude of its effects, we must examine these elements as they relate to organizational change.

Organizations are open systems whose processors are horizontal work units and whose controllers are neutral work units. These work units, or organizational subsystems, are best identified in terms of the functions performed. A systems model of an organization seldom parallels the formal organizational structure. Remember, the objective is not simply to describe the organization, but to understand it (Zmud, 1983).

The prospects for creating more innovative organizations depend on the development of adequate formal organizational models in order to understand the general characteristics of formal organizations and the identification of those variables specifically associated with innovative behavior. The potential design of organizations for innovative roles is problematic because of the tendency of the characteristics of the organization, its managerial systems, and basic organizational processes to promote stability. It is, however, a mistake to assume that this tendency to stability makes an organization incapable of innovative behavior. Organizations are frequently innovative and ingenious in their search for excellence and adaptation to their environmental domains.

According to Rowe and Boise (1973), we know little about the total process of demand, input, conversion, storage, output, and impact which may be relevant to the development implementation of an innovative idea. On the other hand, knowledge is developing rapidly of such broad categories as the environmental factors and processes influencing organizational and technological innovation; the organizational climate needed to induce innovative behavior; the systems required to produce and implement new technical ideas, and the flexible, ad hoc structures that seem likely to evolve as the innovative organizations of the future. It is our position that the systems paradigm can provide some insight into these problems.

4.3 THE GENERAL SYSTEMS CONCEPT

The critical questions to be asked of the systems approach were defined by Scott et al. (1972)

> (1) What are the strategic parts of the system? (2) What is the nature of their mutual dependency: (3) What are the main processes in the system that link the parts together and facilitate their adjustments to each other? (4) What are the goals sought by systems?

The answers to these questions provide the framework for a systems approach to organization theory and design.

The positive features of the systems approach are expressed by a number of organizational theorists, but none have been so superlative in their assessment as Coleman and Palmer (1973). Their evaluation is cogent:

> It is the authors' contention that the systems approach provides a basis for understanding organizations and their problems which may one day produce a revolution in organizations comparable to the one brought about by Taylor with scientific management.

4.4 THE ORGANIZATION AS A SYSTEM

General systems theory focuses on the interdependency of the subsystem components of the system. Most organizations have, at the least, a production, a managerial, and an adaptation or innovation subsystem. Each subsystem has certain goals, and each seeks to contribute to the system's overall objective, thus encouraging the interdependency of the subsystems. This interdependency depends on communication. Since each component has a fairly specialized function, each develops a distinctive nucleus of operating procedures, values, and information-processing requirements. For example, the production component is keyed to efficiency, rationalization, and careful programming of activities. The adaptive component is oriented toward change, innovation, the environment, and the future. The managerial component is oriented toward growth, stability, efficiency, and speed in decision making (Katz and Kahn, 1978).

The characteristics of open systems indicate what principles must be applied in the design of organizations. For instance, organizations must differentiate their functions and tasks to cope with environmental change and complexity. They must design integrative mechanisms to coordinate differentiated tasks and design feedback systems for adaptation, which is also critical for the adoption and diffusion of innovations. Organizations must also incorporate multiple paths to achieve organizational goals, that is, equifinality, in their design and also manage interdependence among their components, and so on. The systemic principles are becoming more critical as we design organizations for change and innovation.

4.5 CHARACTERISTICS OF OPEN ORGANIZATIONAL SYSTEMS

Figure 4.2 shows how the following systems characteristics relate to the new technologies and the tools for managing change.

1. *Suprasystem.* Open systems exchange resources, energy, and information with their environment. Having more or less permeable boundaries, organizations interact with their external environment. Many changes in the organization result from changes in the environment. Organizations must

122   A SYSTEMS APPROACH FOR MANAGING TECHNOLOGICAL CHANGE

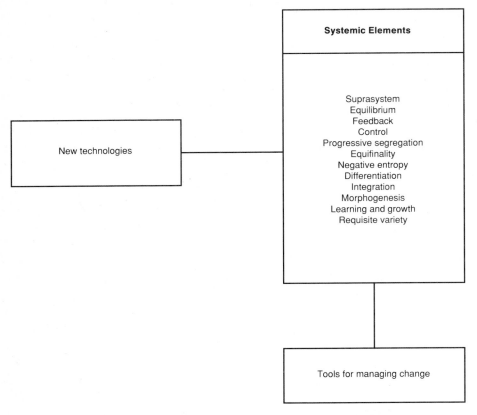

Figure 4.2. Systems characteristics and new technologies.

adapt to their environment if they are to remain viable. The interdependency between organizations and their environments is a fact that managers recognize. The environment (suprasystem) will partially determine the type of structures that organizations adopt to cope with the degree of technological change, complexity, and uncertainty. The structures, in turn, will influence the processes and behavioral patterns within the organization.

2. *Equilibrium.* Open systems tend to maintain themselves in steady states, or equilibrium. A steady state occurs when, given a continuous input to the system, a constant ratio is maintained among its components. Organizations maintain some degree of stability among their components or subsystems through control mechanisms such as rules, regulations, plans, hierarchy, etc. However, organizations as behavioral systems possess characteristics of stability and adaptiveness in a state of dynamic equilibrium. Organizations must maintain some level of stability to engage effectively in adaptive behavior through planning, market research, new product develop-

ment, technological forecasting, and the like. Both maintenance and adaptive activities are required if organizations are to survive.

3. *Feedback.* Open systems maintain their steady states through feedback processes. Organizations have feedback mechanisms that permit various parts of components to adjust to other parts and components. The boundary-spanning units, such as market research and research and development, serve as external scanning or sensing functions to provide information flow between the organization and its environment, which in turn facilitates adaptation. The internal sensing and scanning functions of organizational development units, task forces, liaison roles, etc., are examples of internal feedback mechanisms designed to make internal adaptation easier. At the heart of the open system are the processes, operations, or channels which transform the inputs into outputs. This is where the internal organizational design plays an important role. The transformation process consists of a logical network of subsystems which lend to the output. These subsystems are translated into a complex systems network that transforms the inputs into the desired outputs (Luthans, 1985).

4. *Cycle of Events.* Open systems are a cycle of events. The organization's outputs furnish the means for new inputs that allow for the repetition of the cycle. The organization utilizes raw materials and labor for production, the revenue from sale of the product pays for more materials and labor, and the cycle of activities is perpetuated. The structure of the organization is an interrelated sets of events that completes and renews a cycle of activities. It is events rather than things that are structures, so that social structure is a dynamic rather than a static concept (Katz and Kahn, 1973).

5. *Control.* Open systems maintain their steady states in part through the dynamic interplay of subsystems operating as functional processes. This means that various parts of the system function free of persistent conflicts that cannot be resolved, regulated, nor managed. Organizations have systems with different orientations, cultures, means–ends hierarchies, task designs, and work flows. The potential for conflict is high because of the necessity for integration among these subsystems and because of the different features intrinsic to each subsystem (e.g., goals, values, norms, subcultures, etc.). But organizations have designed integrative mechanisms, control systems, and cooperative procedures to manage conflict and provide unity of activity among the subsystems.

6. *Progressive Segregation.* Open systems display progressive segregation. This occurs when the system divides into a hierarchical order of subordinate systems, which gain a certain independence of each other. In response to change and complexity, organizations become differentiated—they have an elaborate segmentation of jobs, tasks, departments, and functions. As change, complexity, and uncertainty increase, organizations become decentralized; that is, they design relatively autonomous units to which authority for decision making is delegated.

7. *Equifinality.* Open systems display equifinality, that is, identical results obtained from different initial conditions. Organizations can achieve their objectives with varied inputs and transformation processes. This variety is reflected in the different strategies, action plans, and programs that will contribute to the achievement of organizational goals. The multiplicity of paths to organizational outcomes is a cardinal feature of organizations. This variety is also necessary for innovation, change, and creativity.

8. *Negative Entropy.* Open systems display negative entropy. Entropy refers to the propensity of a system to deteriorate or disintegrate. A closed system, because it does not import energy or new inputs from its environment, will deteriorate over time. On the other hand, because of its ability to import more resources, information, and energy than it puts out, an open system can repair itself, maintain its structure, and grow. Organizations in interaction with their environments import information, technology, and resources that are converted into new products and services. The feedback mechanisms perform the functions of sensing, to alert the organization to symptoms of malfunctioning or disintegration. The motivations and efforts of workers are as critical as financial and technical resources to produce a state of negative entropy for organizations. In general, negative entropy—and thus uncertainty—will be high in organizations that foster creativity, innovation, and dynamism.

9. *Differentiation.* Open systems tend to elaborate or differentiate their roles and functions. This tendency to subdivide tasks, programs, workflow, and functions is in response to change, complexity, and uncertainty in the domains of the environment: The problem posed by this tendency is to determine the optimal degree of differentiation for the system.

10. *Integration.* Open systems are faced with a major design problem because of the degree of differentiation of functions and tasks produced by change and innovations, namely, the problem of integration. Organizations must design integrative mechanisms to coordinate their work, resource, and information flow. The optimal degree of integration is a major problem in organizational design and change. A high degree of integration may inhibit the adoption and diffusion of technical and organizational innovations.

11. *Morphogenesis.* Open systems tend to maintain their basic character, attempting to control external contingencies and problematic dependencies with their environmental domains. As growth and expansion occur, basic system characteristics tend to vary incrementally. Under conditions of extreme growth or expansion, a new character, organizational culture, and structure may develop that will serve as a new homeostatic basis. This process of morphogenesis is characteristic of organizations as behavioral systems.

12. *Learning and Growth.* An open system that is to change through learning and innovation must contain very specific feedback mechanisms, a certain variety of information, and particular kinds of input, channel, storage,

cognitive apparatus, and decision-making centers. The capacity to learn and innovate must be institutionalized into the information-processing systems of the organization. The capacity of the system to learn from feedback is essential to its growth.

13. *Law of Requisite Variety*. As the environment of the organization becomes more complicated, it presents more variety with which the organization must deal. Ashby (1968) has described this concept of requisite variety, which argues that if a system faces variety, only by adding variety of its own can the system reduce the effect of the complexity being faced. As the environment becomes more complex, the organization must become structurally and functionally more complex. This complexity may proceed at two levels. At the horizontal level, tasks, programs, work flows, and functions are expanded in size, new jobs are designed, and more complex information systems are devised to handle change and uncertainty. At the vertical level, more levels of management authority are added to the system; the organizational hierarchy expands and grows to coordinate the work flow and its technical complexity.

4.6 THE MANAGER AND THE SYSTEMS APPROACH

Managers today, unlike their predecessors, are managers of change and innovation. This role is reinforced by the expanding technological environment, changes within the organization, and the dynamic interaction between the organization and its environment. The organization is now subjected to increasing pressures to modify classical organizational structure. There are different structural designs (e.g., matrix, project management, "adhocracy") that have emerged to modify established structural lines to cope with complexity, change, and uncertainty. The task of the manager, using the systems concept, is to create an organizational design for change.

Using the systems approach, managers will be involved in integrating the organization as a system. They will design a structure of task and a structure of authority that will maximize the organization's performance within the constraints of the goals of its members. This view is different from that of the past, which tended to emphasize the parts of the structure rather than their interrelationships. The design for managing complexity, interdependence, and uncertainty must be a dynamic one. The increasing use of man–machine interactive systems will demand more and more design effort. The manager must be able to participate in the design and use of responsive, adaptive decision systems. More reliance on factual and empirical information for decisions and new techniques for rational decision making will also propel the manager toward a systems environment. Information will be the means and information systems the structure for implementing the systems approach. The critical element is information management, which requires the proper blending of the remaining elements of the organization into an information systems design. This is the essence of the systems approach (Duncan, 1972).

The importance of the systems perspective to the study of organizations is that it is a useful framework that managers use to conceptualize organizations. The organization is viewed as a behavioral system with degrees of interdependence among its parts—a system composed of subsystems. The systems perspective prevents, or at lease deters, lower-level managers from viewing their jobs as simply managing static, isolated elements of the organization. It encourages managers to identify and understand the environment in which their system operates. It helps managers to see the organization as stable patterns and actions within boundaries, and to give insights into why organizations are resistant to change. Finally, it directs managers' attention to alternative inputs and processes for reaching their goals. As Duncan (1972) continues,

> . . . it is one thing to argue that everything depends on everything else. It is a much different thing to offer suggestions to managers on what precisely will change, and to what degree, if a certain action is taken. Its value, therefore, lies more in its conceptual framework . . .

We expect that our analysis of the systems approach as it relates to change and design will generate guidelines for design and change.

According to Daft (1989), organization managers must be sensitive to the complex nature of the social systems if they are to understand and cope with their organizations. Companies are open systems that must interact continuously with their environment, and they have subsystems to perform specific functions. These systems are also complex because of social characteristics. Organizational dimensions interact so that changing one element may affect the whole system. A single cause does not have a single effect. Organizational systems can be difficult to manage because many important dimensions are intangible.

4.7 A SYSTEMS FRAMEWORK OF THE ORGANIZATION

A model of organizational (Figure 4.3) illustrates our belief that change is contingent on the factors in the external environment and on the response of the internal organizational environment to them. The organizational response is constrained by the structure, processes, and technology of the organization. The major impetus for change comes not only from the external environment—the source of the change may also reside within the internal environment.

The framework in the figure shows that the environment exercises its influence on the organization through the amounts of complexity, uncertainty, and diversity it creates in the organization. The general principles of organization, then, become principles of how to adapt the organization to the complexity, uncertainty, and diversity it faces in its environmental domains.

4.7 A SYSTEMS FRAMEWORK OF THE ORGANIZATION

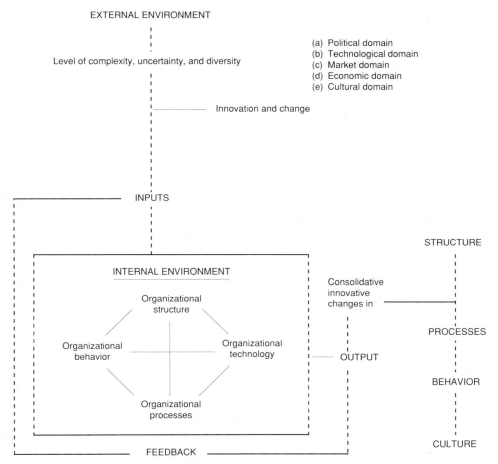

Figure 4.3. A systems framework of the organization.

The framework also emphasizes the relevance of the sociotechnical approach to managing technological change. The technological subsystem impacts on the behavioral, structural, and process subsystems. The interdependence among the subsystems is the reason that a change in one subsystem will produce reactions in others; the reactions must be monitored through feedback mechanisms. The management of interdependence becomes a critical function of the change agent as he attempts to coordinate the changes in work, resource, and information flow that new technologies demand (Sankar, 1988).

Organizational systems consist of sets of people, objects, and activities that are related to one another and can be distinguished from other sets. The precise limits, or boundaries, of a system are drawn by the investigator and depend on the problem being solved. If we focus on the technical subsystem, conceptual boundaries could be drawn at this subsystem level. Other problems, however, might have boundaries that embrace "an entire department, division, company, industry, or total economic system" (Child, 1972). What is important to systems theory is not so much where the boundary is drawn as the interactions and dependencies across boundaries and the nature of the process by which linkage takes place and systems adjust to one another.

An open systems approach to understanding change and innovation requires serious consideration of the organization's environment. The interaction between the organization and its environment is crucial to the innovation process. Many changes in an organization result from changes in its environment. For any organization to be innovative, it must have some degree of system openness—the exchange of information with its environment. New ideas often enter the organization across its boundaries, as does much of the information that creates the need for innovation, so openness to its environment facilitates an organization's innovativeness.

Organization theorists like Drucker, Toffler, Galbraith, Mintzberg, and others emphasize that organizations must adapt to their environment if they are to remain viable. One of the central issues in this process is coping with uncertainty. A major element in the relationship between environment and change is the notion of perceived environmental uncertainty. The existence of such a relationship would appear to follow from the argument that organizational processes, structures, and programs are altered by administrators' perceptions of the organization's environment.

Duncan (1972) argues that organization environments are related to perceived environmental uncertainty, and that perceived environmental uncertainty is related to organizational structure. Our major proposition on this theme is that the internal adjustment process within an organization is not a random one, but can be explained in terms of the perceived uncertainty in the environment and the perceived need to alter strategic properties within the organization, such as goals, structure, strategies and policies, roles, and so on.

The process by which an organization adapts to its environment is a dynamic one consisting of a set of decisions. Miles (1980) suggest that the decision points map will give a definitive picture of organization–environment relations. Their map indicates four major decision points which they consider to be important:

1. The decisions by which the organization selects a portion of the total environment as its particular arena of activity (i.e., its domain) and uses a basic strategy for managing the domain.
2. The decisions by which the organization establishes an appropriate *technology* for implementing its basic operating strategy.
3. The decisions by which the organizaiton creates a structure of roles and relationship to control and coordinate technology and strategy.
4. The decisions made to assure organizational continuity—the capacity to survive, adjust, and grow.

The technique by which organizations respond to their environments depends on the choices made within the organization. Child (1972) calls these the "strategic choices," and argues that the internal politics of organizations determine the structural forms, the manipulation of environment features, and the choice of relevant performance standards selected by the organization. The internal politics in turn depend on the power distributions within the organization, which are themselves a structural condition subject to alteration over time (Luthans, 1985).

Technological change and innovation produce certain effects that necessitate adaptation on the part of the organization. Change increases the complexity of the relationship between the organization and its environment. Change increases the problem of maintaining organizational solidarity and coordination; values and goals become more diverse, pluralistic, and conflicting. Change produces a greater impulse toward rationality in organization decision making and operations through the use of information technology. As a consequence, the management of the change process is a complex problem for the theorist and practitioner, according to Child (1972).

Technological changes in the environment create a situation of stress, or pressure, to which the adoption unit must respond if it is to remain in a relationship of "dynamic equilibrium" with the environment.

4.8 SYSTEMS DIFFERENTIATION, CHANGE, AND INNOVATION

Differentiation refers in part to the idea of specialization of labor, specifically to the degree of departmentalization. But it is broader, and also includes the behavioral attributes of employees of these subsystems, or departments. The process of departmentalizing creates the need for integrating the activities of the departments. The integration of separate, yet interdependent, activities is a familiar problem in the management of organizations. A systems concept that explains this problem is referred to as *system connectedness*—the degree to which the parts or subsystems (behavior, structure, processes) are linked to each other by communication flows. Generally, we expect that the degree of connectedness in a communication network is positively related to the rate of diffusion of innovations (Bobbit et al., 1974).

Again, systems become differentiated to cope with change, complexity, and uncertainty, and then develop mechanisms to coordinate, control, and integrate their functions, work flow, resource flow, and information flow. What happens at this stage is that a consolidation within the system occurs through the use of regulatory feedback systems. These controls inhibit the diffusion of innovations throughout the system. Stability may be sought, through tighter integration and coordination, when flexibility may be the more important requirement. Coordination and control become ends in themselves, desirable states within a closed system, rather than the means of attaining adjustment between the system and its environment. The potential for adoption of a change strategy to make this adjustment possible is remote under such conditions.

The process of achieving unity of effort among the various subsystems in the accomplishment of the organization's task, defined as integration, can be achieved in a variety of ways. The classical writers proposed that the problem could be solved through the creation of rules, procedures, plans, and a hierarchical chain of command that placed managers in the position of integrators, or coordinators. Thompson (1967) observed that this means of integration can be effective only in relatively stable and predictable situations; that rules and procedures lose their appeal as the environment becomes more unstable; and that integration by plans takes on greater significance. But as we approach the highly unstable environment, coordination is achieved by mutual adjustment. Coordination by mutual adjustment in the management of technological change requires a great deal of communication through open channels throughout the organization (Daft, 1989).

It would be expected that highly differentiated organizations would tend to use mutual adjustment as a means of achieving integration or system connectedness. Group-centered decision making, mutual adjustment through network communications, and integrative and sociotechnical project teams are necessary to integrate highly differentiated departments. The distortion, filtering, and uncertainty of information and the time span of feedback are some of the problems that inhibit the diffusion of innovations in highly differentiated systems.

The process of equifinality is a characteristic of open, adaptive, goal-seeking systems, such as formal organizations. It simply means that output (e.g., profit, sales, products, etc.) may be reached from different initial conditions and by a variety of paths; that is, there are more ways than one of producing a given outcome.

In decision making, the search and design processes—the search for innovative paths to achieve a problem solution—usually stop too early to achieve the levels of effectiveness possible if the search had continued or additional creativity been attempted. It is the reduced amount of equifinality in the organization's culture that contributes to premature termination of search and design processes. Little innovation, invention, and creativity in the system as a whole is possible, because of limited search for options.

4.9 NEGATIVE ENTROPY AND CHANGE

The adaptation of organizational systems may be partially explained by the concept of *entropy*. The second law of thermodynamics states that a closed system will, with the passage of time, tend to settle down to a uniform distribution of its energy (Bobbit et al., 1974). That is, a closed system cannot maintain its structure. What is necessary is to arrest the entropic process by importing new inputs, information, and energy into the system (through negative entropy).

The business firms often store negative entropy through such means as (1) a "new venture department" that imports new people, technical ideas, and resources for the organization; (2) selecting and training its members so they can meet present and future demands of the systems; (3) seeking and utilizing technology, the input–output ratio of energy, and the like; and (4) the use of slack resources to reduce the amount of interdependence between units, thus keeping the amount of information within the capacity of the organization to process it (Daft, 1989).

The germane point is that a closed system will tend toward a state of maximum entropy, which will have minimum order or structure. A behavioral system, being open, can import energy and exhibit negative entropic behavior. It can elaborate and make its structure more complex in order to deal with a more complicated technological environment. As the system becomes more complex and moves to counteract entropy, it moves toward growth and expansion. Katz and Kahn (1978) note that large and complex systems operate to acquire some margin of safety beyond the immediate level of existence. The many subsystems within the system, to avoid entropy, tend to import more energy than is required for its output. The basic system does not change directly as a result of expansion. The most common growth pattern is one in which there is merely a multiplication of the same type of cycles or subsystems—more services, departments, and programs rather than new or innovative programs (Katz and Kahn, 1978). The innovative organization must arrest the entropic process by importing new ideas, concepts, and techniques, by creative management, and by designing structures that are adaptive, since a bureaucratic structure is a structure with high entropy.

4.10 THE LAW OF REQUISITE VARIETY AND CHANGE

Information can be thought of as the opposite of uncertainty. As the information increases, the uncertainty related to that system is reduced. Thus, systems with more information have less uncertainty. From the point of view of the behavior of an organization, that behavior will be more predictable as the uncertainty associated with the organization's innovations is reduced. Information is increased in a system if the degree of freedom to choose among signals, symbols, messages, or patterns to be transmitted is increased. More information increases flexibility and thereby the chances that the form of the outcome will be closer to that desired. In many respects, we can think of this as a variation of Ashby's concept of requisite variety (1968).

As the environment of the organization becomes more complex, it presents more variety with which the organization must deal. Ashby (1968) has described a concept of requisite variety, which argues that if a system is faced with variety, only by adding variety of its own can the system reduce the effect of the complexity being faced.

We can extend this argument to the organization–environment interface.

If the relevant environment of the organization has a great deal of variety (heterogeneity, uncertainty, rapid technological change, etc.), then the organization can reduce the effect only by increasing the variety of its actions at the interface. This suggests that buffering units in the organization will increase in number and complexity of functions as the complexity of the environment increases. As the environment becomes more complicated, the organization must become structurally and functionally more complex in order to deal with it successfully. This complexity may proceed at two levels: at the horizontal level, tasks, programs, work flow, and functions are expanded in size, new jobs are designed, and more complex information processing systems are devised to handle change and uncertainty; at the vertical level, more levels of management authority are added to the system, the organizational hierarchy expands and grows to process information and technical complexity, and, at the same time, more sources of discretion develop in the system because of uncertainty and complexity. These sources of discretion are sources of potential power. Uncertainty is a potential source of discretion. Coping with uncertainty is a source of potential power. Those units within organizations that cope with uncertainty and complexity generated by technological change have relatively more power, authority, and influence than other units in the organization.

Ashby (1968) observes that if a system or organization is to adapt to or control its environment, it must contain at least as much variety, or "entropy" or "freedom" of selection of alternatives, as there is in the environment to be controlled. The generality of this concept encompasses information theory. When a system is highly organized, the entropy and hence the information is low. A small entropy figure represents a system capable of little variety. As randomness increases, so does the amount of information needed to restructure the system. However, the regulation and control of systems reduce variety and we notice the change–stability aspects of systems. Technological change invariably calls for more vertical and horizontal expansion of the organization, a degree of structural complexity to cope with technical complexity. This expansion provides a level of requisite variety.

4.11 PROGRESSIVE SEGREGATION AND CHANGE

All systems have some degree of structure. Organizations usually have a relatively high degree of structure and this patterning gives them stability and predictability. The formal structure consists of the patterns of formal relationships and duties, job descriptions, formal rules, operating policies, work procedures, etc. In other words, organizational structure consists of those aspects of the patterns of organizational behavior that are relatively stable and that change only slowly. In fact, one purpose of structure is to supply stability, regularity, and predictability in the organization.

This patterning may inhibit change in the system, perhaps because power, influence, and authority of the hierarchical system have become so structured and crystallized that rigid role expectations and standardized procedures and rules develop. The communication channel is rigidly controlled, which in turn limits the nature and extent of the feedback necessary for the initiation and facilitation of change. But an expansion of the hierarchy is limited by a major characteristic of open systems, namely, *progressive segregation*. This refers to how the system divides into a hierarchical order of subordinate systems, which gain a degree of independence from each other. This intrinsic condition or organizational functioning suggests that organizations are structured for varying degrees of autonomy in its parts. For the effective management of technological change, progressive segregation is essential. That is, some degree of autonomy of the organization's subsystems or level of decentralization of the organizational structure will facilitate the emergence of change initiatives, feedback mechanisms, distributed data bases, and delegation of authority—which can also contribute toward the development of a more participative style of management for managing change.

4.12 INTERDEPENDENCE AND CHANGE

Changes cannot be effected in the technical system without affecting the subsystems of the organization. Since the technical system may influence job design, role relationships, work flow, and information flow, changes in these organizational characteristics can be expected. In the same way, parts of the social system (e.g., personal and role relationships within work groups) can be changed without prior change in technical systems or organizational structure, but the change in organizational behavior is only mild reform, not radical change. Katz and Kahn (1978) view systemic change as the most powerful approach to changing organizations. Systemic change involves changes in inputs (technical, market, economic, cultural) from the environment which generate internal strain, stress, and conflict as well as imbalance among the subsystems of the organization. The internal strain and dissonance are potent causes of the behavioral adaptation of subsystems connected with the changed input.

Although the evidence is not overwhelming, it appears that a properly planned technological change can lead directly to increased job complexity, autonomous and responsible work group processes, more helpful supervision, and high productivity.

> It was found that, at least for some workers, these direct effects can lead to more effective job involvement, improved work group relations, more favourable attitudes to supervision and pride in high productivity (Taylor, 1973).

The implication is that social change can proceed as a direct outcome of certain technological changes, but if this were the only social change that was affected, the outcome would be slow and limited by change factors. Taylor (1973) advances three major hypotheses involving the impact of technology on organizational characteristics:

> (1) sophisticated technology, in and of itself, is associated with more autonomous and participative group process; (2) sophisticated technology will facilitate planned change efforts directed toward increasing participative group process; and (3) the change toward participative group process will be more permanent when the change is facilitated by technology than when it is not.

4.13 EQUILIBRIUM AND CHANGE

The basic concepts of the system approach that attempt to explain organizational change are more oriented toward stability, order, symmetry, and consolidation than toward the sources of change of a system. For example, the concepts of homeostasis and equilibrium as properties of the open system have often been used in organization theory to imply that in its natural state the organization is static. For instance, the critical characteristic of the mechanistic model is static equilibrium, the notion of a state in which there is a constant ratio among the components of the system. Since the model views the system as "analytically closed," external variables are conceived as the major source of change of the system. That is, change is a disturbance of a persistent equilibrium. This concept of equilibrium places a heavy emphasis on the order, continuity, consolidation, and stability of the internal components of the system (Katz and Kahn, 1978).

4.13.1 Dynamic Homeostasis

The organic model of the systemic paradigm avoids the static connotations of equilibrium in the mechanistic model by relying on the concept of homeostasis, a process by which a system regulated itself around a stable state. Its basic feature, self-regulation, requires feedback and control, both operating in such a way as to minimize the adverse effects of change on a system.

This concept in organizational theory, with its notion of the organism as an essentially reactive system, devalues the capacity of the organization to initiate or anticipate change. The major focus of the concept is on the structure-maintaining feature of the system, which assumes that all relevant behavioral alternatives that may emerge in reaction to change fall within the spectrum of the established control machinery. That is, the concept of homeostasis applies to the restoration of the equilibrium through some predetermined internal mechanism. It is the expression of a theme of self-regulation of a system tending toward stability, orderliness, continuity, symmetry, and constancy.

However, under conditions of stability, a system initiates and facilitates changes necessary for the maintenance of stability. It must additionally be noted that certain types of stability are necessary for change—that is, to engage in adaptive responses, some measure of stability in procedures, roles, and structure is necessary. Hence, stability is no guarantee for a state of no change. An organization can adjust its functioning to achieve stability and a variable rate of change simultaneously.

This capacity is particularly relevant for technological change. With technological change a variable rate of change in the subsystems, behavior, structure, and process may be observed. The magnitude of the change may call for either environmental change or strategic (fundamental) change in organizational characteristics. Stability of some organizational characteristics, rules, regulations, procedures, and elements of the management infrastructure may be maintained while changes in job design, strategy, and structure are initiated. A variable rate of change also implies that an organization is essentially a problem-solving and adaptive system. Technology is just a stimulus that may produce such a variable rate of change.

Bennis (1966) has argued that conventional approaches in the study of organizational change are not consistent with the recently emerging view of organizations as adaptive, problem-solving systems. He states that the main challenges confronting today's organization are the response to changing conditions and the adaptation to external stress, and that there is a need for studies which reveal the processes by which the organization, through its members, searches for, adapts to, and solves its changing problems. Bennis (1966) suggests that the methodological rules by which an organization approaches its task and interacts with its environment are the critical determinants of effectiveness, and that without understanding these dynamic processes, knowledge about organizational behavior is woefully inadequate. A number of other theorists according to Bennis have adopted similar views of organizational behavior. This swing to a process emphasis signals a significant new development in ways of thinking about organizations. If the views of these theorists are correct, it would appear that the processes through which organizational adaptation occurs should be a significant subject of analysis, and that it would be important to learn precisely how these processes influence and contribute to the effectiveness of an organization.

4.14 CYBERNETICS AND ORGANIZATIONAL SYSTEMS

The operation of the cybernetic system chiefly entails storing, receiving, transmitting, and modifying information. There is a high level of uncertainty inherent in this operation, since the permutations of the system's data, or fact elements, are enormous. The ordering of the uncertainty in the data, according to Rogers (1975), produces information, and the system becomes more controlled and the end-state or the goal of the system more predictable (Beer, 1959).

A variant of the systems approach is the cybernetic model. Cybernetics as a concept in organizational theory integrates the linking processes in organizational systems and generalizes them to a wide variety of systems. The linking processes are communication, balance, and decisions. It is through them that dynamic and basic interactions are initiated and facilitated in an organization (Scott et al., 1981). Decisions, information (communication), and control (balance or regulation) are indispensable elements of complex systems. Beer (1959) notes that decisions are the events that go on in the network and they are describable—in terms of the information in the system and the structuring of communication. In this sense, then, decisions and information cannot be understood apart from the system's communication pattern, and this pattern in turn is a reflection of the decisions required and the information necessary upon which to base them. Balance, the third linking process, is introduced in the form of control, or regulation, and this constitutes the heart of the cybernetic processes. Regulation of the system network by the information produced in it is the core of cybernetics. One of the requirements of a cybernetic system is self-regulation. Usually identified as feedback, self-regulation controls current activities by adjusting them after comparing their outcomes or performance against some standard or objective (Scott et al. 1981).

4.14.1 Feedback and Change

Katz and Kahn (1978) observe, with reference to feedback, that the reception of inputs into a system is selective. Not all inputs are capable of being absorbed into every system. Systems can react only to those information signals to which they are attuned. The general term for the selective mechanisms of a system by which incoming materials are rejected or accepted and translated for the structure is *coding*. Through the coding process, uncertainty and complexity are simplified into a few meaningful and simplified categories for a given system. The nature of the functions performed by the system determines its coding mechanisms, which in turn perpetuate this type of functioning. With new technologies, the coding and feedback mechanisms are built into the software, and changes are initiated via these mechanisms. Advanced technologies are based on cybernetic principles.

The feedback characteristic of an organization permits managers to determine whether progress is being made toward goal attainment. Without feedback, there would be no means to determine the degree of deviation from or progress toward goals. A second way in which negative feedback differs is that the means, or mechanisms, for obtaining feedback can be *learned* and *created* by organizations.

Feedback-controlled systems in this context are referred to as *goal directed* and not merely goal oriented, since it is the deviations from the goal state itself that direct the behavior of the system, rather than some predetermined internal mechanism that aims blindly.

The significance of feedback control for complex systems, according to Buckley (1967), can be partially expressed by a comparison of "re-cybernetic" machines and modern servomechanisms. In the former, the design must try to anticipate all the contingencies the machine is apt to meet in performing its task, and to build counteracting features into the design; the modern machine, however, uses these very contingencies as information, which, fed into the machine, directs it against them. The concept of feedback in cybernetics as a principle underlying the goal-seeking behavior of complex systems has certain characteristics that distinguish it from the notion of feedback in the organic model. Buckley (1967) observes that

> whereas the concept of equilibrium is restricted to descriptions of steady states, the cybernetic view is based on full dynamics, including change of state as an inherent and necessary aspect of complex operation.

Cybernetics offers (Buckley, 1967)

> to restore the problem of purpose to a fuller share of attention, and even to help make a much needed distinction between the attainment of actual external goals, and the reduction of goal-drive by internal readjustment.

4.14.2 The Systems Approach and Information Processing

From the point of view of cybernetics, any large-scale formal social organization is a communications network. It is assumed that such organizations can display learning and innovative behavior if they process certain necessary facilities, that is, structure, and certain necessary rules of operation, that is, content (Cadwallader, 1959).

First, consider the structure of the system as it might be represented in the language of cybernetics. Any social organization that is to change through learning and innovation—in other words—to be ultrastable, must contain certain very specific feedback mechanisms, a certain variety of information, and certain kinds of input, channel, storage, and decision-making facilities. More specifically, every open system behaving purposefully does so by virtue of a flow of factual and operational information through receptors, channels, selectors, feedback loops, and effectors. Every open system whose purposeful behavior is predictive, and this is essential to ultrastability, must also have mechanisms for the selective storage and recall of information: it must have memory (Cadwallader, 1959).

A cybernetic model would focus the investigator's attention on such factors as (1) the quantity and variety of information stored in the system; (2) the structure of the communication network; (3) the pattern of the subsystems within the whole; (4) the number, location, and function of negative feedback loops in the system and the amount of time lag in them; (5) the nature of the system's memory facility; and (6) the operating rules, or program, determining the system's structure and behavior.

The operating rules of the system and its subsystems are always numerous. Relevant for the present problem are (1) rules or instructions determining range of input; (2) rules responsible for the routing of the information through the network; (3) rules about the identification, analysis, and classification of information; (4) priority rules for input, analysis, storage, and output; (5) rules governing the feedback mechanisms; and (6) instructions for storage (Cadwallader, 1959).

4.15 COMMUNICATIONS, CHANGE, AND INNOVATION

Using the concepts, principles, and rules from the foregoing, Cadwallader (1959) developed a number of propositions on change:

1. The rate of innovation is a function of the rules organizing the problem-solving trails (input) of the system.
2. The capacity for innovation cannot exceed the capacity for variety or available variety of information.
3. The rate of innovation is a function of the quantity and variety of information.
4. A facility, mechanism, or rule for forgetting or disrupting organizing patterns of a high probability must be present.
5. The rate of change for the system will increase with an increase in the rate of change of the environment (input); that is, the changes in the variety of the inputs must force changes in the variety of the outputs or the system will fail to achieve ultrastability.

Since the systems approach recognizes the need to study interactions of the organization's subsystems, it focuses on communication as the key to analyzing and understanding organizations as social systems. Consequently, communicaiton theory and information theory were central in the development of system theory (Kelly, 1980). Communication is the basic process fostering the interdependence of the parts of the total system; it is the mechanism of coordination. In fact, information processing came to be seen by the systems school as the main function performed by all organizations; organizational systems are essentially communication systems. Technological change will therefore have to be managed through the information system of the organization. All of the six subsystems of the information system, namely, the sensor, data-processing, decision, processing, control, and memory subsystems will be activated in the process of managing technological change.

The subsystems are defined by Kelly (1980) as follows (see Figure 4.4):

1. A *sensor* subsystem, which is concerned with the reception and recognition of information.

2. A *data-processing* subsystem, which is concerned with breaking down this information into terms and categories that are meaningful and relevant to the organism.
3. A *decision* subsystem, in which decisions are made. Decisions may involve
   a. Self-regulatory or homeostatic processes.
   b. Adaptive or learning processes.
   c. Integrative processes.
4. A *processing* subsystem, which integrates information, energy, people, and materials to implement the decision, accomplish tasks, and produce output.
5. A *control* subsystem, which ties the whole system together by a set of feedback loops. These loops incorporate the equations of the critical decision variables, which, if not respected, lead to lack of growth and the eventual demise of the organism.
6. A *memory* subsystem, which is concerned with the storage and retrieval of information.

General systems theory is the foundation information systems. It is a very good fit, according to Murdick and Munson (1986). They suggest that to deal with complex systems we must use a methodology for hierarchically decomposing these systems into manageable subsystems (e.g., sensor subsystem, processing subsystem, decision subsystem, etc.). This is the basic approach of general systems theory and systems analysis. An information system itself is a subsystem of a firm. It is a subsystem providing formal information for managing the company from the highest to the lowest levels of decision making.

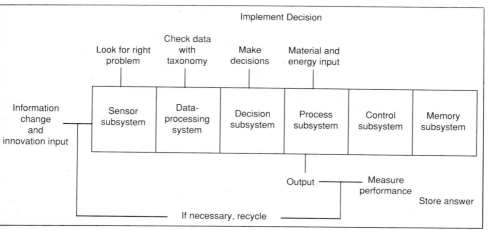

Figure 4.4. The information-processing system. (From Kelly, 1980. Reprinted with permission from Richard D. Irwin, Inc.)

In analyzing and designing an information system, an analyst directly applies many of the concepts from systems theory.

Drawing a boundary around a proposed information system helps an analyst isolate the problem with which he is dealing. Many information systems, especially of the accounting type, are feedback systems for management. These systems produce reports summarizing results of the firm's operations and allow management to take corrective actions in the firm's inputs to change future outputs. In designing an information system, an analyst must always keep in mind the concept of variety and control (Murdick and Munson, 1986).

Feedback concept is very important in information systems. Output from the system is used in decision making. If the output is not relevant to the decisions, then the system is of little use to management. It is important that a feedback loop be incorporated into the system to determine the relevance of output to the decision environment.

The law of requisite variety relates to the control systems. It states that to control a system there must be an available number of countermoves equal to the variety displayed by the output of the system. As a filtering process, the countermoves help us to cope with variety in our environment. Two strategies available to a manager are (1) decreasing variety of outputs from the system and (2) increasing the number of countermoves (Murdick and Munson, 1986).

In managing technological change, the information system of the organization is quite critical. If the system's concepts can improve the design of such a system, it can only lead to more effective management of technological change. As Murdick and Munson (1986) have said, "It would be a mistake to conclude that general systems theory concepts are too abstract and therefore not important to an information systems practitioner."

4.16 CYBERNETIC PRINCIPLES FOR MANAGEMENT OF TECHNOLOGY

Since cybernetics recognizes the need to study interactions of the subsystems in an organization, it focuses on information as the key to analyzing and understanding organizations as social systems. Consequently, communication theory and information theory were central in the development of cybernetic systems and offer the following principles:

1. Communication is the basic process facilitating the interdependence of the parts of the total system; it is the mechanism of coordination.
2. Information processing is the main function performed by all organizations; organizational systems are essentially communication systems.
3. Management in cybernetic language is the processing of information and the conversion of information into decision outcomes.

4. The greater the task uncertainty, the greater the need for information processing.
5. The larger the number of elements in the decision-making process (e.g., the number of departments), the greater the information-processing requirement of the organization.
6. The greater the degree of interdependence among the elements of decision making, the greater the information processing.
7. To increase the capacity of the organization to process information, more flexible integrative mechanisms are needed.
8. Tasks variety, change, and complexity will produce different degrees of uncertainty in the information domain and in decision making.
9. The structural configuration of an organization must be designed with information as the design parameter.
10. Different structural configurations are needed for different information domains.

The argument is made from a systems perspective that an organization must adapt to its environment if it is to remain viable. The greater the degree of technical change, complexity, and uncertainty, the greater the level of *differentiation* of the organization's information systems. This degree of differentiation creates a need for *integration* of the organization's resource, work, and information flow. This degree of differentiation also inhibits the rate of diffusion of innovations in the system. With reference to another major systems concept, the argument is that as organizational systems emphasize control, integration, and regulation of operations, the amount of *equifinality* will be reduced because of the tendency to limit the search for options and for innovative solutions and to codify the organization's information systems. Technological change sometimes decreases equifinality within the system through the process of job simplification and standardization of procedures, programs, and repertoire. As regards *entropy,* the argument is made that organizations fostering creativity, innovation, and dynamism will be high in negative entropy. The ability of organizations to arrest the entropic process is one of the determinants of the viability, richness, and effectiveness of its information-processing capacity. Another systems concept is *requisite variety,* which states that complexity in one system is required to control complexity in another. Therefore, a change in the degree of complexity of the organization's environment will be reflected in the level of vertical and horizontal differentiation of the organization and in the degree of complexity of information systems, decision support systems, and data bases, both integrated and distributed. This is another major principle for organizational design. The more complex the technology, the greater the degree of vertical and horizontal differentiation of the organization. Another design principle states that *equilibrium* is not to be identified with a fixed structure, but must also focus on the capacity of the system to elaborate structure through expansion and growth of its information-processing capabilities. Consequently, an organization can adjust its functioning to achieve stability and a variable rate of change at the same time as it adjusts to technological change. Organizations maintain their equilibrium through *feedback* processes, which permit various parts of the organization and work flow to adjust to other parts and segments of work flow. Feedback mechanisms also facilitate learning and growth. Technological change may require new feedback loops within the work flow and the organizational system. A change in *interdependence* among the parts of a system or segments of the work flow will produce changes in the design of various coordination modes, such as rules, procedures, plans, mutual adjustment, and so on. Technological change invariably alters the type of interdependence (pooled, sequential, reciprocal) and team and coordination modes for managing the interdependence.

4.17 IMPLICATIONS FOR MANAGEMENT

First, it is important to note that organizational change is externally induced. Forces in the environment necessitate organizational change, and it is the responsibility of managers and designers to take such forces of change into account when making organizational plans. Second, if the organization's environment changes, managers must modify organizational design and plans for an appropriate adaptive response. Third, as organizations face increased uncertainty, managers must make use of additional sources of information, and decision making will be more difficult. Finally, an organization's success in adapting to its environment will be determined by whether its managers can learn how to perform in response to changes in its environment and situation. The systems approach, as noted earlier, will provide the tools and perspectives for such an adaptive response (Gerloff, 1985).

The concept of technological strategy will benefit as well from the systems perspective. Technological strategy is an important but often ignored link in the strategic formulation chain. This is not to say that technology is altogether absent from the strategic plans of technology-intensive companies, but it generally appears in a fragmented, piecemeal fashion as part of other functional strategies such as marketing. Functional strategies are the building blocks of business strategy. In turn, functional strategies as themselves defined by a set of interrelated decisions that define the business unit's posture toward financial, manufacturing, technological, and marketing issues (Burgelman and Madique, 1988). The systems concept of interdependence views technological strategy as a critical subsystem of the business strategy that permits feedback between that subsystem and other subsystems, such as marketing, finance, and manufacturing.

The generality of systems theory is one of the elegant features of the sociotechnical approach, since it allows the theory to be adapted with ease to almost any organizational situation. Thus, the theory remains open to continual improvement and revision on the basis of increased experience with it in actual change situations (Hackman and Oldham, 1980). Sociotechnical systems theory offers some clear advantages over both traditional and behaviroal approaches to work design. Traditional approaches often ignore the personal needs of the people who carry out the work (especially social needs that can be fulfilled by group members) and are so oriented toward the efficiency of the technological subsystem that critical aspects of the social system may be ignored. Behavioral or psychological approaches tend to give insufficient attention to the operation of the technical system when work is designed and almost always underestimate the importance of group relations and the organizational environment in affecting what happens in the workplace. By contrast, the sociotechnical approach takes a holistic perspective and emphasizes human needs, values, expectations, and the elements of organizational culture and design in implementing technological change (Hackman and Oldham, 1980).

4.18 LEADERSHIP, SYSTEMS THINKING, AND CHANGE

The systems approach forces the leader to look upon his organization as an information network, with the flow of information providing the decision makers at varying management levels with the information needed to make decisions of all types (Mockler, 1971). Adapting the organization to the information systems needed for effective planning, control, and operational decision making in managing technological change is a primary function of the leader. Systems theory has a revolutionary effect on business operations. Many an executive has found the revolution chaotic (Mockler, 1971).

The systems concept of organizations focuses on the dynamic interaction and intercommunication among components of the system. Systems theory subordinates the separate units or departments of a business to decision-making information and communications networks. Understanding this is critical to leadership excellence in managing change, since the effects of change, whether technological, managerial, or organizational, will produce a dynamic interaction among components of the organizational system. In such a case, both the information networks and decision centers must be activated by the leader to monitor the impact of change.

We will develop the concept of systemic perspective from the major questions posed by the systems approach: (1) What are the strategic parts of the system? (2) What is the nature of their mutual dependency? (3) What are the main processes in the system that link the parts together and facilitate their adjustments to each other? (4) What are the goals sought by the system? (Scott et al., 1981). This conceptual mapping of the organization in terms of its strategic parts and goal structure simply provides some preliminary insights into strategic planning. The first-level systemic perspective of the leader is noted by Andrews (1971), who notes that by considering the relationships between strategy and organizational structure, strategy and organizational processes, and strategy and personal leadership styles, the executive should be able to span a territory crowded with ideas without losing sight of the purpose which he seeks in crossing it.

The second level of the systemic perspective is described by Katz and Kahn (1973). A "systemic" frame of reference involves dealing with the whole gestalt of the enterprise. The analyst, they observe, is obliged to identify his own values, feelings, and perceptual biases as influencing that data he abstracts from the totality. The emphasis is on identifying tendencies and uniformities in the phenomena, and identifying patterns of relationships among the variables comprising the uniformities and tendencies. The *limits* and *constraints* in the situation must also be identified. Within these limits, it is necessary to predict the expected changes in the total pattern, and in the other components, with changes in any one or more of the component variables. Action programs are then thought of as a cyclical single step at a time of (1) changing the characteristics of one or more variables, (2) assessing the outcome on the total system and on other variables, (3) introducing a subsequent multivariable change, (4) testing reactions, (5) reassessing outcomes, and so on (Katz and Kahn, 1973).

To some extent then, the systemic perspective is a matter of processing information and converting information into policy decisions. This process permits forecasting and probable effects of different courses of action and consequently choosing among them. The test of the external perspective is the ability to make appraisals on a predictive basis. This predictive test cannot be rejected by the leadership of the organization. In summary, the systemic perspective is the leader's ability to see, conceptualize, appraise, predict, and understand the demands and opportunities posed to the organization by its environment. Technical, interpersonal, and conceptual skills emphasize some of the aspects of the systems perspective. Katz and Kahn (1973) have proposed that in the upper managerial levels the conceptual abilities of the manager considerably overshadow in importance his technical skills and his skills in human relations. Yet, this conceptual aspect of leadership has been neglected in research.

The approach to leadership, systemic in its emphasis on interactions, has made at least one major contribution, according to Katz and Kahn (1973). For decades, analysts have been seeking, with little success, a key to effective leadership. Such attempts have concentrated upon the leader–follower relationship and their objective frequently has been to develop a "best style" of leadership. The systemic approach provides a basic rationale for putting these efforts to rest. There is no "one best way of leading" because leadership itself is multifaced. Effective leadership at one organizational level does not necessarily mean that the same kind of leadership would be effective at other levels (Katz and Kahn, 1973).

The leader using the systems approach must address the following questions in his search for excellence:

1. How can an organizational system adjust its functioning to achieve stability and a variable rate of change simultaneously?
2. How can dynamic feedback systems be designed to scan the internal and external environments in order to alert the manager to the malfunctions of the system and problematic dependencies in the environment?
3. Can the culture of the organization be predisposed toward high degrees of equifinality in the decision-making process for the search and design of innovative solutions?
4. What are the main processes that link the components of a system together and facilitate their adjustment to each other?
5. Since creative, dynamic, and innovative organizations require high uncertainty and negative entropy, how can the organization's information-processing systems be designed to meet these two requirements?
6. What is the role of strategic planning for the organization? How can strategic planning serve as an instrument of change?
7. What is the optimal degree of differentiation of the organizational system and how can these differentiated units be effectively integrated?
8. How can the organization be designed as an information-processing system with integrated communications networks, instead of as a typical bureaucratic structure?
9. How can the system's management infrastructure be designed to be responsive to the needs of its subsystems?

4.19 THE CYBERNETIC APPROACH AND HUMANISM

The cybernetic view of the organization as a communication network ignores some basic elements in organizational behavior: the notion of organizational culture; the interaction between attitudes, perceptions, values, and norms; the invisible dimensions of culture; and the change behavior of organizations.

Social systems are anchored in the attitudes, perceptions, beliefs, values, motivations, habits, and expectations of human beings. Social structures are essentially contrived systems: The cement which holds them together is essentially psychological rather than biological." Katz and Kahn (1978) The basic conception of the systems approach (homeostasis, entropy, and equilibrium) does not focus on the more complex and subtle aspects of organizations as behavioral systems.

As behavioral systems, the conversion process of the input–output model of organizations does not lend itself to precise mechanical manipulation and design. The conversion process is the motivation of the individual—his perception, attitudes, and outcomes is which he assigns differential values. The conversion of inputs into outputs is not simply a technical process; it is also a behavioral process. This means that the individual is at the center of the decision making process not management information system, or the computer. A computer mystique has developed that attempts to devalue a person's unique attributes, his spark of divinity, intuition, imagination, creative impulse, consciousness, and growth (Sankar, 1991).

The organization is more than a communications network, a homeostatic social unit, or a collection of stimulus–response bonds. This type of conceptualization explicit in the cybernetic system is a return to the mechanistic notions of organizational behavior and management and a return to the scientific management era. It is an expression of the contemporary movement to extend the paradigms, concepts, and models of the physical sciences to the analysis of behavioral systems, not only to enhance the rationality of organizational decision making, but also to lend some pseudorigor or analytical perspective to the social sciences.

The bypass of the more subtle aspects of organizational behavior from the cybernetic standpoint may be traced to the application of the cybernetic concept of the "black box" at the individual and organizational levels as well as the application of the concept of entropy to organizational systems. The black box is a device for converting inputs into outputs. Miller as cited by Scott et al. (1981) finds it necessary to conceptualize the human organism in a communication situation as a black box whose inputs and outputs can be observed and measured. He further cites Broadbent's information flow diagram of the organism with some of the following characteristics: a selective filter, a channel of limited capacity, a "store of conditional probabilities of past events," a "system for varying output until some input is secured" (Scott et al., 1981).

At the organizational level, Scott et al. (1981) note that the black box is particularly relevant to the analysis of complex organizations because they (organizations) (1) convert inputs of productive factors into outputs of goods and services, (2) are often highly complex, and (3) tend to rearrange their structure as a response to the introduction to foreign influences.

> Imagine that a change occurs in an organization. The change is arbitrarily called an input, although some may prefer the word "stimulus." The organization reacts, and an adjustment is the output.

Beer (1959) argues that it is the primary aim of industrial cybernetics to harness the ability of a system to teach itself optimum behavior.

To do it, however, we must know how to design the system in the first place as a machine itself, whose purpose is teaching. There must be exactly the right flow of information in the right places: rich interconnectivity, facilities for growth of feedbacks, many one-transformation circuits, and so on. The exceedingly complex system must be designed as a black box.

This conceptualization of the individual in cybernetic language as a filter, a channel, etc., is a distortion and represents a mechanistic view of his role in the organization and his impact on a variety of organizational variables. This notion is criticized by Blumer (1953):

The human being is not swept along a neutral and indifferent unit by the operation of a system. As an organism capable of self-interaction he forges his actions out of a process of definition involving choice, appraisal, and decision . . . cultural norms, status positions, and role relationships are only frameworks inside of which that process of formative transaction goes on.

The organization as a communications network is only one of such frameworks. However, the cybernetic model conceptualizes this framework as more critical with individual functioning as a reactive system. This is evident in the "cybernetic propositions" on organizational change. The role of the individual and his attitudes, perceptions, motivation patterns, managerial style, and cluster of values are devalued by these propositions in terms of their relevance to the study of organizational change and innovation. March and Simon (1958) argue that organizational theory begins with the nature of human cognitive faculties, that is, that the basic features of organizational structure and function derive from the characteristics of human problem-solving processes and rational human choice. In the cybernetic perspective, the individual is replaced by systems parameters, as exemplified by the cybernetic propositions on organizational change.

REFERENCES

Andrews, K., *The Concept of Corporate Strategy*, Dow Jones–Irwin, Ill., 1971.

Ashby, R., "Variety, Constraint, and the Law of Requisite Variety," in W. Buckley (Ed.), *Modern Systems Research for Behavioral Scientist: A Sourcebook*, Aldine, Chicago, 1968.

Bakke, E., in M. Haire (Ed.), "Concept of the Social Organization," *Modern Organizational Theory*, Wiley, New York, 1959.

Beer, S., *Cybernetics and Management*, Wiley, New York, 1959.

Bennis, W., *Changing Organizations*, McGraw-Hill, New York, 1966.

Blumer, "Psychological Input of the Human Group," in M. Sherif and M. Wilson (Eds.), *Group Relations at the Cross Roads*, Harper & Row, New York, 1953.

Bobbit, H. R. Breinholt, R. Goktor, and J. McNaul, *Organizational Behavior*, Prentice Hall, Englewood Cliffs, N.J. 1974.

Buckley, W., *Sociology and Modern System Theory*, Prentice Hall, Englewood Cliffs, N.J., 1967.

Burgelman, R., and S. Mardique, *Strategic Management of Technological Innovations*, Irwin, Homewood, Ill., 1988.

Cadwallader, M., "The Cybernetic Analysis of Change in Complex Social Organizations," *American Journal of Sociology*, September (1959).

Child, J., "Organizational Structure, Environment, and Performance: The Role of Strategic Choice," *Sociology*, **6** (1972).

Coleman, C., and D. Palmer, "Organizational Applications of Systems Theory," *Business Horizons*, **16** (1973).

Daft, R., *Organization Theory and Design*, West Publishing, St. Paul, Minn., 1989.

Davis, L. "The Coming Crisis for Production Management Technology and Organization" in Design of Jobs, L. Davis and J. Taylor (Eds.), Goodyear, California, 1979.

Duncan, R., "Characteristics of Organizational Environments and Perceived Environmental Uncertainty," *Administrative Science Quarterly*, **17** (1972).

Etzioni, A., *The Active Society: A Theory of Social and Politial Processes*, Collier Macmillan, London, 1968.

Gerloff, E., *Organization Theory and Design*, McGraw-Hill, New York, 1985.

Hackman, J. R., and G. K. Oldham, *Work Redesign*, Addison-Wesley, Reading, Mass., 1980.

Hellrigel, D., and J. Slocum, *Management: A Contingency Approach*, Addison-Wesley, Reading, Mass., 1973.

Hellrigel, D., J. Slocum, and R. Woodman, *Organizational Behavior*, West Publishing, St. Paul, Minn., 1986.

Katz, D., and R. Kahn, *The Social Psychology of Organizations*, Wiley, New York, 1973.

Katz, D., and R. Kahn, *The Social Psychology of Organizations*, Wiley, New York, 1978.

Kelly, J., *Organizational Behavior: Its Data, First Principles and Applications*, Irwin, Homewood, Ill., 1980.

Leavitt, H., "Applied Organizational Change in Industry: Structural Technological and Humanistic Approaches," in J. March (Ed.), *Handbook of Organizations*, Rand McNally, Chicago, 1965.

Luthans, F., *Organizational Behavior*, McGraw-Hill, New York, 1985.

March, J., and H. Simon. *Organizations*, John Wiley, New York, 1958.

Miles, R., *Macro Organizational Behavior*, Scott, Foresman, Glenview, Ill., 1980.

Mockler, R., "Situation Theory of Management," *Harvard Business Review*, 1971.

Murdick, R., and Munson J., *Misconcepts and Design*, Prentice Hall, Englewood Cliffs, N.J., 1986.

Robey, D., *Designing Organizations: A Macro Perspective*, Irwin, Homewood, Ill., 1986.

Rogers, R., *Organizational Theory*, Allyn & Bacon, Boston, 1975.

Rowe, L., and W. Boise, *Organizational and Managerial Innovation*, Goodyear Publishing, Palo Alto, Calif., 1973.

Sankar, Y., "New Technologies and Corporate Culture," *Journal of Systems Management* (April, 1988).

Sankar, Y., *Corporate Culture in Organizational Behavior*, Harcourt Brace Jovanovich, Orlando, Fla., 1991.

Scott, W., and T. Mitchell, *Organization Theory*, Irwin, Homewood, Ill., 1972.

Scott, W., T. Mitchell, and P. Birnbaum. *Organization Theory. A Structural and Behavioral Analysis*, Irwin, Homewood, Illinois, 1981.

Taylor, J., and L. Davis. "Technology and Job Design" *Design of Jobs*, Davis, L. and Taylor (Eds.), Goodyear, California, 1979.

Taylor, J. C., "Some Effects of Technology in Organizational Change," in J. Jun and W. Strom (Eds.), *Tomorrow's Organizations,* Scott, Foresman, Glenview, Ill., 1973.

Taylor, J., P. Gustarson, and W. Carter, "Integrating the Social and Technical Systems of Organizations," in Donald Davis and Associates (Eds.), *Managing Technological Innovations,* Jossey-Bass, San Francisco, 1986.

Thompson, J. 1967. *Organizations in Action,* McGraw-Hill, New York (1967).

Toffler, A., *The Adaptive Corporation,* Bantam, New York, 1985.

Zmud, R., *Information Systems in Organizations,* Scott, Foresman, Glenview, Ill., 1983.

5
AN INFORMATION-PROCESSING MODEL OF ORGANIZATIONAL DESIGN FOR TECHNOLOGICAL CHANGE

5.1 ADVANCE ORGANIZER

The classical design of organizations was based on authority and power as its rationale. With the advent of the information age, a new design based on the information domain is necessary to cope with change. The structure of the organization should be designed to facilitate information inputs to decision centers. Communications is the process by which organizations change. As the elements of the organization change, the information systems must be redesigned to support them.

The principles from classical and contingency schemes of organizational design are no longer able to meet the requirements for more innovative organizational designs. A change in design principles and a strategic shift in focus from the power rationale to the information domain are urgent.

The information-processing principles, based on the systems approach, cybernetics, and information theory, are the new principles that must be blended with the classical and contingency perspectives for an organizational design for technological change (Figure 5.1).

The role of computer technology in organizational design must be considered. The effectiveness of a management information system is greatly enhanced if the structure of the system conforms to that of the organization of the firm. The geometry of the organization is also affected by computer technology by virtue of its impact on centralization/decentralization, span of control, job design, and decision making.

An information-processing model of organizational design is developed with a call for changes in classical and contingency design principles. The

5.1 ADVANCE ORGANIZER **151**

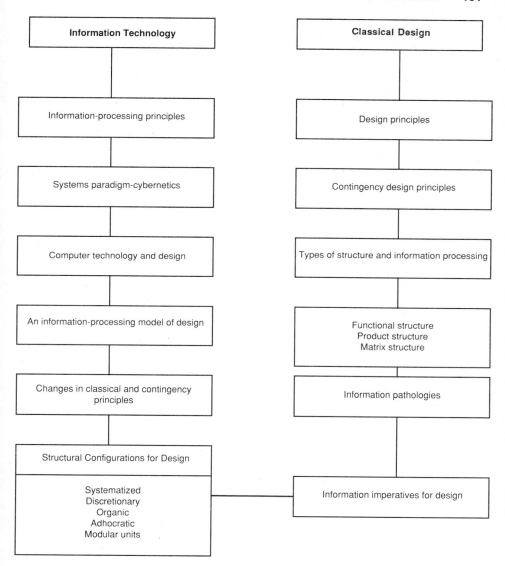

Figure 5.1. Information processing and organizational design.

information domain and its effects on the functional, product, and matrix structures are also examined. The information pathologies in the classical and contingency design make a change in design essential for the information age. The information imperatives for this new age also reinforce this need for a new design. A variety of structural configurations are developed from the information-processing model.

Different structural configurations within a complex organization are needed because of the variety, complexity, and uncertainty of the information domain within it. Changes in the information domain are generally produced by technological change. The systematized modular unit, the discretionary modular unit, the organic modular unit, and the adhocratic modular unit are different structural configurations used for different stages of the change process.

5.2 INFORMATION TECHNOLOGY AND CLASSICAL DESIGN

In organizing the components of a business to achieve its objectives, traditional business organization theory has emphasized the relationships between people focusing on the tasks to be performed, the job positions related to performing these tasks, and the appropriate authority and responsibility for each job position.

McDonough (1969) gives some of the principles of organization that show the traditional approach to organization:

1. Be sure that adequate provision is made for all activities.
2. Group (departmentalize) activities on some logical basis.
3. Limit the number of subordinates reporting to each executive.
4. Define the responsibilities of each department, division, and subdivision.
5. Delegate authority to subordinates wherever practicable.
6. Make authority and responsibility equal.
7. Provide for controls over those to whom authority is delegated.
8. Avoid dual subordination.
9. Distinguish clearly between line authority, functional authority, and staff relationships.
10. Develop methods of coordination.

These principles clearly focus on the role relationships within an organization, and on the physical and functional departmentalization of the business unit. The commonplace block-diagram organization chart reflects this concept. Such relationships are, of course, important in thinking about organizations, but overemphasis of these relationships can obscure the information and communication links so vital to effective decision making within the corporation.

When changes are introduced within the traditional organization structure, according to Mockler (1971), new departments or units are normally added or new responsibilities given to existing departments. Sometimes these additions or changes are made to meet new business needs, sometimes to take maximum advantage of an individual executive's particular combination of talents, and other merely to adjust to the personalities of individual executives. Such a fragmented development process almost invariably leads to some decrease in the effectiveness of the decision-making processes within an organization.

The systems concept of organization attempts to avoid this problem by focusing on the dynamic interaction and intercommunication among components of the system. Systems theory subordinates the separate units or departments of a business to decision-making information and communication networks. Understanding this difference is fundamental to understanding how systems theory has affected business organization and decision making.

The initial chart picturing a business organization restructured around the information flows, instead of around the authority and responsibility units, does not look substantially different, for during the first phases of the changeover only a few departments have been added and some job responsibilities shifted. The change in basic organization philosophy has a profound effect, however, for it creates major changes in the way an organization functions—changes that affect the lives of all the individuals operating within the business system, and changes that after a period of time produce major adjustments in the structure of the business organization. Both in theory and in practice, therefore, the theory of systems is revolutionary for an established business (Mockler, 1971).

Information also affects structure by the manner in which information systems are designed. These should conform to the organizational structure and the delegation of authority within the company. Only then can each organizational unit's objective be established and its contribution to company-wide goals be measured. This means that organizations must be designed around information flow and those factors of information chosen to plan and control performance. Frequently, organizational structure and performance reporting do not coincide. In these cases, information systems cannot truly reflect plans and results of operations. Another major cause of organizational and information mismatch is the lag between organizational changes and the information systems to facilitate them. As needs, structure, and managers change, the information system should be changed to support them. Rarely does one find a change in informational systems matching that in organizational responsibilities and the needs of managers. The result is often an "information lag" (Mockler, 1971).

The structure of the organization should be designed to facilitate information inputs to decision centers. Communication is the process by which organizations change. As the needs, structure, and processes of the organization change, the information system should be redesigned to support them. These changes are necessary to make the adoption and diffusion of innovations within the organization easier.

The definition of communication found in classical organizational theory is the sending of information from superior to worker. The information specifies the method an employee is to use to perform certain tasks. All the decisions about how work is performed are made by the designers. This is centralized decision making in its purest form. The classical theorists equated formal structure with the organization and ignored the roles of communication in coordination; coordination is the counterpart of the division of effort. Coordination was the arranging of separate efforts to produce a predetermined definite end. Classical organizational structure can be visualized as a formalization of communications channels. The objective of formalization is control over organizational activities by making the patterns of information flow uniform and predictable (Mockler, 1971). Two classical principles—scalar chain and centralization—are the means by which control, predictability, and consistency were maintained. These outcomes were the organizational imperatives in classical design. Communication is the critical process in organizing. It links individuals, groups, machines, tasks, authority levels, and functional specializations. This classical design perspective inhibits the adoption and diffusion of innovations because it emphasizes the imperatives of bureaucracy, such as predictability, consistency, standardization, control, and stability.

5.3 INFORMATION, STRUCTURE, AND CHANGE

The availability of information or its accessibility to a decision-making unit, the ease of obtaining information, and the speed with which information flows in the organization are determined primarily by the form of the organization's communication structure as well as the extent to which the organization has an open, trusting climate. In addition, the quality of electronic data processing, retrieval capabilities, input–output, and display features are highly important.

The information process is central to an organization system. It affects control and decision making, influence and power, interpersonal relationships, leadership, and so on. Dorsey (1957), in emphasizing the significance of communications, states that power consists of the extent to which a given communication influences the generation and flow of later communications.

Furthermore, the communication net consists physically of a complex of decision centers and channels which seek, receive, transmit, subdivide, store, select, recall, recombine, and retransmit information. Power and communication are found in all organizational relations: among organizational subsystems, between them and the suprasystem, and between the managerial overlayers and the action underlayers. Power and communication are the two main implementing factors which "transmit" the signals of the managerial centers to the performing units and carry their "feedbacks." Organization networks and consensus—formation structures—are, above all, institutionalized power and communication pathways (Dorsey, 1957).

March and Simon (1962) refer to this aspect of communications. They contend that in highly specialized organizations, information enters the organization at certain specified points. Here, the communications are screened and categorized in terms of a conceptual scheme that has been approved by the organization. There is one place of entry for most communications and one way of categorizing them. All other parts of the organization depend on these technological gatekeepers for the degree to which the data reflect reality (March and Simon, 1958):

> The person who summarizes then assesses his own direct perceptions and transmits them to the rest of the organization becomes an important source of information premises for action . . . by the very nature and limits of the communication system, a great deal of discretion and influence is exercised by those persons who are in direct contact with some part of the "reality" that is of concern to the organization."

The important consideration from the standpoint of organization theory is that there is a network or a grid of personal power centers, though sometimes latent and not expressed. They may or may not coincide with the official structure of authority. Power is not institutionalized in the sense that one can look in the organization manual and find out where it resides. For example, a person of comparatively low status may be a power center because he has been around so long that only he knows the intricate rules and regulations well enough to make immediate decisions.

5.4 POWER, INFORMATION TECHNOLOGY, AND CHANGE

We noted earlier that an organization can be described as a pattern for the distribution of power and also as a pattern for the flow of information. The quickest way to change the distribution of power in an organization is to change the characteristics of the information flow. Computers change the flow of information, a strategy was noted also by Pettigrew (1972) in his study of the computer installation decision in a British company. Control over information is a critical resource for mobilizing power in a decision-taking situation. The ability to absorb uncertainty is a major source of power because it creates a dependency on information for effective decision making. The extent to which information technology channels information to some groups and not others affects the power distribution in an organization (Hussain and Hussain, 1982).

When computers were first used for payroll and other accounting purposes, the controller or financial management department often gained control over the information system. Later, as applications in production and inventory control became more widespread, ownership by staff analysts reporting to production management became more common. Now, with virtually all phases of business affected by the computer, it is common to find a separate functional department housing the computer resource. Because of the technical knowledge needed to understand computing and its applications, these departments have increased their organizational power relative to operating areas.

Robey (1986) further argues that the pursuit and use of power is always couched in legitimate terms because blatant admission of power-seeking violates political norms and undermines one's basis of influence. Therefore, a more acceptable rationale for controlling information must be voiced by information system departments in order to get the cooperation typically required to collect the information from other departments. More rational criteria, such as increasing cost-effectiveness or efficiency, usually justify this data collection. For example, production planners often use computer systems to analyze production and inventory costs, sales forecasts, manpower availability, and other operations data. By doing so, they can produce more efficient production schedules and reduce total costs substantially. As a consequence, their power in the organization is commensurately enhanced, their salaries and staffs are increased, and they enjoy higher status.

It is hard to deal openly with organizational politics, especially when information technology is involved. However, Robey (1986) contends that managers must not naively accept proposals for computer centralization or decentralization without considering resulting power distribution and the impact on other areas of organization performance. Technologically, decentralizing the computer resource by giving ownership to operating departments is quite feasible with distributed processing, as noted earlier. Operating departments often jusitfy this decentralization in the name of efficiency, greater responsiveness to local problems, and so forth. Underlying these reasons is also a quest to regain power through control of information. Likewise, arguments that favor centralized data processing also stress rational criteria like reduction of wasted hardware, more efficient use of system analysts and programmers, sharing between departments, and so on. Again, implicit in these arguments is also a quest for power and control (Robey, 1986).

Organizational structure lends predictability and stability to interpersonal communication and thus facilitates the accomplishment of administrative tasks. A critical reason for studying organizational communication is that it occurs in a highly structured context. An organization's structure tends to affect the communication process.

5.5 SOME DESIGN PRINCIPLES

5.5.1 Classical Design

The organizational chart designed by both classical and contingency principles depict an organization as being structured around the authority and power variable rather than around information-processing systems, information flows, and information links so vital to effective decision making.

In the design of organizations, two major perspectives have emerged, the classical and the contingency perspectives. The latter involves a set of principles that produced the bureaucratic design. The major organizational design principles are (1) scalar chain, which establishes a chain of command or a hierarchy of authority; (2) unity of command, which enforces no bypassing of the chain of command or communications channel; (3) unity of direction, which stipulates the clustering of activities, programs, and functions with similar objectives into departments; (4) division of labor, which subdivides and fragments jobs and the work flow into component elements; (5) specialization, the utilization of specialized expertise, information, and knowledge for each job component; (6) formalization, the use of rules, regulations, and procedures to coordinate the information, resource, and work flow; (7) span of control, which identifies the number of hierarchical levels and supervisory span; and (8) centralization, the locus of authority of decision making in the strategic apex of the organization.

The design of the bureaucratic structure using these design principles was aimed at enhancing predictability, consistency, control, order, stability, standardization, and efficiency. These were the organizational imperatives in classical design. The critical question for design in the information age is whether these classical design principles are relevant for organizational design and the management of technological change.

5.5.2 Contingency Design

The contingency perspective provides another set of principles for organizational design. The design of an effective organizations is contingent on the environment and the type of technology adopted by the organization. From the environmental perspective emerges the concepts of differentiation and integration. Because of technological and environmental change, complexity, and uncertainty, organizations must have a high degree of differentiation among its functions, tasks, and programs and a high or low integration among these elements. If the environment is stable, then the organizational design with a low degree of differentiation and either high or low integration will be effective.

The major design problems from the contingency perspective for a manager are: (1) What is the optimal degree of differentiation among organizational units? (2) What is the optimal degree of integration among these units? (3) What type of structural configuration is appropriate for each unit because of differences in perceived uncertainty in their subenvironments?

The contingency design principles from the technological imperative are that (1) the interdependence among the work flow (pool, sequential, reciprocal) will determine the type of structure for an organization; (2) the degree of technical complexity in mass, continuous, and unit technology will influence the type of structure adopted by an organization; and (3) the degree of variety and analyzability of the work flow will determine the type of structural configuration for each unit of an organization.

5.6 A CHANGE IN ORGANIZATIONAL DESIGN

If the organization is to be designed as an information-processing system, a strategic change in focus on organizational parameters, principles, and paradigms is imperative.

Modern organizational theory is far from unified. Each theorist and researcher in organization and management gives a different emphasis and provides a different breakdown of the elements. Lawrence and Lorsch (1967), for example, identifies the following five elements: structures, planning, measurements and evaluation schemes, rewards, selection criteria and training. Most recently, organizational theorists have focused not only on the internal structuring of organizations, but also on the contextual dimensions of the whole organization, such as size, technology, and environment (Daft, 1989).

According to Luthans (1986), despite divergent viewpoints, the systems approach, more than any other conceptual approach, has led organizational theorists to take a unified view of the organization as a whole made up of interrelated and interdependent parts. This view of organizations reflects the concept of synergism. The synergistic effect means that the whole organization is greater than the sum of its parts. This is an outgrowth of and closely related to the gestalt school of psychological thought. We will examine some major concepts of the systems paradigm that have relevance for the design of the organization as an information-processing system in subsequent sections of this chapter.

The recent view of organizations as information-processing systems facing uncertainty serves as a transition between systems theory and cybernetics information theory. The information-processing view makes three major assumptions about organizations. First, organizations are open systems that face external environmental uncertainty and internal, work-related task uncertainty. Galbraith (1973) defines task uncertainty as the difference between the amount of information required to perform the task and the amount of information already possessed by the organization. The organization must have mechanisms and be structured in order to diagnose and cope with this environmental and task uncertainty. In particular, the organization must be able to gather, interpret, and process appropriate information to reduce the uncertainty. The second assumption is as follows. Given the various sources of uncertainty, a basic function of the organization's structure is to create the most appropriate configuration of work units as well as linkages between these units to facilitate the effective gathering, processing, and distribution of information. The final major assumption involves the major organizational units of the system. Because the subunits have different degrees of differentiation, it is necessary to examine the information subsystems (e.g., the sensor, data-processing, decision, control, and memory subsystems) of these subunits. Here, we are concerned with the structural mechanisms that will facilitate effective coordination among differentiated yet interdependent subunits.

Tushman and Nadler (1978) formulate the following propositions about an information-processing theory of organizations:

1. The tasks of organizational subunits vary in their degree of uncertainty.
2. As work-related uncertainty increases, so does the need for increased amounts of information, and thus the need for increased information-processing capacity.
3. Different organizational structures have different capacities for effective information processing.
4. An organization will be more effective when there is a match between the information-processing requirements facing the organization and the information-processing capacity of the organization's structure.
5. If organizations (or subunits) face different conditions over time, more effective units will adapt their structures to meet the changed information processing requirements.

The above propositions summarize the current state of knowledge concerning the information-processing view of organizations. Although the focal point of this approach is the interface between the environmental uncertainty—both external and internal—and information processing, it is closely related to systems and contingency theories.

In Chapter 4, we noted that the systems and cybernetic approaches recognize the need to study interactions of the subsystems in an organization; they focus on information as the key to analyzing and understanding organizations as social systems. Consequently, communication theory and information theory are central in the development of cybernetic systems and offer some insights for the new field of organizational science. Communication is the basic process facilitating the interdependence of the parts of the total system; it is the mechanism of coordination. Management in cybernetic language is the processing of information and the conversion of information into decision outcomes.

Some Systemic–Cybernetic Principles for the New Corporate Design: A Summary

1. The greater the task uncertainty, the greater the need for information processing within the organization.
2. The larger the number of elements in the decision-making process, such as the number of departments and product lines, the greater the information-processing requirement of the organization.
3. The greater the degree of interdependence among the elements of decision making, the greater the information-processing requirement of the constellation of tasks.
4. To increase the capacity of the organization to process information, more flexible integrative mechanisms are needed.
5. Task variety, change, and complexity will produce different degrees of

uncertainty in the information domain and the decision-making centers.

6. The structural configurations of an organization must be designed with information as the design parameter.
7. The rate of innovation is a function of the rules organizing the problem-solving trails (output) of the system.
8. The capacity for innovation cannot exceed the capacity of the information-processing systems of the system.
9. The rate of innovation is a function of the quantity and variety of information.
10. A facility, mechanism, or rule for changing organizing patterns of a high probability must be present to facilitate flexibility of processing.
11. The rate of change for the system will increase with an increase in the rate of change of the environment (input). That is, the changes in the variety of the inputs must force changes in the variety of the outputs or the system will fail to achieve ultrastability.
12. The speedy conversion of information into decision outcomes is facilitated by a variety of feedback loops, information channels, coordination modes, and data-base configurations.
13. The information-processing capacity of an organization is determined by U, N, and C (task uncertainty, number of decision elements, and degree of connectedness respectively). Any change in U, N, or C will change I (information-processing requirement).
14. Within each structural configuration of the complex organization, the relevant information subsystem (sensor, processing, etc.) must be activated for optimal utilization or rational choice processes (Chaffe's seven information criteria).
15. At the three levels of the organization (strategic, tactical, and operational), uncertainty in (1) cause–effect relationship, (2) decision objectives, (3) means–end chain, and (4) decision outcomes means that information-processing requirements for each problem set will be different at each level.
16. At the nonprogrammed decision level, the semistructured decision level, and at the programmed decision level of the organization, different communications networks, channels, connectivities, configurations, and feedback loops will necessitate different modular units and structural frameworks.
17. The flow of information, communication networks, and channels do not have to coincide with the flow of authority in the organization.
18. The degree of centralization and decentralization will be determined by the information systems architecture and parameters, the DDP configurations, and characteristics of the information domain rather than the power rationale.

19. The greater the rate of change and the degree of complexity of the environment, the more the organization must differentiate its information systems, work flows, and coordinative modes.
20. The degree of integration of the organization's subsystems will be partially determined by U, N, C.
21. The degree of complexity, variety, and uncertainty of the information technologies adopted by the organization will partially determine the type of modular units adopted by the organization. With advanced manufacturing technologies, the technological imperative in organization design emerges as the critical design parameter.
22. A culture audit of the communications artifacts, symbols, rituals and rites, value systems, and patterns of basic assumptions is a requisite in the design of the new organizational paradigm.
23. A joint optimization of information system attributes and individual attributes (intuition, imagination, creativity, vision, etc.) will involve, among other factors, user-system compatibility, a variety of cognitive styles for information processing, conditions for curiosity, fantasy, and challenge, intrinsic motivators (e.g., cognitive drive, self-actualization, variety, etc.), and the redesign of work as the medium through which the individual develops his value potentialities.

5.7 THE IMPACT OF COMPUTER TECHNOLOGY ON ORGANIZATIONAL DESIGN

Computer technology affects major elements of organizational structure, decision making, job design, and strategic managerial functions, and hence organizational architects must redesign the organization around information technologies.

It is important to examine the changes that the computer generates in organizations in which it is implemented because managers need some guidelines in adapting organizations to computer technology. The impact of computer is compared to the effects of the industrial revolution in changing hierarchies, realigning political power structures, radicalizing labor, and disrupting social and psychological patterns.

The effect of computer technology on *organizational structure* is demonstrated by (1) a decline in the span of control which changes the organization's geometry, (2) a decline in the number of levels in the organizational hierarchy, (3) a consolidation of departments, (4) a shift from parallel departmental configurations to functional forms, and (5) an increase in the centralization of control (Whisler, 1970; Daft, 1989; Child, 1984). The impact of the computer on *decision making* is demonstrated by (1) an integration and consolidation of previously separate decision systems, (2) a shift upward in the locus of decision making in the organization, (3) the rationalization and quantification of decision making, (4) changes in decision-making functions at middle and lower management levels in the organizational hierarchy (Hussain and Hussain, 1982; Robey, 1986). The impact of computer technology on *job design* is demonstrated by (1) a tendency to routinize and narrow job content at clerical and supervisory levels, (2) the expansion of job content at upper management levels, (3) the growth of information-processing jobs, and (4) the need for job enrichment programs. Finally, computer technology affects the *major functions of management,* namely, planning, organizing, coordinating, and controlling, through its impact on report generation, inquiry processing, and data analysis (Kroeber, 1982).

A major variable that computer technology affects is the *power distribution* within the organization. Since power and authority are variables that provide the rationale for organizational design, this design parameter will be examined separately. We had noted earlier that an organization can be described as a pattern for the distribution of power and also as a pattern for the flow of information. The quickest way to change the distribution of power is to change the characteristics of the information flow. Computers change the flow of information in an organization by affecting the information channels, networks, locus of feedback loops, and configurations for distributed data bases. Control over information is a critical resource for mobilizing power in a decision-making situation. The ability to absorb uncertainty through decision support systems, strategic information systems, and distributed data processing is a significant source of potential power for information analysts and managers. Issues of centralization and decentralization of information systems and data bases are also linked to the power variable. Organizational politics as it relates to information technology is a critical subarea for studying the impact of computer technology on organizational design (Sankar, 1991). Since computer technology affects most major design parameters, organizational architects must develop guidelines for adapting the organization to information-processing systems and designing the organization around the information domain.

Another way to look at the impacts of computer technology on the organization is to study the changes that MIS (management information system) produces after it is implemented.

Murdick and Munson (1986) developed a "before MIS" and "after MIS" picture from various studies to illustrate the changes produced by information technology.

Before the MIS is installed, the following situation is typical:

1. The organization is departmentalized by function.
2. Communication is primarily up and down the hierarchy, with some committees for integration.
3. Management style tends to be autocratic or paternalistic.
4. Leadership is based more on power than on professional competence.
5. The role of the manager is very important because of ambiguity of situations caused by lack of information. The emphasis is on managers who can forge ahead in ignorance using low levels of conceptualization and management sophistication.

6. Nonmanagers are limited in their scope of action because of ambiguity of objectives and lack of information about the business system as a whole.
7. Aspirations and motivations of nonmanagers are heavily influenced by goals set by their managers.
8. Motivation of nonmanagers is directed more toward filling social needs on the job than toward self-fulfillment through control of one's own work.
9. Group conformity, group norms, and group cohesion are sought above individual job performance.

After the MIS is installed, changes that may occur because of its effect on the organization include:

1. The organization becomes more sophisticated in structure as a complex cross-functional structure, or a product/investment center/functional/geographical structure.
2. More communication is formalized into networks represented by the MIS. Emphasis is on getting information directly to the individual responsible for a task rather than having it pass through his manager.
3. Leadership style must become flexible and participative, or else it fails.
4. Management must be based more on competence than power. Those managers lacking technical competence become very insecure.
5. The old role of the manager changes. Emphasis is on analytical skills and a high level of conceptualization. Some of the pre-MIS managers become obsolete.
6. Nonmanagers share in developing clear-cut objectives based on information about the total business system.
7. By setting their own goals, the nonmanagers develop their aspirations and motivations based on their performance and beliefs.
8. Motivation is based on a climate for growth and control over one's own situation.
9. Group cooperation rather than group conformity becomes an objective.

The changes in the nine organizational characteristics make the organization more conducive to technological change. The adoption and diffusion of innovations are facilitated by such characteristics as flexible communications networks and leadership style, a high level of conceptual skills, a climate of growth, and group cooperation.

5.8 THE COMPUTER-BASED INFORMATION SYSTEM AND ORGANIZATIONAL DESIGN

The effectiveness of a management information system is greatly enhanced if the structure of the system conforms to that of the firm's organization.

Computer control resembles a spectrum, parallel to a company's organizational chart. Information from lower operational levels is communicated to higher levels, which summarize the information, send it still higher, and also make control decisions to be sent back to lower levels (Robey, 1986).

Information is the essential factor within each organizational level. At the strategic apex of the organizational hierarchy, higher managements need information to formulate strategic plans and to evaluate them; this is at the policy level. At the planning level, information is required to convert strategy into tactics (detailed plans and schedules and their evaluation). At the operational level, information is required to carry out production or refining or marketing plans. Each level requires a different category of information to achieve the types of decisions for which it is responsible. Generally, this information will contain less detail as it rises through the levels of the hierarchy. The differences in control objectives from level to level within the organization have different influences on the type of computer hardware and software required to implement control at each management level (Robey, 1986).

A fundamental question confronting the system designer as he begins to think about the data-base aspect of management information systems is: How many levels of data base are there going to be in the organization? There is the organizational hierarchy or management hierarchy with different information needs at its various levels, and so the question arises whether the information system can serve all these levels in an organization with a single data base.

5.9 INFORMATION TECHNOLOGY, CENTRALIZATION, AND CONTROL: CHILD'S PERSPECTIVES

With the potential of new information technology for making precise data speedily available to any level of the organization and for integrating information and offering analytical facilities such as trend analysis, the questions of centralization and the role of middle management become highly relevant (Child, 1981). They present some significant organizational design choices. The case for centralization is based on its efficiency: centralized decision making involves fewer people, less formalization, and less investment in control and integration systems. The case for delegation is based on flexibility: when decisions are delegated to managers who are closer to the action or problem, they can respond to changing circumstances without delay, instead of waiting for the information to be passed up the organization's hierarchy. The use of information technology to enhance information systems such as those monitoring production processes, retail sales, or distribution offers an opportunity to resolve some of the tension between efficiency and flexibility in decision making. It creates the option of recentralizing decision-making discretion because of programs to integrate data and simplify its presentation. This recentralization aims at consolidating a small elite of senior managers and associated specialist and technical experts within a primary sector. There is no overwhelming technical reason, Child (1981) further argues, why information technology cannot be used to facilitate more effective delegation. This can be done in two main respects. The first is by plugging each local unit of an organization into a common file system, thereby allowing a unit to be immediately aware of the situation of other units and of wider consequences of the decisions it might make. By feeding its proposed courses of action into a common information system and signaling any departure these would entail from rules or precedents, a local decision unit could inform the center and other units and elicit rapid feedback. The balanced view and awareness of the whole, which are advantages claimed for centralization, could therefore be taken into account within a delegated system. Second, the improved analytical facilities of new information technology, such as programs for sensitivity analysis and financial modeling, when combined with greatly improved data could be used to enhance the capacity of local units to make sound judgments in their decision making (Child, 1981).

The application of modern information technology permits substantial changes in the context of the manager's work and in the structure of management. Excessive time is spent by managers on problems and incidents of short-term character. Many of these problems are caused by the lack or inadequacy of information and absence of routines that adequately utilize the information available. Interpersonal communication also requires a substantial segment of managerial time. Some management and clerical jobs may be almost wholly devoted to information processing. New technology has the potential to improve information deficiencies and reduce the time devoted to communication between physically separated locations. This can free top management for strategic issues and also economize on middle managerial and clerical staffing requirements (Child, 1981).

The introduction of superior technological facilities for communication has some relevance for the grouping of activities that is adopted. It makes it possible to place less reliance upon roles, such as those of liasion officers or integrators, designed to ensure that the necessary degree of integration takes place within an organization, and instead to substitute a use of new communications technology which distributes information more effectively and permits ready interaction over spatial distances. Structural arrangements intended to improve integration are costly in terms of staffing and can blur the allocation of responsibility. If, with the application of new technology, the linkage and overlays between different activities can be secured with reduced investment in coordination roles, then a simplification of management structure should be possible (Child, 1981).

Robey (1986) reviewed a number of studies that show the computer to be associated with centralization and concluded that the studies have three things in common. They report cases of organizations that have computerized systems and centralized structures. There are also reports of organizations that exist in relatively stable or simple task environments. Though the studies largely differ in methodology and sometimes in basic definitions, they all support the proposition that task environment moderates the impact of computers on organization structure. Another set of studies involving a gas company and international oil companies with computerized systems, as well as a large manufacturer of heavy industrial equipment, points to an increase in decentralization. In the firms showing a high degree of decentralization, the task environments were complex, dynamic, and uncertain—factors that encourage decentralization. To conclude, it appears that conditions in organizational task environments do affect the degree of centralization or decentralization. Since computers were evident in all the organizations, it may safely be assumed that computerization does not explain the variance in structure. Instead, we may regard computerized, information technology as a flexible mechanism that can facilitate either form of structure depending on the more basic requirements that the task environment imposes (Robey, 1986).

5.10 THE INFORMATION-PROCESSING MODEL OF ORGANIZATIONAL DESIGN

Computer-based information system is organizational design. Power and authority are conventional variables in organizational design. Information processing and conversion must emerge as the critical variables for the design of organizations for the information age. Any design for change, according to Toffler (1985), must relate to the information-processing system of the organization. This system must be designed with some degree of flexibility to cater to the different information-processing requirements of each level of the organizational hierarchy, that is, the operational, tactical, and strategic levels (Murdick and Munson, 1986).

The systems view of organizations considers the integrative nature of information flows. The structure of the organization should be designed to facilitate information inputs to decision centers. The pattern of structured–unstructured decisions varies with levels of the organizational hierarchy and so must the information systems. Organizations, whatever their mainstream transformations, are information-processing systems. In order to serve as an information processor, the organization must facilitate the acquisition, interpretation, and synthesis of information in the context of organizational decision making (Tushman and Nadler, 1978). The ability of an organization to process information effectively is limited by the quality and fit of its structure and processes, the capacities of its communications channels, and the behavior of its people. Depending on the manner in which these factors combine for a given organization, it has two broad types of information systems through which its members can disseminate, gather, and synthesize information. One of these systems is mechanistic, formal, and documented, and it interconnects positions of the formal structure. The second system is organic, informal, undocumented, and it interconnects people rather than positions (Robey 1986).

The role of information technology as supporting bureaucratic forms potentially blinds us from considering its role in more organic designs (Robey, 1986). Information technology does not always result in more bureaucracy. Information systems and distributed data bases can give project managers the information needed to coordinate project activities across functional boundaries. The information system in this context supports an organic structure. Various organizational forms are compatible with information technology. Some information systems centralize bureaucratic decision making; others support more organic designs and matrix designs with dual authority command hierarchies. Information technology is a flexible tool that can be used to support many design alternatives (Robey, 1986).

The critical nature of the information-processing perspective to organizational design can be considered in terms of corporate strategy as well. Different elements of an organization's strategy can be seen as posing different requirements for information processing. If strategies can be measured in terms of the kind and amount of information processing required to implement them, then we can create a general framework for hypothesizing fit or congruence between strategy and structure. There is a good fit between structure and strategy when the information-processing requirements of a firm's strategy are satisfied by the information-processing capacities of its structure.

5.11 AN INFORMATION-PROCESSING MODEL

The systems and cybernetic concepts of organizations focus on the dynamic interaction and intercommunication among components of the system. Systems theory subordinates the separate units or departments of an organization to decision-making and communications networks. The cybernetic view of organizations is concerned with decisions, control, self-regulation, and feedback operating in dynamic equilibrium. Both take into account the integrative nature of information flow. This concept is demonstrated in Figure 5.2, where each organizational entity is seen as an information system with the components of input, processor, and output. Each is connected to the others through information and communication channels, and each organizational entity becomes a decision point. Thus, the communication process is dynamic, because as the objectives of the organization change, the content of communication must also change in order to alter or reinforce the actions of various segments of the organization. In the language of the systems model and cybernetics, communication is a linking activity of organizations. This means that organizations must be designed around the information domain

168 AN INFORMATION-PROCESSING MODEL OF ORGANIZATIONAL DESIGN

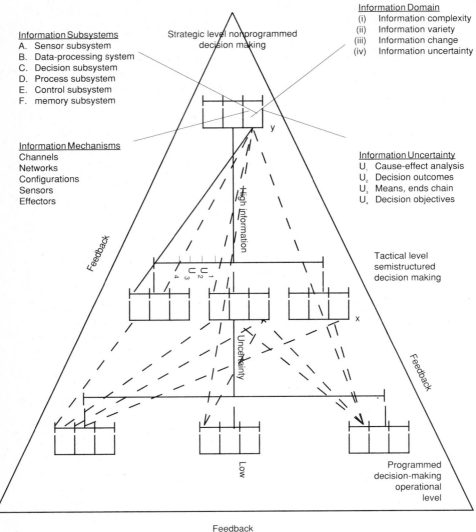

Figure 5.2. The organization as an information-processing system.

and those parameters of information (e.g., interdependence, complexity, uncertainty, feedback, variety, information subsystems, etc.) that are relevant to plan and control performance.

At a practical level, designing the organization as an information-processing system means that the information subsystems (e.g., the sensor, data-processing, decision, process, control, and memory subsystems) may be integrated into each structural configuration within the organization and along levels in the organizational hierarchy. At the strategic level of the organization, where there is high uncertainty in decision outcomes, objectives, and the means–ends chain, more information processing is critical for strategic planning. The sensor subsystem must be activated through a variety of information systems: the strategic information system (SIS), decision support systems, and distributed data bases (DDB). Planning goals and action goals at the tactical and operational levels of the organizational hierarchy, respectively, means that the other information subsystems, which use other varieties of information systems and data bases, may be more effective for semistructured and structured decision making.

At each level, information mechanisms such as channels, networks, feedback loops, and distributed data-processing (DDP) configurations will vary with elements of the information domain. At the strategic level, the mechanisms will require some degree of flexibility, with information channels, networks, and configurations linked to decision centers. At each organizational level—the strategic, tactical, and operational—information needs will vary because of the elements of the task and the information domain. Linking information channels and networks to decision centers is more effective than the traditional linkage to power centers. Information, rather than power and authority, emerges as the critical contingency variable in organizational design. The structural configuration will also vary with the information domain. Designing the organization as an information-processing system also means that strategic changes in classical and contingency principles for organizational design must be considered.

5.12 CHANGES IN CLASSICAL AND CONTINGENCY DESIGN PRINCIPLES

The information-processing model violates some of the basic classical concepts of organizational design. Unity of command, which stipulates no bypassing of communication channels, is outmoded since access to data bases varies along all levels of the organizational hierarchy. The localization of feedback loops within a subsystem is redundant, since feedback loops are now determined by the type of data bases, such as distributed data-base configurations, including ring, star, wheel, etc. (Murdick and Munson, 1986). The principle of centralization, which calls for the locus of decision making in the strategic apex, is neutralized because of the conjunction of decision centers with information centers and data bases at lower levels of the hierarchy. The scalar chain principle, which establishes a chain of command based on authority and power rather than on an information domain organized around the need for information in managing work flow complexity, variety, change, and uncertainty, must be reexamined. The unity of direction, defined as the grouping of similar functions and tasks into a structural framework rather than into information configurations, must be further explored.

We will examine one classical principle in more detail, since it affects the geometry of the organization, namely, the span of control. This classical principle determines the number of hierarchical levels and the supervisory span of control. The Lockheed system identifies five factors that determine the optimal span of control of a supervisor: geographic contiguity, similarity of functions, amount of planning, type of coordination, and complexity of functions (Luthans, 1986). With computer technology, these factors become marginal in determining levels in the hierarchy and supervisory span. The information domain is more critical. Such information parameters as the following will determine the optimal span of control: (1) coordination of information flows; (2) number of feedback loops; (3) types of data bases (integrated versus distributed); (4) complexity of data bases and information taxonomies; (5) types of DDP configurations; (6) degree of interdependence among information systems; (7) types of communications networks; (8) degree of integration among the information subsystems (sensor, data processing, etc.); (9) the number of information channels in the work flow; and (10) the connectivity among the information systems.

The information-processing model also shift the emphasis from the contingency design parameters. The optimal degree of differentiation and integration is determined not only by the degree of environmental change, but by the information domain, variety, complexity, change, and uncertainty of information as well. Environmental determinism is still critical, but so is the type of information domain that must be designed to cope with environmental change. With reference to the technological imperative, the organizational structure must cope with technical complexity in Woodward's conventional paradigm and with types of interdependence (pooled, sequential, and reciprocal, á la Thompson's menu). Now the structure must also cope with computer technology, integrated versus distributed data bases, information taxonomies based on degree of complexity and variety of information inputs, information-processing capacity of information subsystems, varieties of DDP configurations, interdependence among the information subsystems, feedback loops within each structural configuration, elements of task uncertainty, information-processing needs, and types of information channels. Principles from cybernetics, the systems paradigm, and information theory, described in Chapter 4, are more relevant and informative as design principles for an organization in the information age. The contingency model of organizational information processing is more complex and dynamic for organizational design. This model can incorporate classical and contingency principles that survive the imperatives of the information age. To emphasize the cogency of these imperatives, we will examine briefly the effectiveness of a variety of structures in the context of information-processing requirements (Figure 5.3).

5.13 INFORMATION SYSTEMS AND STRUCTURE

Information systems operate within organizations and must be tailored to fit the form of organization. There are three organizational forms that will relate to information systems.

5.13.1 Functional Structure

Information systems for functional-form organizations tend to be hierarchical in nature. Information, subdivided by function and data, that is reported at lower-level organizational functions is hierarchically summed up for presentation at the high-level functions (Lucas, 1986). The information systems is generally designed for managerial functions, and therefore summary information by product is problematic. The functional form tends to centralize decision-making authority in the strategic apex. The information systems have predetermined rules for actions and are less flexible for problem solving. Information filtration, distortion, and overload are the major information pathologies in this structure. Because of the emphasis on rules, procedures, operations manuals, and codified information taxonomies, the functional structure cannot cope effectively with information change, complexity, variety, and uncertainty. In addition, the differentiation of organizational units

Parameters	Functional	Matrix	Product
1. Complexity of information	Low	Medium	High
2. Variety of information	Low	High	Medium
3. Uncertainty of information	Low	High	Medium
4. Diffentiation of information subsystems	Low	High	Medium
5. Changes in environment	Low	High	Medium
6. Product changes	Low	Moderate	High
7. Technology	Standard	Complicated	New
8. Product portfolios	Small	Diverse	Several
9. Size of organization	Small	Large	Large
10. Integration of information subsystems	Low	Medium	High
11. Information-processing capacity	Low	High	Medium

Figure 5.3. Types of structure and information processing.

and information-processing systems is limited in this structure. Furthermore, the functional form is slow to respond changes in both the external environment and product characteristics because of its hierarchical bureaucratic decision-making and information-processing systems.

5.13.2 Product Structure

With the product-form organization, activities are grouped by outputs or products of the firm. Each of the product divisions is organized internally by managerial functions such as engineering, marketing, and manufacturing. The performance of the product line can be monitored effectively, since information systems and data are organized around the product line. The managerial functions within the product line can be better coordinated through an integrated data base. There is, however, a duplication of information systems for these managerial functions (sales, finance, accounting, marketing) across product lines. An integrative mechanism, such as an information steering committee, is necessary to avoid this. A distributed data base configuration may also work. The product structure, because of its decentralization of decision making and its communications channels, can cope more effectively with change, complexity, variety, and uncertainty of information. The differentiation and integration of information systems are more problematic for this type of structure. The product form is swift to respond to changes in the environment, product characteristics, technology, and organizational size because of its decentralized decision-making apparatus and information systems.

5.13.3 Matrix Structure

The matrix structure combines elements of the functional and product forms. Its primary feature is the dual nature of its chain of command, information–reporting relationship, and management systems such as performance appraisal, control, and budgetary systems. The information system for the matrix structure is more complex than the other types of organizational forms. It is imperative that the information system be designed to facilitate the dual nature of the system—the functional arm and the product arm of the matrix, which necessitate dual reporting, budgeting, evaluation, and control information systems. The management information system is more complex and must be integrated for the two arms of the matrix. The matrix has a high capacity for information processing and speeds up the decision-making process. Both integrated and distributed data bases can be satisfactorily managed in the matrix form. The matrix can cope with information changes, complexity, variety, and uncertainty because of its integrated product and functional data bases and its strategic suspension of the classical design principles, scalar chain and unity of command. The matrix structure can respond adequately to changes in the environment, product portfolios, technology, and organizational size because of its vertical information systems and distributed data-base configurations. The dual command hierarchies may be problematic because of conflicting role expectations, power bases, and information priorities. However, the high information-processing and information-conversion capabilities of the matrix negate these problems.

These types of structures have many negative and positive features in terms of information processing. The information-processing capacity is limited in all of the structures because they were not designed with the information domain as the primary parameter. Moreover, there are major information pathologies in these structural configurations which impact on the performance of the organization.

5.14 INFORMATION PATHOLOGIES AND ORGANIZATIONAL DESIGN

5.14.1 Classical Design

The major information pathologies of the classical organizational design are information filtration, distortion, overload, and lags in feedback. The main cause of the organizational and information mismatch is the lag between organizational changes and information systems to facilitate them.

Authority and power provide the rationale for organizational design in the classical perspective. In the hierarchical functional structure of the classical organization, the major design parameter is the formal authority at the strategic apex and a chain of command. The functional form has always provided for the routine exchange of information across the chain of command. The coordination of the work flow is provided by rules, regulations, procedures, and operations manuals that regulate the information flow. The coordination of information flow across the organizational units is problematic. The vertical flow of information down the organizational hierarchy through prescribed communications channels is an organizational imperative. The centralization of authority for decision making at the strategic apex legitimizes these communications channels. Information filtration in the classical design is a major pathology of the system. Information overload in the organizational hierarchy is a common feature of the system. Information distortion because of bureaucratic codes, symbols, operations manuals, and specialized information taxonomies is another pathology of the classical design. Feedback on change initiative at lower levels of the hierarchy is quite limited and occurs with extensive lags. Categorization as a decision-making technique is common because of the ritual of categorizing problems into files, codes, and information taxonomies. The main bureaupathology of the classical design lies in the information-processing field. Classical organizational structure can be visualized as a formalization of communications channels. The objective of formalization is control over organizational activities by making the patterns of information flow predictable and uniform. Two classical principles, scalar chain and centralization, are the means by which control, predictability, and consistency are maintained.

5.14.2 Contingency Design

The major information pathologies in the contingency organizational design are the lack of variety of information, communications networks, and configurations of distributed data bases. The organizational structure does not conform to the information-processing systems of the organization.

In the classical contingency design, there are some information pathologies, but they are not as problematic and dysfunctional as in the classical bureaucratic design (Davis, 1986). The chain of command, the hierarchical structure of authority, is still the predominant mode. To cope with environmental and technological change, a high degree of differentiation among organizational units of the product structure is necessary. To coordinate the work flow among these differentiated units, interdependence must be managed through the information flow. In managing interdependence (pooled, sequential, reciprocal, and team), certain integrative mechanisms must be designed. Unlike the classical design, which relied heavily on rules, regulations, and manuals as integrative mechanisms, the contingency design uses liaison roles, task team overlays, and integrating units that may cut across the chain of command in the product structure. These integrative mechanisms are more flexible for the processing and conversion of information into decision outcomes. However, the decision centers are not necessarily aligned with the communications channels, communications networks, or information centers. Power and authority are still the critical criteria that provide the rationale for organizational design rather than information processing and conversion. There are problems in the contingency design of information filtration, information overload, information distortion, and lag in feedback on change initiatives, but these pathologies are not as severe as in the classical design because of more flexible coordination modes for information flow in the vertical and horizontal dimensions of the organization. However, the organizational design does not enhance the flow of information or optimize the use of a variety of communications networks and channels. The processing and effective conversion of information are still problematic.

5.15 INFORMATION IMPERATIVES IN ORGANIZATIONAL DESIGN

The information-processing capacity of an organization must match the information-processing requirements of the network of tasks. A variety of integrative mechanisms must be institutionalized within the structural configurations of the organization to enhance its information-processing capacity, a capacity critical in managing technological change.

5.15 INFORMATION IMPERATIVES IN ORGANIZATIONAL DESIGN

The limitations of both the classical and contingency design of the organization as an information-processing system emphasize the urgency of articulating the information imperatives in organizational design. These guidelines are as follows. There are four criteria for communications effectiveness: speed, accuracy, variety, and richness. Speed is essentially the rapid flow of information to decision centers along the vertical *and* horizontal dimensions of the organization. Accuracy is the minimization of errors in information distortion and filtration. The information taxonomies within the system must be concise, lucid, and logical to permit understanding of the information at various levels of the organizational hierarchy. The use of codes, symbols, and jargon must be kept to a minimum. Information filtration along the organizational hierarchy must be minimized through a change in the organizational culture and its value system, code of ethics, and reward systems. The other important design variable is information richness. Richness pertains to the information-carrying capacity of data (Daft, 1989). Information richness is related to the medium or channel through which it is communicated. Some media are richer because they provide more information to managers. Information richness is important because it relates to the ambiguity of management problems. Rich media provide multiple cues and feedback. Variety as a design variable is as critical as speed and accuracy. When task variety is high, problems are frequent and unpredictable. Uncertainty is greater, so the amount of information needed will also be greater. Managers spend more time processing information and need access to larger data bases. When variety is low, the amount of information to be processed is low (Davis and Olson, 1985).

In addition to these criteria of communications effectiveness, other information imperatives are guidelines. (1) Decision centers must be aligned with the information subsystems. (2) A *network* structure of control, authority, and information versus a hierarchical structure is mandatory. (3) A variety of communications networks must be experimented with and instituted in the design. (4) Integrated and distributed data bases in a variety of configurations must be linked to the computer system. (5) Vertical and horizontal linkage mechanisms for coordinating the information flow must be designed. (6) The organization must be designed around the information subsystems (the sensor, data-processing, decision, process, control, and memory subsystems) rather than around authority and power. (7) Information is the power variable to be used in organizational design, since management is redefined as the processing of information and the conversion of information into decision outcomes. (8) The *culture* of the organization, namely, its symbols, basic assumptions, artifacts, and norms must be managed to provide information for strategic change. (9) Organizational design is the structuring of *task* and *authority* to achieve organizational goals. The structuring of both must be aligned through information-processing and -conversion mechanisms. (10) Because of change, complexity, and uncertainty, the information-processing capacity of the organization must be increased through all the information subsystems for effective strategic planning.

All of the foregoing information imperatives play a role in the effective management of technological change because they interact to produce a degree of flexibility in the structure which facilitates the adoption and diffusion of innovations within the system.

5.16 DESIGN OF STRUCTURAL CONFIGURATIONS FOR INFORMATION PROCESSING

Different structural configurations within a complex organization are necessary because of the variety, complexity, change, and uncertainty of the information domains of the organization. These changes in the information domain are generally produced by technological change. The immediate impact of technological change is on the information subsystems. The degree of variety, complexity, and uncertainty of information will be increased owing to technological change. These changes in the information domain will be reflected in a variety of structural configurations. Different structural configurations are appropriate for different stages of the change process.

The need for the organization to process information arises from the division of work. Organizations are designed when activities are clustered to form jobs, which in turn, are clustered into departments. This creates a work flow that necessitates information flow and resource flow and a need to manage interdependence among these three flows. The greater the level of work subdivision and specialization, the greater the level of interdependence and the greater the need for high levels of coordination. Many coordination modes have been identified, among them standards, plans, and mutual adjustment (Thompson, 1973) plan and feedback; and hierarchical, administrative, or voluntary modes (Kast and Rosenzweig, 1979). These modes are aids to information processing. They facilitate the interpretation, analysis, synthesis, and evaluation of information for decision making at the organization's operational, tactical, and strategic levels. For effective coordination, communications channels must be established along the vertical and horizontal dimensions of the organization to manage the cluster of tasks.

The tasks within an organization may be categorized as containing degrees of complexity, change, variety which produce degrees of uncertainty. With reference to high complexity, tasks are not analyzable because of ambiguity in information categories. Complexity of information ranges from interpretation, analysis, and synthesis to evaluation of information for problem solving. With low complexity, standardization of information categories and the use of operations manuals are adequate for routine problem solving. High change means various elements in the environment necessitate changes in task parameters. Information systems have to be redesigned to cope with these changes in parameters. Low change indicates that there is some level of stability and therefore procedures can be routinized because of predictability of decision outcomes. High task variety shows that problems are frequent and unpredictable. The amount of information to cope with unexpected and novel situations is high because of the need for variety of procedures, techniques, plans, and feedback to manage the work flow. When variety is low, tasks are simple and repetitive and therefore standardized procedures may be adequate for problem solving of limited elements of the task and work flow. High degrees of change, complexity, and variety produce high levels of uncertainty in cause–effect analysis, decision outcomes, and decision objectives. Performance expectations are not clear and procedures are vague and ambiguous. More information as well as access to larger data bases is needed to cope with high uncertainty. Low degrees of change, complexity, and variety produce low levels of uncertainty. Cause–effect analysis is well understood, decision objectives are simple, and performance expectations and procedures are routinized. A simple data base is adequately integrated for problem solving because of predictability of the outcomes of the task.

Change in complexity, variety, change will produce a high or low degree of uncertainty and the need for more or less information and for various types of integrative mechanisms. The more information processing required for task execution, the greater the need for flexible integrators of the organic types because of high uncertainty in cause–effect analysis and decision objectives. The more routinized the tasks, the lower the need for more information and the greater the need for standardized integrators of the bureaucratic types because of predictability of decision outcomes and objectives. A mix of bureaucratic and organic integrators is generally needed for coordination of the work flow and information flow.

The four dimensions of the information domain—change, complexity, variety, and uncertainty—produce four types of structural configurations: the systematized modular unit, the discretionary modular unit, the organic modular unit, and the adhocratic modular unit (Figure 5.4). Each has a distinct role in the management of technological change. At the evaluation stage of the change process, an organic modular unit is needed because of its degree of flexibility and the activation of the sensor subsystem, and its strategic information systems. At the initiation stage of the change process, the adhocratic unit is necessary because of the temporary problem-solving status of this unit. Its flexible integrative mechanisms and its variety of data base configurations are helpful at this stage of the change process. At the implementation stage of the change process, the discretionary modular unit is needed because there is some degree of uncertainty about the effects of the implementation process. Conflict, role ambiguity, and vague performance expectations are prevalent at this stage. At the routinization stage of the change process, the systematized modular unit is effective because of the need for rules, regulations, procedures, and a management infrastructure to integrate the innovation into the management systems.

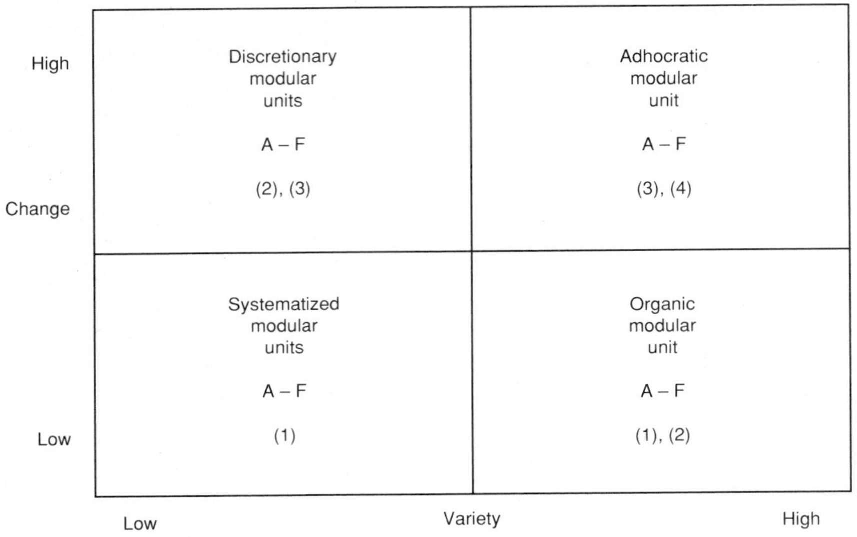

	Low	Variety	High
High Change		Discretionary modular units A – F (2), (3)	Adhocratic modular unit A – F (3), (4)
Low		Systematized modular units A – F (1)	Organic modular unit A – F (1), (2)

Information Domain
(1) Integrated data base
(2) Distributed data base
(3) Decision support systems
(4) Strategic information systems

Information Subsystems
A. Sensor subsystem
B. Data-processing subsystem
C. Decision subsystem
D. Process subsystem
E. Control subsystem
F. Memory subsystem

Figure 5.4. Information domain and structural configuration.

5.16.1 The Systematized Modular Unit

This unit, or structural configuration, is characterized by low complexity, variety, and change. There is high predictability of cause–effect outcomes, goals, and means–ends analysis. The use of rules, regulations, operations manuals, and program repertoire can serve as integrators of the work, information, and resource flows. Centralization of decision making in the strategic apex, prescribed channels of communication, and localization of feedback loops in the subsystem are adequate design strategies. The use of integrated data bases is effective for problem solving. The amount of information can be small—directed toward a limited set of routinized applications. A computerized inventory control system is an example of information support used for a routinized constellation of tasks in this configuration. An available store of knowledge, information systems, and operations manuals can be assembled to handle problems in this information domain. A tight integration among the information subsystems (sensor, data-processing, decision, process, control, and memory subsystems) can be designed because of uniformity, standardization, and predictability of decision outcomes, decision objectives, and means–ends analysis. All of the subsystems can be regulated, controlled, and systematized. For example, the sensor subsystem is not active in this unit because of limited changes in the environmental domain. The data-processing subsystem will use standardized computer applications and integrated data bases because of low complexity, variety, and uncertainty of decision outcomes and objectives. The decision subsystem programs means–ends analysis with a high predictability of outcomes in the task. The control subsystem uses mostly rules, regulations, programs, and hierarchy to coordinate the work flow and information flow. The memory subsystem, which focuses on the storage and retrieval of information, can function effectively because of the simplicity of data bases and low complexity and variety of information inputs.

5.16.2 The Discretionary Modular Unit

This unit is characterized by high change and complexity and low variety of information. Task variety is not high, but problems are more ambiguous and complex and it is difficult to analyze cause–effect outcomes, decision objectives, and means–ends analysis. Problems are solved on the basis of judgment and analytical techniques. There is moderate uncertainty because of high change and complexity, and therefore decision outcomes and performance expectations are not clear. The program repertoire for coordinating the work flow and information flow requires some degree of flexibility to cope with elements changing in the environmental domain. The information-processing and -conversion mechanisms cannot be fully integrated into a data base. Decision support systems must be designed to enhance the effectiveness of strategic planning. The delegation of authority to lower levels of the organizational hierarchy must be initiated because of complexity of inputs in the decision-making process. The sensor information subsystem must be activated for strategic planning to scan the environment; the data-processing system will call for distributed data bases; the decision subsystem is complex because of uncertainty in cause–effect outcomes; the process subsystem, which implements decisions, must be flexible; the control subsystem will be of the clan or market category rather than bureaucratic because of a need for flexibility; and in the memory subsystem, storage and retrieval of information must be characterized by high speed, clarity, and synthesis of information categories.

5.16.3 The Organic Modular Unit

This unit has low complexity and change and high variety, which produce some degree of uncertainty. The variety in the task demands some variations in information-processing capabilities. Problems in this configuration have some degree of uncertainty in means–ends analysis, but not in decision outcomes and decision objectives. Managers need access to large data bases and some decision support systems. Problem solving can be managed by formulas, techniques, and available knowledge bases. The program repertoire to coordinate the information flow and work flow can be managed through plans, MIS, and the hierarchy. The information-processing and information-conversion mechanisms can be integrated data bases. The sensor subsystem is inactive; the data-processing system is integrated; the decision subsystem is enhanced by plans and programs; the process subsystem is activated for variety in data bases; and the control subsystem requires many mechanisms to coordinate the diverse elements of the task and work flow interdependence.

5.16.4 The Adhocratic Modular Unit

This is the most problematic unit for managers because of the high degree of uncertainty produced by high change, complexity, and variety. The computation of goals, cause–effect outcomes, and means–ends analysis is complex. Problems are ill-structured or nonprogrammable, and performance expectations are ambiguous and vague. There is no programmed repertoire to coordinate the work flow and information flow. Organic integrators such as lateral relations, MIS, and decision support systems are essential. There must be decentralization of decision making and communications channels to cope with high uncertainty. Managers need to interact directly with data bases to construct conditional probabilities of outcomes. The use of distributed database configurations must be experimented with to produce variety of information. Flexible integrating mechanisms must be employed because of complexity of work flow. The sensor subsystem is actively scanning the environment through strategic information systems (SISs); data processing is breaking down information into complex categories; the decision subsystem is adaptive, integrative, cybernetic, and enhanced through decision support systems; the processing subsystem integrates information and task elements in complex clusters to implement decisions; the control subsystem uses mostly clan control; and the memory subsystem has to institutionalize some methodology for learning and adaptation. These information subsystems of a complex organization are difficult to manage well because of the volatility of the information domain. Strategic planning and basic research departments are examples of adhocratic modular units. They may be conceptualized as temporary problem-solving units.

The importance of this information-processing view of organizational design is that it determines the organization's pattern of problems and information needs. The information support systems and organizational structure should provide information to managers based on the pattern of decisions to be made. When the organization is designed to provide the correct amount and type of information to managers, decision processes work well. When information systems are poorly designed, problem-solving and decision processes will be ineffective and managers may not be able to explain the problem. The processing of information and the conversion of information into decision outcomes can be effectively executed if the structural configuration matches the elements of the information domain or the information needs of the task, and its degree of complexity, variety, and change. Information becomes the contingency variable, rather than authority and power, for the design of a modular unit. It is possible to have all four modular units in an organization because of the different degrees of task complexity, variety, change, and uncertainty. The network of tasks and variety of information-processing requirements of these tasks make it crucial to design a variety of structural configurations for the organization. This information contingency perspective has the following objectives: (1) to increase the information-processing capacity of the organization; (2) to provide flexibility in program repertoire for coordinating the information flow and work flow; (3) the use of a variety of data bases for information processing; (4) the conjunction of an information/communications channel with decision centers; (5) the use of a variety of communications networks; (6) the locus and integration of information subsystems within a modular unit; (7) the speeding up of the decision-making process; (8) the use of cybernetic principles of feedback, control, and self-regulation to design information systems; (9) the use of the information domain rather than authority as the design parameter; (10) to assist in the design of adaptive organizations for the information age; (11) to facilitate the management of technological change; (12) to design modular units for each stage of the change process, namely, evaluation, initiation, implementation, and routinization.

5.17 THE CREATIVE INDIVIDUAL IN THE NEW CORPORATE DESIGN

A new corporate design based on the systems and cybernetic principles can be as dysfunctional as the bureaucratic machine if a culture audit is not undertaken concurrently to blend the technical imperatives of the new technologies with the humanistic imperatives of the new corporate person. Our view of design is consciously oriented toward improving the quality and effectiveness of organizational life, not just providing computer support for current practices. All innovative technology leads to new practices which cause social and organizational changes whether anticipated or not.

The information processing perspective in the previous sections captures many important aspects of coordinating the activities of people in organizations, but it leaves out some of the most important factors about why people are there in the first place, how hard they work, and whether they find their activities satisfying or alienating (March and Simon, 1958).

According to Fromm (1968) if we are only concerned with input-output figures, a system may give the impression of efficiency. If we take into account what the given methods do to the human beings in the system, we may discover that they are bored, anxious, depressed, tense and so on. The result would be a two fold one. (1) Their imagination would be hobbled by their psychic pathology, they would be uncreative, their thinking would be routinized and bureaucratic and hence there would be inertia in the system (2) They would suffer from tension, stress, and alienation which will reduce their creative potential.

Argyris (1973) not only comments on the conditions that inhibit individual growth and self actualization but also identifies some of the major requirements of tomorrow's organization. In order to meet the challenges of complexity, change, and uncertainty, modern organizations need (1) much more creative planning, (2) the development of valid and useful knowledge (3) increased concerted and cooperative action with internalized long-range commitment and (4) increased understanding of criteria for effectiveness that meet the challenges of complexity. These requirements depend upon (1) continuous and open access between individuals and groups; (2) free reliable information where (3) interdependence is the foundation for individual and departmental cohesiveness and (4) trust, risk taking, and keeping each other informed is prevalent, so that (5) conflict is identified and managed in such a way that destructive win-lose stances are minimized and effective problem solving is maximized.

The culture of the new design will be based on joint optimization, a socio-technical paradigm which is an ethical imperative. Ethical ground rules are the heart of organizational culture. Ethics is the fulcrum in culture for producing change. Pastin (1986). A new design that neutralizes the basic information pathologies of both the classical and contingency design and that stress the information imperatives noted earlier has a greater chance of developing an entrepreneurial culture than the typical bureaucratic culture. Within such a culture, the ethos of the organization facilitates risk taking behaviour, the exercise of discretion and automony in decision making, the prevalence of organizational "slack" for absorbing the consequences of errors in risk taking behaviours. The value system of the entrepreneurial culture rewards the exercise of initiative, judgement, creativity and ethics in decision making. This is the type of culture that organizations must design and nurture to facilitate the emergence of new technologies and their integration into all the domains of the organization Sankar (1991). Designing the new corporation must go hand in hand with a culture audit of its norms, dominant values, philosophy, rules of the game, climate, and patterns of assumptions, organizational symbols and communications rituals, artifacts, and symbols. Virtually every successful new model for re-inventing the corporation was based on the mutual interest of corporations and the people within them. One of the most fundamental shifts is movement away from the authoritarianism-hierarchy to the new lateral structures, lattices, networks, and small teams where people manage themselves. The new corporate design will, therefore, create a new corporate culture that is more humane, value-based, and dynamic.

5.17 INFORMATION, CREATIVITY, AND INNOVATION

The information domain around which the organization is designed is necessary for creativity. The role of communications networks, information channels, and data bases in promoting innovation is rather self-evident, but yet managers engage in behaviors and adopt strategies that lead to information pathologies. The organization must not only promote those structural requirements for innovation, but also those general requirements for innovative creative behavior.

A general practice in organizing work is to constantly remove elements of creativity (involving an element of risk or uncertainty) and group work by dividing and subdividing tasks to the point where no judgment or interpersonal contact remains or is required. Workers and technicians are by no means insensitive to this process. Their frustration is often perceptive and articulate, and comments such as "We are human" and "The work is not fit for human beings" are not uncommon. Information processing and its efficiency preoccupation may be costly in individual and social terms. (Fromm, 1968).

Creative ability is often associated with related concepts such as intuition, inspiration and imagination. These qualities share a common aura of mystique and serve to lend a magical flavour to notions of creativity. Historically, creative behaviour has been studied in terms of nature rather than nurture. It was assumed to be innate, and only the most gifted could lay claim to its power. This belief is no longer upheld and most theorists agree that creative behaviour can be directly stimulated and developed through particular methods. Every individual has creative potential, although it does exist in varying degrees. The significance of this discovery relates as much to the corporate world and the business community as to the realm of psychology. Confronted with an accelerating rate of environmental and technological change, an increasing sense of economic uncertainty, and a sharper competitive posture, most present day firms require constant innovation. Logical analysis is not a reliable means of problem solving in a dynamic environment. Constant change makes it difficult to employ traditional trend analysis and static planning strategies. Common sense and experience is not enough to guarantee success. Given these new developments, it is not surprising that companies are beginning to realize the importance of creative management. The concept extends beyond a creative leader, and collectively refers to the culture within the organization. All levels, not only the vertical hierarchy of management, require the freedom of imagination.

Whatever is flexible and flowing will tend to grow. Whatever is rigid and blocked will atrophy and die (Heider, 1985).

There has been an appalling neglect of the human capacity to create, and this is particularly obvious in the business world (Whiting, 1987). Power is given to facts and figures that are the result of machines, while ignoring the abundant supply of imagination which is a uniquely human product. Technology is moving in new directions and behind each step is an innovative idea. Machines can only create what we have programmed them to create and it is a misconception to think they will provide answers to questions of the future. Their abilities are necessarily rooted to the past. Companies must pursue new dimensions in management practices, strategies, and approaches. The prosperous business is now exploring the benefits of quality circles, enlightened teamwork, corporate visioneering, and an entrepreneurial attitude. As Whiting (1987) points out, productive efforts must be based upon the abilities of many individuals so there can be a blend of divergent and convergent thinking patterns. He maintains that alternative thinking is the essence of the creative act.

Jung (1921) defines creativity as a psychological function that explores the unknown and senses possibilities and indications which may not be readily apparent. There is a strong link to intuitive ability. The leader who is successful at playing "hunches" is seen as visionary. The intuitive thinker sees beyond the boundary of space and time and is not limited by logic or reason.

Innovative corporate cultures are a necessity for the future. They are creative work environments where risk-taking is the norm. Stimulation is a constant feature of the climate that can raise the potential for employee burn-out and stress related ailments. Pressure is often experienced by those expected to produce. The stress is unavoidable, but it can be used in such a way that it results in positive benefits. The structure of the organization should be flexible enough that it can change as the environment dictates and can act as a buffer against stress. Team support and participative leadership will allow the individuals to exercise authority over the creative process and allow them to seek diversion when it becomes necessary. It should also be recognized that an organization cannot function if all its employees are creators. There are routine jobs that are no less necessary because of their simplicity. Individuals must perform a variety of tasks that do not involve stimulation and intellectual challenge. The successful company can maneuver around this obstacle by rotating the mundane tasks amongst employees. On the other hand, some individuals do not desire a challenging career and are satisfied with simpler tasks. As long as the company is matching the employee's goals with assigned work requirements they are likely to avoid an inconsistency that would adversely affect production.

> Creative imagination is worth more than mere book knowledge. Education and intelligence are merely the ways by which we facilitate the liberation of this creative energy (R. Simpson, 1927).

Creativity is a quality that exists in every person and it is one of the features that distinguishes humans from other species. It has allowed mankind to widen the gap in the evolutionary race. It may not always be apparent, however in the right climate, it will flourish. The organizations of the future must invest substantial time and effort in developing this trait because it will distinguish the successes from the failures. The impetus is underway for present firms to begin the transition that will allow for this metamorphosis to occur. The barriers that are now in place are substantial, but they are not impregnable. The goal is to broaden the spectrum of creativity and to incorporate it into every organization. In this way organizations will forever move forward and avoid the crippling effects of stagnation, as as they move towards human and corporate excellence.

REFERENCES

Argyris, C. "Today's Problems with Tomorrow's Organization" in J. Gun and W. Strom, *Tomorrow's Organization,* Scott Foresman, Glenview, Illinois, 1973.

Child, J., *Organizations,* Harper & Row, New York, 1981.

Child, J., *Organizations,* Harper & Row, New York, 1984.

Daft, R., *Organizational Theory and Design,* West Publishing, St. Paul, Minn., 1989.

Davis, G. B., and M. H. Olson, *Management Information Systems: Conceptual Foundations, Structure, and Development,* 2nd ed., McGraw-Hill, New York, 1985.

Davis, D., *Managing Technological Innovation: Organization; Strategies for Implementing Advanced Technologies,* Jossey-Bass, San Francisco, 1986.

Dorsey, J., "A Communication Model for Administration," *Administrative Science Quarterly,* December (1957).

Fromm, Erich, *The Revolution of Hope: Toward a Humanized Technology.* Frederick Ungar, New York, 1968.

Galbraith, J., *Designing Complex Organizations,* Addison-Wesley, Reading, Mass., 1973.

Gerloff, E. W., *Organizational Theory and Design: A Strategic Approach to Management,* McGraw-Hill, New York, 1985.

Gorry, G., and Scott Morton, "A Framework for Management Information Systems," *Sloan Management Review,* Fall (1971).

Hall, R., *Organizations: Structure and Process,* Prentice Hall, Englewood Cliffs, N.J., 1977.

Heider, G., *The Tao of Leadership,* Bantam, New York, 1985.

Hofer, C., "Emerging EDP Pattern," *Harvard Business Review,* March–April (1970).

Hussain, D., and K. Hussain, *Information Processing Systems for Management,* Irwin, Homewood, Ill., 1982.

Jung, C., Psychological Types in H. Read, M. Fordham, and G. Asler (Eds.), *Collected Works of C. G. Jung,* Vol. 6 Princeton University (1921).

Kanter, J., *Management-Oriented Information Systems,* 2nd ed., Prentice Hall, Englewood Cliffs, N.J., 1977.

Kast, E. and Rosenzweig, J., Organization and Management: A Systems and Contingency Approach, McGraw-Hill, New York, 1979.

Kelly, J., *Organizational Behavior: Its Data, First Principles and Applications,* Irwin, Homewood, Ill., 1980.

Kroeber, D., *Management Information Systems: A Handbook for Modern Managers.* Free Press and Collier Macmillan, New York, 1982.

Lawrence, P., and J. Lorsch, *Organization and Environment,* Harvard Business School, Division of Research, Cambridge, Mass., 1967.

Lucas, H., *Information Systems Concepts for Management,* McGraw-Hill, New York, 1986.

Luthans, F., *Organizational Behavior,* McGraw-Hill, New York, 1986.

McDonough, A., *Centralized Systems—Planning and Control,* Thompson, Wayne, Pa., 1969.

March, J. and H. Simon, 1962. *Organizations,* John Wiley, New York, 1958.

Mockler, R., "Situation Theory of Management," *Harvard Business Review,* July 1971.

Murdick, R., and Munson, J., *MIS Concepts and Design,* Prentice Hall, Englewood Cliffs, N.J., 1986.

Pettigrew, Information Control as a Power Resource, *Sociology,* 6, 1972.

Robey, D., *Designing Organizations,* Irwin, Homewood, Ill., 1986.

Sankar, Y., "Organizational Culture and New Technologies," *Administrative Sciences Association of Canada (ASAC). Canada Conference,* 1987.

Sankar, Y., *Corporate Culture in Organizational Behavior,* Harcourt Brace Jovanovich, Orlando, Fla., 1991.

Simon, H., *The Shape of Automation for Man and Management,* Harper & Row, New York, 1965.

Simpson, R., "Creative Imagination," *American Journal of Psychology,* Vol. 33 (1927).

Thompson, V., "Bureaucracy and Innovation" in L. Rowe and B. Boise, *Organizational and Managerial Innovation,* Goodyear, California, 1973.

Toffler, A., *The Adaptive Corporation,* Bantam, New York, 1985.

Tushman, M., and D. Nadler, "Information Processing as an Interacting Concept in Organizational Design," *Academy of Management Review,* July (1978).

Voich, D., Jr., H. Mottice, and W. Shrode, *Information Systems for Operations and Management,* South-Western Publishing, Dallas, 1975.

Whisler, J., *The Impact of Computers on Organizations,* Praeger, New York, 1970.

Whiting, B., "Entrepreneurial Creativity Needed More Than Ever," *The Journal of Creative Behavior,* 21(2) 93–107 (1987).

6
INFORMATION TECHNOLOGY AND CHANGE

6.1 ADVANCE ORGANIZER

The new information society will be more turbulent and complex than the previous industrial society. If companies must change, employees will also have to adapt to these changes. Flexible, well-designed computer-based systems can help workers adapt to change by way of such things as "job aids" built into the systems. Computer technology facilitates the unification of previously fragmented control systems and assists integration through its enhancement of communication. These possibilities present opportunities for change within management with regard to the hierarchical location of decision making, the complexity of coordinative mechanisms, and the size of middle management. The adoption and diffusion of innovations and change can also be made easier by the organization's information system. The management of change, a crucial function for the contemporary manager, must also be linked to computer technology.

A major factor in understanding the innovation process is an understanding of the organization's information system. The strategic objectives of the information technology (e.g., control and integration versus strategic planning and change) will determine the effectiveness and innovation potential of this system. As an instrument for strategic planning, environmental scanning, and the management of change, MIS (management information system) can be effective in the management of technological change.

An understanding of managerial decision making is a prerequisite for good systems design. Four types of decision environments are examined. Since change is a consequence of decision making, it is important to consider MIS in these decision environments. In the management of technological change, a more systematic effect must be made to diagnose (1) the decision environment in terms of goals and technical uncertainty produced by the innovation; and (2) the types of decisions (structured, semistructured, and unstructured) at various levels of the organizational hierarchy. This will help identify the information systems and type of data bases that must be developed and used.

Four stages of the change process are identified: diagnosis, initiation, implementation, and integration. Two types of data bases, integrated and distributed, and their role in the management of technological change are reviewed. Next, we discuss the effects of information technology on organizations in terms of middle management, organizational structure, and power. Some factors limiting the applications of MIS in the management of change are reviewed from the perspective of Kanter (1977).

Finally, it is observed that the power of human information processing must not be devalued by the mystique of computer technology. The power of intuition, creative visualization, lateral thinking, and perception, and a variety of decision styles are noted from the viewpoints of Robey (1986) and De Bono (1971).

Figure 6.1 summarizes the points to be touched on.

6.2 THE IMPORTANCE OF INFORMATION TECHNOLOGY AND CHANGE

To understand the innovation process is a principal factor in understanding the organization's information system. The design of a management information system will partially determine the change orientation of the organization: incremental or innovative. If the design objective stresses planning over control, the probability of innovative change is greater. The major design objectives of a management information system are the reduction of uncertainty in information and the enhancement of the critical functions of the manager in planning and controlling the organization's operations. Planning provides valuable information for both the evaluation and the initiation stages of the change process. In the management of change, information for planning, operations control, and managerial control is vital for the satisfactory integration of the innovation with management systems.

The change potential of a management information system (MIS) is associated with the feedback feature of the system. A large-scale formal organization is a communications system. It is assumed that it can display learning and innovative behavior if it contains specific feedback mechanisms, a certain variety of information, certain kinds of inputs, channels, and storage and decision-making facilities. Therefore, a change in an organization will first manifest itself in changes in the information network, channels, feedback mechanisms, and so forth. The management information system must cope with a changing environment. To adapt to change, the information system relies on feedback, evaluation, and adjustment of data input. The design of

6.2 THE IMPORTANCE OF INFORMATION TECHNOLOGY AND CHANGE 189

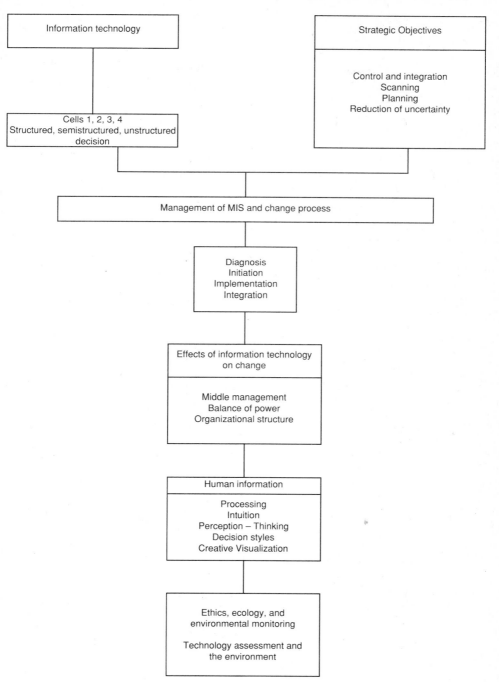

Figure 6.1. Information technology and change.

the data base must be adaptive to the organization's information requirements for innovation.

The role of a management information system in managing change is quite critical. At the evaluation stage, for example, the management information system, by providing information on strategic planning, can facilitate organization diagnosis. Various scenarios may be developed as potential methods for bridging the performance gap indicated by organizational diagnosis. On the basis of this organizational diagnosis, a decision to innovate may be made. At the initiation stage, a management information system assists in the formulation of alternatives. Simulations permit managers to visualize each alternative and project the changes into the future. At the implementation stage, the control function of the management information system becomes more relevant. The control function will monitor the systems to determine if the objectives of the innovation are being achieved.

A management information system also affects the main functions of management, namely, planning, organizing, controlling, coordinating, and staffing through its impact on report generation, inquiry processing, and data analysis—the major roles of an information system. Some additional changes postulated as a result of information technology are the shrinking of middle management, shifts in the balance of power, and changes in organizational structure. A number of conditions also limit the application of MIS to the innovation process.

While information technology plays a large role in the management of technological change, the role of human information processing, with its elements of intuition, creative visualization, lateral thinking, and variety of cognitive styles, is critical. The human decision maker is still at the center of the organizational stage in the management of the complex process of change.

In this chapter, we examine a strategic role that information systems are assuming as a catalyst to foster innovation. Can information systems executives do anything to facilitate the diffusion of innovation in their organizations? Can installing group communications systems help employees be more innovative? Peters and Waterman (1982), in their best-selling book *In Search of Excellence,* identified eight characteristics of excellently managed companies that are typically not found in less successful companies. "Creativity" is getting an idea. "Innovation" not only includes getting an idea but implementing it. Successful companies stress innovation over creativity because they expect the people who get the ideas to implement them. Daft (1989) views innovation often as a group activity, so creating an environment in which people can share ideas easily can provide an atmosphere where innovation is more likely to happen. Peters and Waterman (1982) note that "the nature and uses of communications in excellent companies are remarkably different from those of their nonexcellent peers:

> "The excellent companies are a vast network of informal, open communications. The intensity of communications is unmistakable. . . . It usually starts with an insistence on informality.

For any one project, most communication will be among the members of a small group or task force. Usually, to facilitate the project, these people have been brought together into one work space. With such new developments as computer networks, computer messaging, voice store-and-forward messaging, and computer conferencing, this constraint can be lifted. Members of the group can be located almost anywhere—at different geographic location, in a different organizational unit, and at different level of management."

Peters and Waterman also observe that effective companies almost universally exhibit a "bias for action." If they see something that needs doing—a new marketing method or innovation, a special product for a market niche, or just about anything new—they organize people to achieve this goal almost immediately. They examine the problem until a feasible task emerges with which they can initiate a few useful steps. Peters and Waterman (1982) describe two other related characteristics:

Task Forces. Often, ad hoc task forces are assigned with mangers (not staff) who are just as senior as the problem justifies. These task forces go to work, do the job assigned, then disband. Often, a task force of this kind is given no more than ninety days to accomplish its mission. So the overall project is divided into a series of ninety-day subprojects.

Quick In, Quick Out. As results occur, if the early results are negative, the whole idea is abandoned quickly. If the results are positive, the task force gathers suggestions on how best to take the next steps.

Murdick and Munson (1986) argue that information technology directly supports a "bias for action" through the use of new fourth-generation software tools and software prototyping (iterative development). The general procedure is to identify a portion of the overall application system that will be immediately useful to the users, analyze the needs, and quickly set up the new application using a fourth-generation language. The users start employing it and soon see some things they want changed, added, or deleted. These changes are made, and then the next version of the application system is used. The process is repeated until the users have something they are happy with. So, encouraging prototyping—by professional staff and end users—is another way to foster innovation.

6.3 STRATEGIC OBJECTIVES AND INFORMATION TECHNOLOGY: CHILD'S PERSPECTIVES

Adapting the organization to the information needed for effective planning, control, and operational decision making enables the organization to provide the information infrastructure for change and innovation.

Some strategic objectives that managers have in mind when introducing information technology are identified by Child (1984): (1) reduction in operating costs, (2) increased flexibility, (3) improvement in quality of the product or service, and (4) increased control and integration. We will examine two of these strategic objectives only, increased flexibility and increased control and integration. Child's observations on these objectives are discussed below.

6.3.1 Increased Flexibility

From a manager's viewpoint, one of the most attractive features offered by new technology is the prospect of being able to run a range of production items through a single facility with a minimum of cost and upheaval when changing from one type to another. The computer programming of equipment permits its rapid adjustment to suit changes in production, given the generic limits within which it is designed to perform (Child, 1984). The development of computer-aided design (CAD) adds to this a far greater speed of product changes and redesign, and research is under way to link the CAD/CAM (computer-aided manufacture) stages directly together. In this field, the availability of relatively inexpensive microelectronics is permitting a synthesis between areas that were previously segregated because each was tied to costly, large computers; such areas include computer graphical design, numerical-control machine tools, production schedules, and industrial robotics. The new integration between these areas greatly reduces the cost of product change and significantly improves the economics of flexibility. Rumvelt as cited in Child (1984) suggests that the use of CAD/CAM avoids the requirement of having large production units, for once computers begin to specify parts fabrication needs in a standardized language, and once automated facilities exist that can turn these specifications into a part, the need to have them within the same enterprise diminishes. This points to smaller organizational units. This degree of flexibility is a necessary condition for technological innovation (Child, 1984).

6.3.2 Increased Control and Integration

Control and integration are critical requirements for the successful functioning of an organization. In both cases, it is the ability to transmit information directly across distances and the capacity to apply computational or synthesizing routines, when called for, which are the crucial advantages of the new technology. Information can be passed directly from the scene of operations to management. The feedback loop in the control cycle is significantly strengthened. Integration is enhanced by electronics through the ability (1) to bring together a range of information into one location; and (2) to link physically separated individuals in an interactive mode via telex, communicating word processors, and audiovisual teleconferencing. Modern information technology permits a physical as well as an analytical unification of control information. For example, information on the status of a production plant's work flow and stocks can be displayed in the same location as information on the condition of its equipment from a maintenance perspective. If required, data on temperature, humidity, etc., in the production environment can also be added. An integration of this type of information takes away much of the rationale for inherited patterns of job specialization between, say, industrial engineers, maintenance and production workers, and their respective supervisors (Child, 1984).

Information technology increases management control in three respects. First, it provides faster and more precise knowledge of operating conditions and results via feedback. Second, it reduces the scope of uncertainty and indeterminacy about the work to be done and how to do it. There are a growing number of electronic applications that, with appropriate software, can capture the special knowledge held by experts and subsequently exercise this repeatedly as a routine. Once this stage is reached, the mystique of special expertise is dissipated as a basis for resisting the imposition of managerial standards for cost, efficiency, and output. Third, information technology unified previously segmented control systems and thereby increases the potential for a comprehensive and balanced assessment of performance and change. In manufacturing, the dream (now being translated into reality with flexible manufacturing systems) is of a unified control system complementing integrated computer-aided design, manufacture, and production systems. Such a system, argues Child (1984), would combine controls in regard to physical movements, condition of the plant, stocks, wastage, energy consumption, unit costs, and deployment of personnel. Integrated control of this kind can optimize the overall balance of production activities and maximize flexibility of adjustment.

6.4 CONTROL, INTEGRATION, AND DESIGN CHANGES

Child observes that the unification of control through modern information technology encourages a corresponding integration and simplication in organizational structures. As the interdependencies between financial activities are made more visible through integrated control systems and shared data services, so the logic of team working and networks emerges as the "natural" basis of organization, rather than patterns of work and communication defined predominantly by departmental boundaries. Structural barriers of a vertical kind are likely to weaken as well. The integration of control information, its ready access to top management, and the availability of powerful analytical models renders redundant such roles of consolidating, interpreting, and transmitting information, which was the role of middle management and some staff specialists' jobs. A shrinkage of the middle management level, a reduction in hierarchical levels, and other staff and support roles below the strategic level are some of the design changes in organizations (Child, 1981).

Some of these changes are in turn producing changes in the information-processing requirements at various levels of the organizational hierarchy and, therefore, in terms of the use of information technology. Increased control and integration through information technology and manufacturing technologies will help in the implementation and routinization stages of the change process. Control and integration are crucial for organizational imperatives such as predictability of outcomes, consistency in work flow, standardization of procedures, and evaluation of performance outcomes.

Information is a fundamental resource of an organization. It is so essential to the operations and management of an organization that it is often referred to as the "fabric of an organization" or the "agent for sustaining organizational stability." Voich et al. (1975) observes that managers need information to define goals and to guide the operations of the organization for effective performance. Managers also use information to formulate plans, transmit plans to subordinates, coordinate execution of plans, and monitor the performance of those plans. In addition, managers need a continual flow of information about their customers, their suppliers, and the environment in order to redefine goals and alter plans when a change in the environment necessitates such action.

6.5 STABILITY, CHANGE, AND PLANNING

As a means of coping with information uncertainty and change, managers apply planning through two extreme profiles. Since the planning process can assist in integrating the physical and manpower inputs of an organization, it may contribute to the maintenance of stability. According to Hellriegel and Slocum (1974), plans, by providing predetermined courses of action, create means for partially predicting and controlling the future actions of an organization's human and physical resources. Within organizations, plans enable one unit to anticipate the actions of other units with which it is interdependent.

The attempt by organizations to adapt and innovate, on the other hand, is usually expressed in their long-range plans with their specifications of goals, strategies, and action programs aimed at identifying new markets, products, services, and so on. Hellriegel and Slocum further note that most organizations that would be considered adaptive seem to expend considerable effort on planning activities. These organizations are continuously doing two things: (1) they are developing alternative strategies and tactics in anticipation of different changes in the environment, and (2) they are gathering and analyzing information concerning environmental change and reactions to the organization's present outputs. Information processing and conversion can be geared toward stability or innovative change.

6.6 MIS AND THE REDUCTION OF UNCERTAINTY

The information process can be viewed as the transformation of raw data into information. This conceptualization of information as a process emphasizes two basic concepts: data and transformation of data into information. Data are generally interpreted as unstructured facts and represent potential information. Only when data have been transformed and restructured via the information process do they become information—information, then, is the output of the conversion process.

Information can be defined as data which, after transformation via the information process, reduces uncertainty related to the outcome of a particular problem, event, or activity, Murdick and Munson (1986). This definition of information points to a primary function of MIS, which is to reduce uncertainty encountered in the performance of the various functions that comprise the process of management.

Information can be thought of as the opposite of uncertainty. As the information increases, the uncertainty related to that system is reduced. Thus, systems with more information have less uncertainty. From the point of view of the behavior of an organization, that behavior will be more predictable as the uncertainty associated with the organization's operations is reduced. Information is increased in a system when the degree of freedom to choose among signals, symbols, messages, or patterns to be transmitted is increased. Increasing information increases flexibility and hence increases the predictability that the form of the outcome will be closer to that desired. Any change in an organization will produce uncertainty in information. To the extent that an organization's information systems can reduce this uncertainty, information plays an effective role in the management of change. As noted earlier, the change decisions emphasize the potential role of MIS in the innovation process.

6.7 MIS, THE MANAGEMENT PROCESS, AND CHANGE

In the management of change, information for planning, operations control, and managerial control are vital to effective integration of the innovation with management systems. To the extent that a management information system can meet these information requirements, it will contribute to the effective management of change.

A classification of management functions that is closely related to three major groupings, which identifies the types of major problems organizations regularly face, is provided by Anthony (1965):

1. *Strategic planning* is the process of deciding on objectives, on changes in these objectives, on the resources and policies used to attain these objectives, and on the policies that are to govern the acquisition, use and disposition of these resources.
2. *Management control* is the process by which managers ensure that resources are obtained and used effectively and efficiently in the accomplishment of the organization's objectives.
3. *Operational control* is the process of ensuring that specific tasks are carried out efficiently and effectively.

Specifically, information is required to make possible the fundamental functions of planning, operational control, and managerial/financial control. These are also major functions of management information systems. Data are transformed into information to facilitate the performance of these managerial functions. The organization is an information network, with the flow of information providing the decision makers at varying management levels with the data needed to make decisions of all types. As a result, business information systems, which were once considered mainly accounting record-keeping systems, are now being called upon to support all the planning, control, and operational decision-making processes within an organization. As such, these systems are required to store and process not only information concerning internal operations, but also data concerning competitors' plans, strategies, programs, and performance as well as significant information on the economic, technological, and market domains. The computer is indeed a catalyst for change and innovation.

With reference to technological change, strategic planning can be associated with the organizational diagnosis and initiation stages of the change process. The decision to adopt the innovation is made at this stage with the information and data base at this level. The management control and operations control levels are associated with the implementation and routinization stages of the change process. The information and data base here are more oriented toward applications and the design of management systems to monitor change.

6.8 PLANNING CHANGE

For any organization facing a complex, uncertain environment to be successful, planning must be handled well. However, planning may be the most difficult task performed in organizations. Not only does it require such skills as creativity and synthesis, which many individuals have not fully developed, but it is often preempted by current operating problems because of the tendency of many organizational reward systems to emphasize short-term performance (Zmud, 1983).

The following activities of the planning process are identified by Zmud (1983). Planning premises are those facts, trends, and assumptions that provide a context for planning. Planning goals evolve from a consideration of hierarchically communicated organizational objectives in light of the work unit's current planning premises. Planning strategies might reflect current actions, or completely new actions. Strategy selection is the formal appraisal of expected outcomes of each strategy, resulting in the selection of a plan. Plan implementation involves the allocation of necessary resources, delegation of the necessary authority, and communication of the plan to members. Plan monitoring includes periodic assessment of whether the plan is being carried out as designed, whether progress has been made toward the planning

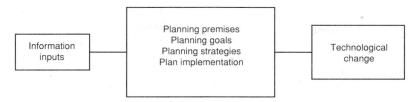

Figure 6.2. Planning and technological change.

goals, and, if the plan is appropriate, given a continuously changing organizational context (Figure 6.2).

The MIS, according to Murdick and Munson (1986), plays an important role in the development of corporate strategy in that it provides a continuous, formalized, and structured gathering of information, both internal and external. The MIS uses organized inputs from managers given specific responsibilities in certain areas, as well as information based on manipulation of transaction data and models.

6.9 ENVIRONMENTAL SCANNING

Since technological change is directly influenced by the environmental domain, the scanning of the environment must be incorporated into planning systems of the organization. One set of planning premises is linked to the technological strategy of the firm.

Murdick and Munson (1986) developed a comprehensive model that is relevant to planning and change. The environmental monitoring system consists of executives who are assigned specific responsibilities for monitoring external factors and developing files on various topics, including economics, ecological issues, demographics, resources, and technology. The predicting and forecasting system consists of line and staff executives who are assisted by the modeling systems when needed. This predictive and forecasting system attempts to develop alternative scenarios of the future through such methods as scenario building, Delphi techniques, or trend analysis.

Murdick and Munson (1986) further observe that the most valuable application of MIS is to assist managers in planning. This need for planning is constantly increasing, since the time in which the firm must adapt to its environment is growing shorter. In addition, changes in the environment are occurring more rapidly. Strategic planning must incorporate the technological change as one of its components. Planning for technological change, using a variety of data bases and information systems, is crucial to the organization's viability and growth.

In this context, information is required to define and evaluate these alternative objectives, resource allocation patterns, and policies. Only information can provide the means for comparing the expected relative costs and benefits of various operating alternatives and serve as the basis for optimal selection. The future direction of the organization and the supporting structure, facilities, and operations is defined on the basis of planning information. The information here is designed to cope with change by reducing uncertainty. Murdick and Munson (1986) argue that one of the major priorities for MIS executives is in the area of strategic uses of information systems. The main challenge in the types of new applications will be the increased emphasis on strategic information systems. In the late 1980s, the managers of most organizations are likely to be demanding information systems that enhance their company's competitive positions. Information systems executives will then be judged on how quickly they can get new strategic systems built and installed and the system's role in improving their company's strategic planning capabilities. This emphasis on strategic information systems can require a change in orientation for many application systems developers. At this strategic level, the information systems can have great impacts on the internal workings or organizations. They can lead to changes in organizational structure and they can be used to foster innovation. Such impacts are strategic in that they can significantly alter the rate and direction of growth for the organization. Another strategic role is that information systems can influence and are influencing new products, new markets, acquisitions, and innovations. They thus have an impact on the organization's external environment.

6.10 INFORMATION FOR OPERATIONS CONTROL

The use of information for operational control is communicative in nature. Information is a means of conveying objectives, plans, policies, and procedures to all segments of the organization. This type of information facilitates the coordination of operating performance in a current day-to-day, ongoing process. Operational control involves a continual ensuring that the organization's operations are being carried out as planned. It includes coordination of the individual efforts of those involved in group tasks and the regulation of resource and work flows among the different work groups. In its operational control context, information serves as both a process activator by communication requirements to each group and as a feedback mechanism by providing a basis upon which to determine whether requirements were received, understood, and carried out. This type of information is designed for operational stability. The emphasis is on standardization, uniformity, and consistency—all factors that interact to produce stability among the components of the system.

Information also provides a link between the organization and its environment. Consumers' use of and performance data on the organization's product and service must be processed so that information on these elements can be defined, translated, and communicated to the organization's internal segments responsible for design and production. Similarly, organizational requirements for resources must be communicated to appropriate sources of supply. In the event of changes in product lines or portfolios, the demand for information is increased.

The control system design requires the following actions, according to Zmud (1983):

Selection of Control Targets: the elements by which performance will be monitored.

Derivation of control indicators: the actual measures used to represent control targets.

Establishment of control standards: performance levels, stated in terms of the control indicators, that are appropriate to goal attainment and acceptable to organizational members.

Specification of corrective actions: adjustments required in behavior that return performance to a path consistent with that planned.

The information required for monitoring the change process is identified in Figure 6.3.

While criteria for productivity are tangible and can be programmed to monitor the objectives of the new technology adopted, other elements of the change process do not lend themselves to such a degree of objectivity. For example, the changes in the degree of interdependence among subsystems of the organization produced by technological change, the changes in the type of mechanisms for coordinating the work flow, information flow, and resource flow, and the changes in behavioral–structural–process outcomes must be monitored as the organization goes through a period of strategic change.

Murdick and Munson (1986) make the following observations on the dynamic nature of feedback control systems. Because of the long-range implication of decisions and the many feedback systems in which the strategic level of the organization is involved, the time lag in adjusting the business to a change in the environment is usually a relatively long one. Most management information systems are based on feedback control that involves a time delay between the occurrence of events and corrective action. A system may be kept right on standard, however, if the response to disturbances or changes can be anticipated so that the system may be changed in anticipation. This type of control is called steering control. It requires a sophisticated MIS that measures not only change but rates and directions of change. The MIS must provide information not only to the manager of the system, but to all elements of the system that perform operations.

The principal design objectives of a management information system, then, are the reduction of uncertainty in information and the enhancement of the critical functions of the manager in planning and controlling the operations of the organization. Both objectives are designed to facilitate more effective decision making, so MIS is essentially a tool for optimizing the decision-making process of the organization. The objectives of a manage-

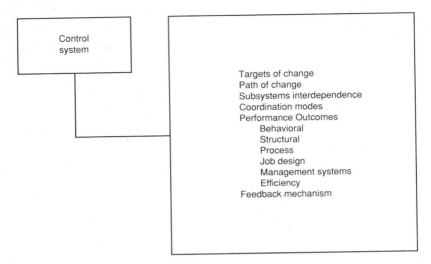

Figure 6.3. Elements of change.

ment information system are more critical today because the role of organizational information, and the formalized use and processing of information, have expanded at a phenomenal rate. The reasons for such growth include (1) rapid growth in size and complexity of organizations and their environments, (2) proliferation of staff groups within large organizations, (3) improvements in planning and control techniques, and (4) improvements in information technology.

6.11 MIS, FEEDBACK, AND CHANGE

The information processes can be viewed conceptually as an input–output transformation unit (Figure 6.4).

The MIS must cope with its changing environment. To adapt to change, the MIS relies on feedback, evaluation, and adjustment of data input. The design of the data base must be adaptive to the organization's information requirements for innovation.

The change potential of a management information system is associated with the feedback feature of the system. The objective of a management information system is to collect data and produce information so that sooner or later the input to the process or, less frequently, the structure of the process can be altered. Viewed in this light, the management information system is part of the feedback link controlling a main process. The feedback is completed, according to Donald (1979), when

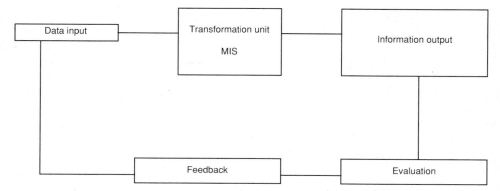

Figure 6.4. Data input and feedback. (From Donald, 1979. Reprinted with permission from Pergamon Press, Inc.)

1. The managers take decisions on the basis of the information provided.
2. The mangers take action to initiate change
   a. in the input parameters in the existing process system.
   b. in the structure of the process system.
   c. decide on the basis of incoming information that no change is necessary.

It is perhaps unfortunate that we have no word in common use to describe the information/decision/feedback link. Such a description might emphasize that this link is an extrinsic one that has to be constantly monitored to see what is working. Information by itself is only of potential use. It is "information in action" that creates results and either (Donald, 1979)

1. Controls the main process to its existing limits.
2. Creates new limits by taking divergent action to alter the existing goals.
3. Alters the process structure, the better to achieve new or existing goals.

Considered as a system by itself, an MIS is frequently an open-loop system; its end product is information, not information in action, and it is extremely difficult to place a value on the information it produces.

A key to understanding organizational search and innovation processes is to understand the organization's information system. From the point of view of cybernetics, any large-scale formal social organization is a communications network. It is assumed that it can display learning and innovative behavior if it possesses certain necessary facilities, that is, structure, and certain necessary rules of operation, namely, content (Cadwallader, 1959).

6.12 DECISION MAKING, CHANGE, AND INFORMATION TECHNOLOGY

Since change is a consequence of decision making, it is necessary to examine the decision-making environments within the organization, the information requirements for these environments, and the role of the organization's information systems in facilitating the change potential of these decision environments.

6.12.1 MIS, Decision Making, Design, and Change

According to Keen and Scott Morton (1978), information systems should exist only to support decisions and therefore organizational activity should be characterized in terms of the types of decision making involved. An understanding of managerial decision making is a prerequisite for effective systems design and implementation. Most MIS groups become involved in system development and implementation without a prior analysis of the variety of managerial activities. This has prevented these groups from developing a sufficiently broad definition of their purpose and has resulted in generally inefficient allocation of resources.

MIS is defined, in part, as an organized set of processes to support decision making within an organization. The key to providing this support is an understanding of how decisions are made. Some of the major issues relevant to decision making are identified by Kroeber (1982). Do managers pick the best solution or merely one that works? To what extent do they rely on intuition? By what means are data manipulated to answer critical questions in the decision process? And, for the MIS manager, how can decision making be enhanced by a computer-based information system? Here, the most important aspects of modern decision making will be surveyed, omitting details of actual computational procedures.

Understanding managerial decision making is necessary for effective systems design. Four types of decision environments are examined. Since change is a consequence of decision making, then it is important to consider MIS in these decision environments. In the management of technological change, a more systematic effect must be made to diagnose (1) the decision environment in terms of goals and technical uncertainty produced by the innovation; and (2) the types of decisions (i.e., structured, semistructured, unstructured) at various levels of the organizational hierarchy. This diagnosis will help identify the information systems and type of data bases that must be developed and used.

Decision Environments and MIS. The model of diagnosing decision environments in organizations is the one developed by Thompson (1967); refer to Figure 6.5. The use of a model is contingent on the organization setting. Two characteristics of organizations that determine the use of decision models are (1) preferences about goals (goal uncertainty) and (2) beliefs about cause–

6.12 DECISION MAKING, CHANGE, AND INFORMATION TECHNOLOGY

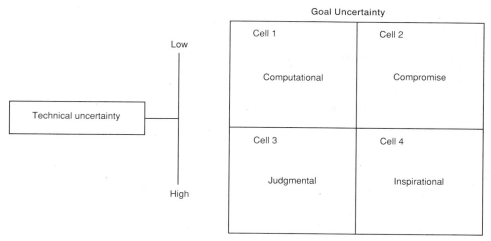

Figure 6.5. Decision environments. (From Thompson, 1967. Reprinted with permission from McGraw-Hill.)

effect relationships (technical uncertainty). Analyzing organizations along these two dimensions suggests which model will be used to make decisions. The definition of these dimensions of the model is taken from Daft (1989).

Goal uncertainty refers to the agreement, among managers about which organizational goals to pursue. This variable ranges from complete agreement to complete disagreement. When managers agree, the goals of the organization are clear and so are performance standards. When managers disagree, organization direction and performance expectations are in dispute.

Agreement about goals is important for the problem identification stage of decision making. When goals are clear and agreed upon, they provide clear standards and expectations for performance. Problems can be identified. When goals are not agreed upon, problem identification is uncertain and more attention has to be focused on defining goals and problem priorities.

Technical uncertainty refers to the knowledge, understanding, and agreement about how to reach organizational goals. This variable can range from complete agreement and certainty to complete disagreement and uncertainty about cause–effect relationships. Technical uncertainty is important to the problem solution stage of decision making. When means are well understood, the appropriate alternatives can be identified and calculated with some degree of certainty. When means are poorly understood, rational alternatives are ill defined and uncertain. Intuition, judgment, and trial and error become decision criteria.

Cell I. When managers have a definite preference for outcomes, few production defects, high output, and so on—and they have complete knowledge of the factors that cause these outcomes—they can literally compute the best course of action to follow. This is the ideal environment for computers and quantitative methods. There are many computer programs to compute the order-quantity to minimize inventory costs, the product mix to maximize profits, the allocation of resources to minimize project time, and solutions to many other similar problems (Kroeber, 1982). Rational decision procedures are used because the goals are agreed upon and cause–effect relationships well understood. Decisions are made in a computational manner. Alternatives can be identified and the best solution adopted through analysis and calculation (Daft, 1989).

Kroeber (1982) further observes that more recently there have been attempts to make judgmental decisions directly by computers. The computer programs involved simulate rather than replicate human judgment; that is, the computer model will make the same decision a human might, although it obviously cannot duplicate the free-form, associative thought processes of the human mind. Initiation of decision making by computer is called heuristic programming. Decisions have been made heuristically by computers with success in such areas as graduate school applications, bank loan applications, and large-scale personnel matters. Heuristic programming holds great promise in the support of managerial decision making.

Cell II. Another decision environment exists, according to Thompson (1967), when managers know exactly how to achieve certain outcomes, but their preference for any given outcome is unclear. Such decision situations call for compromise. Compromise decisions are quite common in the public sector, where conflicting or unclear preferences are present in almost every situation. The dispute over preferences cannot be resolved by MIS, although MIS may help gather and analyze data on public opinion. But once preferences are established, the decision environment shifts back to the computational quadrant and MIS becomes a major element in the decision process.

Cell III. In other decision environments, the preference for an outcome may be equally clear, but the knowledge of how to achieve such an outcome is less than complete. Many decisions in human resource management fall in this category. The objective to hire a production manager may be clear, but the choice of the person to fill that position cannot be made by computer. Instead, decision makers exercise judgment; they review the information available on candidates and make a decision subjectively. However, MISs are not excluded completely from such judgmental decision processes. Files are designed on existing staffing resources and classified in terms of attributes and other managerial and professional criteria; these attributes and criteria ranked and rated in terms of numerical scores or values (Kroeber, 1982). Daft (1989) makes the following observations on this decision environment. Goals and standards are certain, but technical alternatives are vague and uncertain. Techniques to solve the problem are ill-defined and poorly understood. When an individual manager faces this situation, intuition will be the decision guideline. Rational, analytical approaches are not effective because the alternative cannot be identified and calculated in a logical way. Hard facts and information are not available. MIS can be used to map the goals but not cause–effect outcomes. Since technical uncertainty is high, scenarios must be developed to assess outcomes. Some categories of technological change will have this high degree of technical uncertainty in cause–effect outcomes.

Cell IV. In the final decision environment, the manager is unsure of what he prefers, and even if he wasn't, he would not know how to achieve the desired outcome. In such extreme states of uncertainty, an inspirational approach is the only mode. Such situations are more common in the public sector. This environment poses dilemmas for MIS and the manager, since the politics of decision making may emerge as more central to information processing. In situations of uncertainty and conflict, political behavior is apparent, goals are not certain, and means are not clear. Inspiration refers to an innovative, creative solution that is not reached by logical means. In a manager experiencing this high level of uncertainty, intuition and creativity are important for both the problem identification and problem solution states of decision making (Daft, 1989). This is also the decision environment for many high-tech firms operating in dynamic, complex, and uncertain environments. Information systems should exist only to support decisions. Understanding managerial decision making is essential for effective systems design and implementation. Most MIS groups become involved in system development and implementation, according to Keen and Scott Morton (1978), without a prior analysis of the variety of managerial activities. This has prevented them from developing a sufficiently broad definition of their purpose and has resulted in generally inefficient allocation of resources.

In the management of technological change, a more systematic effort must be made to diagnose the decision environment in terms of the goals of the innovation and the cause–effect outcomes or technical uncertainty that the innovation generates. This diagnosis will help identify the information systems and data bases that must be developed and used. While uncertainty in decision goals and decision outcomes may be high at the strategic level, at the implementation level the goals are more concrete, but the technical uncertainty of the innovation may still be high until the routinization stage is reached.

6.12.2 Types of Decisions and Information Systems

Another framework for classifying the types of decisions within an organization is suggested by Simon (1965), whose distinction between "programmed" and "nonprogrammed" decisions is quite useful (Kroeber, 1982). Simon (1965) states that

> Decisions are programmed to the extent that they are repetitive and routine, to the extent that a definite procedure has been worked out for handling them so that they don't have to be treated de novo each time they occur . . . Decisions are nonprogrammed to the extent that they are novel, unstructured, and conse-

quential. There is no cut-and-dried method of handling the problem because it hasn't arisen before, or because its precise nature and structure are elusive or complex or because it is so important that it deserves a custom tailored treatment . . . By nonprogrammed I mean a response where the system has no specific procedure to deal with situations like the one at hand, but must fall back on whatever general capacity it has for intelligent, adaptive, problem oriented action.

Keen and Scott Morton (1978) substitute the terms "structured" and "unstructured" for programmed and nonprogrammed because they imply less dependence on the computer and more dependence on the basic character of the problem-solving activity in question. The procedures and the kinds of computation and information vary depending on the extent to which the problem in question is unstructured. The basis for these differences is that, in the unstructured case the human decision maker must provide judgment and evaluation as well as insight into problem definition. On the other hand, it is argued that in a very structured situation, much if not all of the decision-making process can be automated. Systems built to support structured decision making will be significantly different from those designed to assist managers in dealing with unstructured problems. A fully structured problem is one in which intelligence, design, and choice are structured. That is, we can specify algorithms, or decision rules, that will allow us to find the problem, design alternative solutions, and select the best solution. An unstructured problem is one in which none of these three phases is structured. Another class of decisions is identified as semi structured—decisions with one or two of the intelligence, design, and choice phases unstructured.

Information systems, it is observed, should be centered around the important decisions of the organization, many of which are relatively unstructured. It is therefore essential that models be built of the decision process involved. As Keen and Scott Morton (1978) state,

> Model development is fundamental because it is a prerequisite for the analysis of the value of information, and because it is the key to understanding which portions of the decision process can be supported or automated. Both the success and failures in the current use of computers can be understood largely in terms of the difficulty of this model development.

Information systems designed to support structured decision making will be significantly different from those designed to assist managers with unstructured decision making. The pattern of structured–unstructured decisions varies with levels in the organizational hierarchy.

6.13 MIS AND THE MANAGEMENT OF CHANGE

Can MIS assist in the management of change in an organization? This section, based on the information provided in the previous part, will indicate how MIS can assist management in the implementation of change. Change may be

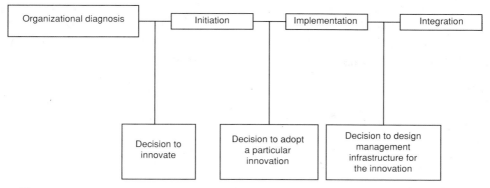

Figure 6.6. Stages in the change process. (Adapted from Robey, 1986. Reprinted with permission from Richard D. Irwin, Inc.)

brought about as a result of a managerial decision based on either internal factors, such as an attempt to streamline operations, or external factors, such as technical, political, economic, or competitive change. Organizational diagnosis is the first step in the management of change.

To study the effects of MIS on the management of change, we will use the model for change developed earlier. The model indicates a four-step approach to change (Figure 6.6).

> The innovation process involves four stages: organizational diagnosis, initiation, implementation, and integration. The stages are separated by three decision points which mark movement to the next stage. Between organizational diagnosis and initiation comes the decision to innovate, as one possible response to the need. Between initiation and implementation comes the decision to adopt a particular innovation generated in the initiation stage. Finally, between the implementation and integration stages comes the decision to design the management infrastructure (e.g., programs and operating procedures, information systems, and control and feedback systems) to support the innovation and facilitate its integration with management systems. While the stages are conceptually distinct, they are not so clearly delineated in practice. Often, initiation uncovers recognition of more needs, and problems encountered during implementation may require more work in initiation. In other words, most successful innovations involve team or reciprocal interdependence rather than sequential or pooled interdependence (Robey, 1986).

6.13.1 Organizational Diagnosis: The Decision to Innovate

If used efficiently, management information systems permit the accumulation of large amounts of data leading to earlier detection of change. They are used by many organizations to detect changes in the markets, and allow corporations to continuously monitor consumer preferences to detect changes in purchasing habits or seasonal changes. The information used may be drawn from various sources: periodic surveys, sales, intelligence gathering about competitors, etc. MIS permits all of this data to be accumulated in an easily presentable manner in order to attract the manager's attention. Trend analyses shown in a graphic manner and/or estimations and statistics are used by organizations to indicate their position vis-à-vis competitors and changes in consumer preference and expenditure. Intelligence gathering may be used to keep an eye on technological and sociopolitical change. Intelligence gathering was performed prior to MIS; however, today's management information systems communicate these changes throughout the organization at a rate never before expected and in quantities of information never previously matched. This increased information flow becomes more critical as technological, economic, and political changes increase. MIS, by providing information on strategic planning, can facilitate organizational diagnosis. Various scenarios may be developed as potential methods for bridging the performance gap indicated in the organizational diagnosis. In any event, the database management systems will be used to provide inputs in the design of various scenarios.

Organizational diagnosis may also focus on organizational processes. Process forces for change include decision-making and communications breakdown. Such breakdowns are reflected sooner or later in the production statistics of the organization or of a subsection. Changes in the behavioral factors will be reflected in morale and in other areas such as absenteeism and turnover. These factors will show up in the production statistics as well. The role of MIS is to assist management in detecting a problem and pinpointing its cause. Such tools as trend analysis and reports are used by managers to detect that a problem does exist. By manipulating the data, the manager can find the cause. On the basis of this diagnosis, a decision may be taken to innovate.

6.13.2 Initiation: The Decision to Adopt an Innovation

This step combines diagnosis of the problem and identification of alternative techniques or solutions. MIS assists the manager in evaluating the problem as a result of the massive information-storing capability and the quality and rapidity of processing. To evaluate the problem, the manager will interact with the computers, asking questions or requesting additional information, and wanting rapid responses on alternative solutions to the problem. MIS permits very rapid responses, especially if video display terminals are available. MIS decreases the total time required to evaluate a problem and increases the accuracy of the evaluation by extensive information processing.

MIS also assists in the formulation of alternatives. As noted in the section on decision making, the amount of solutions generated after an MIS is installed increases. This is the result of increased time available and the application of simulation. The amount of time available to develop more alternatives results from the decreased time required to develop each alternative. Simulation, on the other hand, permits managers to visualize each alternative. It is easier to understand alternatives and their impacts if one can visualize the alternatives instead of relying on imagination and memory. In addition, the quality of the alternatives increases as a result of simulation and manager–computer interaction in the data-base management systems.

Another important factor in evaluating the alternatives is the participation of other managers and subordinates. The increased communication capability of MIS permits more persons to be involved in the evaluation of alternatives to solve the problem. Computer simulation is the key to obtaining the appropriate solution. It permits the manager to project the proposed change(s) into the future. From this, the manager can choose the best solution. In addition to simulation, the manager can also use other models, such as linear programming and decision tree analysis, which are designed to indicate the optimal solution.

6.13.3 Integration: The Decision to Design Systems

The manager, after having decided which solution is to be used, must then detect the limiting factors, that is, the factors that may inhibit the implementation of the innovation, and counteract them. Some of the limiting conditions are leadership climate, formal organization, and organization culture.

Depending on the leadership climate, the lower or middle manager must convince the executives that the proposed change will be beneficial to the organization. MIS can be used to assist him in this task. If the middle manager has access to an MIS, by use of simulation he can build himself a very good case showing the problem, the proposed solution, and the impact in the long run. He can then place all of this in a language (e.g., graphs) that will be easily understood by the executives.

The formal organization is seen as including policies of top management, organizational structure, control systems, etc. This may in itself be a restricting factor on change. MIS cannot be used in this case to overcome resistance emerging from the formal organizational structure. If the change creates shifts in the power sources of the organization, there will be resistance. If it reduces the number of levels in the management hierarchy, there will also be resistance, unless the organizational culture emphasizes change as a major theme in the organization.

Organization culture includes group norms, values, and information activities. These are main factors in resisting change. The means of overcoming such resistance include participation and increased communication. Participation can increase as a result of MIS. Video display terminals, if liberally located, can be used by a manager to increase the amount of personnel and levels involved in a decision and a solution. In addition, by placing information in an easily legible format or making it visual, MIS permits more people to be involved in the diagnosis, analysis, and evaluation of the problem. The feelings resulting from being left out of the decision-making process and being uncertain about the effects of a change can create resistance to change.

6.13.4 Implementing the Change

This step is almost entirely executed by the manager. MIS can only be used as a means of communicating instructions in an easily understood format. This phase is similar in many ways to the control function. When implementing a change, management must set a standard that is to be maintained, or a goal to be reached. The control functions of MIS can be used to compare the actual results of the change(s) with the standard. The report-generating function of MIS will indicate the variances. It is the manager's decision to then react based on the size and acceptability of the variances. The control function will monitor the systems to determine if the objectives of the innovation or change program have been achieved.

Thus, management information systems can be used quite effectively by managers in implementing change. They are instrumental in detecting and diagnosing the problem. Once the problem is diagnosed, the manager uses MIS to develop and select alternatives. Finally, MIS can be used to decrease the resistance to change and implement and evaluate the solution.

What information is important—and who has it—may vary at different stages of the implementation process, but someone must coordinate the iterative work of gathering it, and that someone is the implementation manager. The illustration that Barton and Yogan used emphasized the need for information. A computer numerically controlled (CNC) system was implemented and the need for information by the project managers led to the following recommendations (Barton and Yogan, 1986):

- Observe the current job routine. System designers visited the floor several times and each time interviewed eight to 10 operators about their work procedures.
- Pay special attention to those parts of the work that required users to make decisions or see information about which tools or materials to use, which sequence of steps to follow in machining, and which jobs operators ought to run first.
- Discuss with workers what they found especially frustrating or rewarding about their work. In this case it turned out that they liked some flexibility in the sequencing of jobs, felt that choice of materials should be theirs, and were often frustrated by the difficulty of finding tools.

6.14 THE BUREAUCRATIC FRAMEWORK, MIS, AND CHANGE

The context in which an MIS operates may be bureaucratic. Where the organizational structure is bureaucratized, certain conditions prevail in the organization that inhibit communications effectiveness by constraining information search, processing, and conversion. For example, an emphasis on hierarchical authority, procedural specifications, behavioral rules, and impersonality may lead to a crystallized authority system with rigid role expectations, a communications system with limited feedback processes, and standardization of organizational activities. These structural characteristics will constrain the decision-making process and hence the change orientation of the decision outcome. That is, the manner in which the decision issue or problem is defined and its parameters delimited and alternatives surveyed will be constrained by an inventory of rules and procedures in the organizational structure.

These structural characteristics of an organization influence the problem definition process and its outcome. The actual outcome is determined by the manner in which the decision-making model is constrained. The decision-making model may be constrained by the degree of specific rules and procedures (formalization), the hierarchy of authority (centralization), and the degree of specialization (complexity) in decision making. When these dimensions are highly structured, channels of information and amount of information available within the unit are restricted. When dealing with high environmental uncertainty and change, a very high degree of emphasis on the hierarchy of authority can cause decision-unit members to adhere to the specified channels of communication and selectively to feed back only positive information. Strict emphasis on rigid rules and procedures may hinder the unit from seeking new sources of information when new information inputs (which may not have been anticipated when the rules and procedures were initially developed) are required to adapt to the uncertainty of the environment.

To manage change effectively, MIS must be designed around distributed data-processing configurations that facilitate the decentralization of decision making. A delegation of authority for decision making to lower levels in the organizational hierarchy will promote a culture for risk taking, exercise of initiative, and search for options. the vertical flow of information through the chain of command will also be complemented by a lateral flow which will expand the information-processing capacity of the system to cope with change and uncertainty. The information pathologies of the bureaucracy, such as information distortion, information overload, information filtration, and limited feedback cycles will be neutralized by distributed data-processing configurations.

6.15 MIS, INTEGRATED DATA BASES, AND CHANGE

An integrated data base of an MIS poses design problems of managing conflict because of (1) the high degree of interdependence among information systems; and (2) the degree of coordination of information inputs for the strategic, tactical, and operational levels of the organization.

The central feature of the MIS concept is its system perspective; that is, design, operation, and monitoring of information processing are integrated into a comprehensive system, according to Voich (1975). The MIS as a system deals with all information subsystems in combination, rather than with several individually. A total system implies, then, that all the functional systems have been designed and implemented. Although many companies can state that they have made inroads into the major functional areas, few would boast that they have implemented all elements of each of the functional systems. Studies of medium and large companies have shown that there is a relatively low saturation ratio of computerized applications. Many companies have not yet tackled complete functional system areas like market research, or major subsystems of other functional areas, like production scheduling and control. An integrated system implies that all functional systems are linked together into one entity. Individual systems must be designed to meet the needs of a restricted area of the company's operation, according to Kanter (1977), but with the needs of the whole organization in mind. Functional areas are not tackled in isolation, but only after realization of their relationship with the total system. This is a most desirable goal and represents a characteristic of MIS; however, an integrated system and a management information system are not synonymous. As Kanter (1977) has noted,

> It is possible to develop an integrated system aimed not at the management level but at the clerical and operating levels. The result could be a completely integrated system but not MIS.

The need for integration and coordination of segments of an information system is noted by McDonough (1969):

> It is clear that after more than twenty years of experience with computer applications in business, the major problem remains one of coordination . . . We should look ahead by moving out beyond the analysis of single or isolated systems and attack some of the problems of coordination among the variety of systems and among the varied content of positions in the organization . . . To provide for coordination of these situations is no easy matter. The starting point, however, is the recognition that neither systems content nor job content can be designed separately. The search for better management must bring jobs and systems together in an integrated framework.

A management information system is defined as a group of information systems that interconnect in their design, operations, and management to serve operations and facilitate the performance of the management process (Hussain and Hussain, 1981). The MIS thus requires clustering the information needs of various operating segments of functional systems into an integrated data base which is flexible enough to provide all types of information. The complete MIS comprises special purpose information systems integrated into a single unified framework. This means that the individual information systems, such as payroll, personnel, inventory, sales receivable, purchase payable, and forecasting, are combined. Since each is a system, the resulting MIS is a system of systems.

According to the MIS concept, the design of the data base, data coding scheme, and data processor must be comprehensive, yet flexible enough to service a variety of needs using a minimum of data input.

Design of the MIS processes requires a high degree of integration of information needs and data sources. Specific information elements on reports must be defined as the basis for determining the data needed to process the information for the reports. The management of MIS, Voich (1975) concludes, directly involves the design and monitoring of information uses and the MIS processes required. Once users' requirements for information are determined and the reports and MIS processes are designed and implemented, monitoring of information processing ensures performance according to plan. The interdependence among the inputs of information of the functional systems (e.g., marketing, finance, production, personnel) in the data base, the levels of management hierarchy (strategic, tactical, operational), and the uncertainty of information inputs in the event of change predispose an organization toward a high incidence of conflict. In other words, the greater the interdependence among the units and levels using the MIS, the greater the potential for conflict.

Integrated data bases are important to the effective management of technological change. The interdependence among functional units of information must be monitored. Changes in one subsystem will effect changes in other subsystems of the information system. The integrated data base provides a unified view of the enterprise, consistency among subsystems, linkages in the subsystems (vertical and horizontal), and control of information flow. As the system changes, the data base will also change. An integrated data base can facilitate and/or inhibit change because of all these systems characteristics.

6.15.1 MIS, Distributed Data Processing (DDP), and Change

DDP can serve as a mechanism for initiating and facilitating change by (1) distributing processing capabilities and intelligence, (2) relocating computing power from a large centralized computer to dispersed sites where processing demand is generated, and (3) adapting to various organizational structures and changing patterns in the flow of information.

Distributed data processing is one way of organizing equipment and personnel to implement a management information system. Generally, a decision to utilize this processing structure is made during the development stages of a new information system, but sometimes an ongoing system is converted to DDP. In either case, operations shift from a centralized DDP center to dispersed locations (Hussain and Hussain, 1982). DDP combines features of both centralized and decentralized processing.

Distributed data processing is the linking of two or more processing centers (nodes) within a single organization, each center having facilities for program execution and data storage. The linkage provided by DDP, according to Hussain and Hussain (1982), facilitates centralized control over policies and processing while the system retains the flexibility of decentraliza-

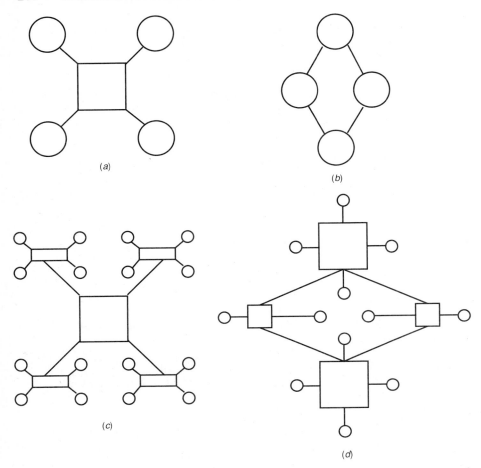

Figure 6.7. DDP configurations: (a) star, (b) ring, (c) star–star, (d) ring–star. (From Hussain and Hussain, 1982. Reprinted with permission of Richard D. Irwin, Inc.)

tion. The facilities at sites are tailored to local needs and controlled by local management, avoiding the rigidity of centralized hardware and personnel, while integration of the sites minimizes duplication of effort.

DDP involves a network of processors. Figure 6.7 shows a sample of DDP configurations. In the star network, failure of the central computer impairs the entire system. The ring structure overcomes this problem, for rerouting is possible should one processing center or its link fail. The ring allows interaction and off-loading without dependence on a central host. But star and ring configurations are essentially horizontal systems, that is, each processor is an equal. However, many firms prefer the hierarchical distribution configuration, as it requires the least reorganization, corresponding to the hierarchical structure that is already in existence within the corporation. This system has a central host computer and common data base with minis and micros at dispersed sites.

6.16 INFORMATION TECHNOLOGY AND ORGANIZATIONAL CHANGE

Business Week (1983) conducted a survey on the impact of computers on management and concluded that there had been a shrinking of middle management and changes in both structure and power distribution in the organization. The impact of computers on management is compared to the effects of the industrial revolution in changing hierarchies, radicalizing labor, realigning political forces, and disrupting social and psychological patterns (Figure 6.8). Both economic necessity and the technological revolution have combined forces to keep a permanent crunch on middle managers. A partial list of the effects on management follows:

1. Corporate structure is changing to accommodate broader information gathering and to let data flow from shop floor to executive suite without the editing, monitoring, and second-guessing that have been the middle manager's functions.
2. Middle managers who survive find their roles expanded and their functions changed. Generalists, not specialists, are needed, as companies demand solutions to interdisciplinary problems.
3. As corporate pyramids are flattened, with fewer levels, there are lowered expectations and more lateral moves.
4. Marketplace and manufacturing decisions are made by first-line managers whose power has been eroded by staff. Foremen now serve in pivotal roles, managing better-educated, more demanding workers, and integrating maintenance, engineering, and personnel managers into operating systems.

These changes create a receptivity for innovation. The design of the organization with fewer levels in the hierarchy, more distributed data-processing configurations, and more delegation of decision-making authority to lower levels in the organizational hierarchy, as well as the integration of functional units through information systems, and the shifts in balance of power, are all conducive to the adoption and diffusion of innovations.

6.16.1 The Shrinking of Middle Management

To make good decisions today, managers must deal with several issues, including the internationalization of their markets, complex financial transactions, and the impact of technology on their products. Technology is also changing the very nature and need for middle management jobs. Sales staff in industries from brokerages to pharmaceuticals are consulting their computers rather than their managers for pricing, inventory, and market information. With less paperwork, sales managers can cover larger territories and get faster feedback on performance and problems. Similarly, computer-aided design and manufacturing allowed Chrysler Corp. to halve its engineering group to 4000 without sacrificing its product-development programs.

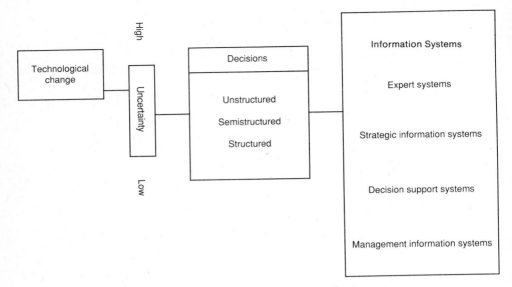

Figure 6.8. Technological change and information systems.

Business Week (1983) further noted that the middle ranks of the largest corporations were undergoing a revolution in job security. More than 40% of the 1200 major companies surveyed cut their middle management staffs in 1982. In nearly half of those companies, the cutbacks were substantial. In more than half, cuts were across the board, affecting every corporate department. However, with virtual unanimity, U.S. middle managers retained confidence in the goals and viewpoints of top management. Nearly two-thirds of the executives agreed that the staff reductions increased productivity in their companies or departments. One factor that seemed to mitigate the adverse effects was the availability of some kind of retraining program. Nearly two-thirds of such retraining involved work with computers. This is especially significant considering the respect that computer-based information elicits from most U.S. middle managers. No fewer than 85% used computer-generated data that they considered essential to their jobs, and 91% believed that this computer access increased their own productivity. Almost as many believed that computer access could promote careers. Some 84% believed that computers could "increase the number of variety and responsibilities you can handle." Furthermore, 41% believed that the continued integration of the computer into managerial operations was likely to lead to "further consolidations of departments and functions." Lastly, two-thirds of the middle managers questioned reported that their companies were moving ahead with office automation.

6.16.2 Shifts in the Balance of Power

In industries as diverse as those producing automobiles, computers, aluminum, oil, and machine tools, companies are discarding their organizational charts and simplifying their chains of command. This gives chief executives more flexibility, keener and quicker assessment of performance, clearer and faster lines of action, and a better grasp of information needed to run their companies. But as the new structures are put into place, the balance of power in corporations is shifting. Marketing, strategic, and financial planners are being deposed, and operations managers, who not only know how to devise plans but also how to implement them, are taking over. Authority is now being delegated to plant managers, salespeople, and engineers, and bypassing staffs completely. Rigid hierarchical structures have begun to crumble at the best-managed companies. Replacing them are leaner, more fluid organizations, with fewer levels of management and more direct lines of communication between the top and the bottom. Instead of relying on entrenched specialist bureaucracies, companies are pulling together a few key managers on an ad hoc basis to solve immediate problems, and disbanding them just as quickly. In place of corporate cultures in which titles determine power, companies are creating environments in which compensation depends on performance, and freedom to improve performance provides the psychic reward.

6.16.3 Changes in Organizational Structure

Many of the changes noted in this section are seen as only an evolutionary step toward more flexible, decentralized, and undefined structures. Organizational experts envision a fluid organizational structure based not on formal hierarchy, but on the changing needs of the corporation. Such a structure places a high priority on consistent retraining to develop more flexible managers who can be deployed as needed. International Business Machines (IBM) Corp. has long had this type of structure in place. Futurists believe that the bureaucratic, multilayered organization is a dinosaur. They predict that a growing number of companies will turn to special units to attack markets, create ad hoc task forces to solve problems, and form new relationships that will enable groups of managers to work "inside and outside" the corporation at the same time.

These kinds of approaches are already being applied. IBM formed seven "independent business units" to explore such high-potential markets as robotics, remote computing services, and analytical instruments. General Electric Co. created a task force that reorganized its corporate staff and then disbanded after its job was done. Xerox Ltd. and Connecticut Mutual Life Insurance Co. have spun off groups of managers and then put them under contract as outside consultants. In their book on corporate cultures, Deal and Kennedy (1982) predict the coming of the "atomized organization." Telecommunications networks and common cultures—not an organizational chart—will link the "atoms." The new organizational style will have the most significant impact on the traditional cultures on which many corporations are built. As the structures change, the distribution of power will change as well. It is essential that it flows to the best performers and not to their monitors.

6.17 SOME FACTORS LIMITING MIS APPLICATIONS IN CHANGE AND INNOVATION

These factors are identified and discussed by Kanter (1977). Our discussion is derived from his analysis.

6.17.1 Design of the Data Base

The design of the data base for a management information system should not be left to computer technicians, either by delegation or default, for technicians lack the expertise, judgment, systemic perspective, and motivation of managers, the ultimate users of the system. Managers should be active in the design of the data base. They should be involved in both its creation and maintenance, since the quality of information derived is directly related to the quality of the data base itself. There is a phrase in data processing, "GIGO": garbage in, garbage out. Computer power is too potent to be relegated to technical experts who know little about organizational structure, design, and innovation. Information systems that affect management require a greater degree of interaction and involvement between the systems function and the management function. Although it is lessening, this management–systems gap is still a major cause of the slow evolution of systems aimed at improving the management decision-making process. Decision-making information systems are critical to the decision to innovation. A flexible data-base system is therefore critical to the innovation process.

6.17.2 Unstructured Nature of Data

The data required by top management is unstructured, nonprogrammed, future oriented, inexact, and external. The type of information required by top management is the most difficult to acquire, update, and process, and therefore few computerized systems exist in this area. Another factor restricting the development of MIS for top management and strategic policy-making is the lack of definite cause-and-effect relationships of data. The problem of obtaining the type of data required for strategic decisions, when combined with the questionable decision rules of how to analyze the data meaningfully, presents a formidable obstacle for computerized information systems directed at the strategic planning area (Kanter, 1977). Managerial activities are unstructured and constantly changing. Information, to be valuable, must change too. Yet, according to Daft (1989), computer information systems often fail to support management because they are not flexible enough to keep pace. Technology induces standardization and requires procedures to govern data handling into and out of the computer. Formal information systems, once in place, have a life of their own and do not adapt to the changing needs of users.

6.17.3 Management Science Techniques

Although there are many cases in which management science or operations research (OR) techniques have been used with a good deal of success in facilitating top management decision making, the companies that utilize management science are still in the minority. Assuming that the proper technique has been selected for tackling a particular problem, there are still some major barriers to overcome. A communication gap exists between management and the operations research analyst who is responsible for preparing the set of linear equations if, for example, the problem is one that can be handled by linear programming. Management is skeptical of making a decision based on a set of complex equations it cannot interpret, and generally the OR specialist finds it difficult to explain in management terms what the mathematical processing will achieve. Furthermore, if a manager realizes that the data used as input to the model are not valid, reliable, or accurate, he is suspicious of the results. Some of these techniques, methods, and models are also still at the theoretical level and may provide an architectural design that is elegant but not feasible for innovation.

6.17.4 The Element of Intuition

Many top managers, it is further argued, feel that good strategic decisions are made more by intuition than by quantitative analysis of the available data. They do not ignore the data completely, but believe that there is a strong intuitive factor, particularly when little or no data is available. Studies suggest that some managers have more "precognitive" ability than others and that this ability gives them a better batting average in making decisions intuitively as opposed to logically. This notion is reinforced by De Bono (1971), who observes that in an age where computers perform remarkable and diverse functions and often replace people in the workplace, it's easy to forget that even the most sophisticated computer cannot generate a single new idea. Only the human mind is capable of both processing information and creating something new. It is these amazing abilities that are critical to the innovation process.

6.17.5 Problem Definition and Diagnosis

Effective tools to define problems are still lacking. Data processing has not focused on the area of problem definition. Despite the advances in hardware and software, there have not been substantial advances in basic systems methodology. We talk about the "systems approach" to solving problems, but we really haven't developed the tools for defining the problem to be solved. For example, how does one go about analyzing the determinants of profitability so that the resultant formula can be modeled on a computer? This is not a trivial job when one has to consider both the short and long range. It seems obvious that, before a computer can help, the problem must be defined and the cause and effect relationships determined.

For well-structured situations, the situation is defined and its main characteristics are defined. A model is then constructed to provide the basis for estimating possible outcomes of the decision over a range of possible conditions. Certain aspects of ill-structured problems, according to Bass (1983), can also be modeled. For example, people tend to be consistent when diagnosing cause–effect relationships. They seem to have preferences from among relational "operators," which systematically bias their diagnoses and subsequent search efforts. Programs can be constructed to model these operations. On the other hand, we must be willing to accept the inability to completely define the problem in advance, since we lack complete information. So we move on to search and to make a trial choice. However, we may return to the definition phase several more times as additional information becomes available during the search and trial choices. MIS can make these phases of the decision-making process easier.

6.17.6 Changing Conditions

Many MIS designers have discovered that innovative and turbulent changes in business strategy, as well as changes in top management and the structure of the organization, have compromised the systems approach they had planned. While computer systems handling the administrative functions, such as order processing and inventory control, remain relatively stable in times of organizational change, management temperament and style can be affected dramatically by changes of managers at the strategic apex of the organization. Additionally, the uncertainty generated by the changes at the strategic level cannot be programmed. The cause–effect relationships may be unclear and the outcomes difficult to estimate.

6.17.7 Long-Range Planning of MIS

Because it is so sensitive to organizational and business strategy changes, the MIS planning focus must be more long-range in nature than that for more conventional information systems. This long-range planning is particularly difficult in the data-processing field, where technological change has seen so many computer installations go through four generations of hardware in less than 10 years. Most computer manufacturers have been preoccupied with hardware and software planning at the expense of systems planning or decision support systems for management and innovation. In an era of megatrends, long-range planning and information processing cannot handle the variety and complexity.

6.18 HUMAN INFORMATION PROCESSING AND CHANGE

Information technology contains three distinct elements, according to Robey (1986): (1) man's symbolic systems of communication, (2) computer and telecommunications hardware, and (3) software. Only the second of these is directly involved with the physical computer.

6.18.1 Man's Symbolic Systems

The human nervous system, or biocomputer, remains the most complex system on this planet, with information-processing capabilities that surpass those of any electronic computer. Because of human limitations on memory, consistency, and speed, however, electronic computers have been developed to complement human information processing. But man is the author of the symbolic systems which form the basis for information technology. It is this contribution in design and programming that makes the computer feasible as an information-processing mechanism. Man's imagination, intelligence, intuition, insight, and logic cannot be supplanted by the computer. The marvel or miracle is the biocomputer (Robey, 1986).

The interface between a person and the computer is critical to its efficiency. The computer is not the central element of the information system. As seen in Figure 6.9, he observes the physical system and applies his intelligence, his ability to judge situations rather than respond to them deter-

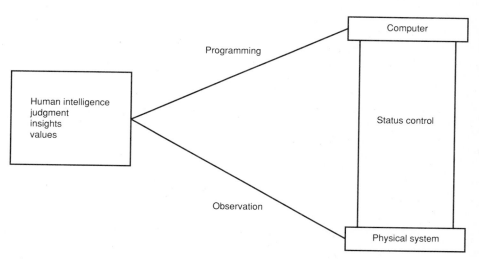

Figure 6.9. Interface between man and computer. (Adapted from Robey, 1986. Reprinted with permission from Richard D. Irwin, Inc.)

ministically, and his set of values. He establishes the criteria and the goals to be achieved by the system and communicates his conclusions to the computer via programming. This is a critical input to the computer; it cannot be absent, as might be inferred from most models or diagrams that depict elements and operations of the computer.

This powerful combination of man, computer, and physical system multiplies the resources of all three. Automatic control combines the benefits of the industrial revolution with the information revolution. The digital computer makes this system quite distinct from the man–engine combination alone. It handles great quantities of data at tremendous speeds, extending the realm of operations that can be performed. Programming techniques also combine decision making with mathematical calculations, and this permits discontinuous or discrete control afforded by analog devices (Robey, 1986).

Rather than belittling the activities of man's brain, the limitations of thought by computers show man's superior intelligence more clearly. The programming of artificial perception calls not only for the same perception on man's part, but also a transcendent awareness of how that perception works.

De Bono (1971) makes the following observations on the role of human information processing and the creative process. Logic, mathematics, and computers are, in his view, second-order information-processing systems. They go to work on information that has already been selected out and parceled up. "New think" is more concerned with the first stage, for herein lie the new ideas. New think usually involves lateral thinking. Lateral thinking is based on biological information-processing principles and seeks to get away from the patterns that are leading one in a definite direction and to move sideways by re-forming the patterns. Lateral thinking is necessary precisely because the brain functions so well in the vertical fashion.

In this age of computer technology, the creative function assumes greater importance for change. The capacity of the mind to generate new ideas and new ways at looking at things is the basis of progress. New ideas provide the impetus in science, technology, and art (De Bono, 1971).

A great deal of time and effort is spent in learning how to get the best out of electronic computer systems. Much less attention is spent on trying to get the best out of the brain system. Unfortunately, the brain system seems to be so very different from computer systems that what is learned with these may be more misleading than helpful.

In an age where computers perform remarkable and diverse functions, often replacing people in the workplace, it is easy to forget that even the most advanced computer cannot generate a single new idea. It is only the human mind that is capable of both processing information and creating something new. Thinking is regarded as only a tool for assimilating, classifying, storing, and retrieving information. Information is much easier to teach than thinking. Thinking may even seem to be mere guessing. The relationship between thinking and information can be considered in this situation, according to De Bono. Data become information only when they are looked at through the spectacles of an idea: "Einstein looked at the data that had been seen through the Newtonian idea and by looking at them in a different way came to a different conclusion" (De Bono, 1971). The constant play between information and ideas cannot be neglected. Ideas are generated by the application of thinking to data. When we collect information, we collect data that have been organized by the old ideas. To improve these ideas, we need thinking, not just more information.

When they think of new ideas, many people think of technical inventions or scientific theories. In both instances, the appropriate technical knowledge seems to be required before any new ideas are possible. This is perfectly true, but clearly technical knowledge is not enough, for even the people who possess it do not automatically come up with new ideas. Technology by itself does not generate new ideas.

De Bono (1971) further observes that it is important to be quite clear about the distinction between information and perception.

> The function of perception is to direct attention to the information, to put different bits of information together, to abstract certain principles, to make predictions, and so on. In real-life situations, information is never provided in such a neatly packaged form. Information is obtained by exploring experience, asking questions, knowing where to look for it, and making assumptions and developing scenarios. To say that a person needs more information before he can start thinking is pointless, because thinking is concerned precisely with extracting that information from experience. Perception is the processing of information for use. Thinking is the exploring of experience for a purpose. The role of perception in thinking is seen through the operations that have been carried out: organizing, challenging concepts, conducting queries, and observing cues. All these take place in the perception area. They are the devices and frameworks for direction attention.

The role of perception, intuition, and lateral thinking in managing technological change can therefore be considered as critical as information technology in managing such change.

6.18.2 Cognitive Decision Styles

This section is based on Robey's model of cognitive decision styles. He argues that human information processing is essential to organizational information systems because human actors are the ultimate recipients of information and decision makers (Robey, 1986).

> Descriptive studies on human information processing can be divided into three groups: model fitting, process tracing, and cognitive decision styles. Model fitting is essentially a statistical exercise intended to describe a human decision maker's internal model for weighing criteria, estimating probabilities of unknown events, and trading off potential outcomes against one another. Process

tracing attempts to describe information processing by getting decision makers to "think out loud" while they make a decision. By following one's verbalized thought process and mapping recurring tendencies, a researcher can derive the structure of mental processes. Both of these first two approaches require analyses of data generated by decision makers who themselves may be unaware of their information-processing tendencies. The approaches also reveal that human beings make predictable errors in perception and judgment that may be corrected by using an appropriate organizational information system or information technology. The cognitive decision style approach focuses on general patterns of gather and using information in a variety of situations (Robey, 1986).

A useful typology of psychological types developed by Carl Jung is used by Robey to describe cognitive decision styles. According to this theory, information processing consists of two processes: perception and judgment. Each process can be carried out in two qualitatively different ways. Perception can be achieved by sensing or by intuition. Judgment is accomplished through thinking or feeling. Each of us develops preferences for a way of perceiving and a way of judging, and the combination of these preferences defines our cognitive style. Figure 6.10 summarizes Jung's typology by defining each of the terms and by showing the four possible combinations.

1. *The Sensing–Thinking (ST) Style.* People who prefer sensing as a way of collecting information place great faith in their five senses to define what is real. The sensing person uses sight, hearing, taste, smell, and touch to understand the world. This factual, detailed information is then analyzed logically and objectively with the thinking mode of judgment. Sensation combined with thinking means a person prefers situations where concrete facts can be analyzed logically. In business situations, people with ST style often feel comfortable in jobs like accounting or data processing.

2. *The Sensing–Feeling (SF) Style.* People who combine sensing with feeling prefer facts and detail. They deal with things that can be verified by the senses, just like the ST style. However, people with SF prefer a subjective approach to judgment where personal values and "gut feelings" enter into decisions. Especially when they are dealing with other people, feelings are more prized than thinking. The SF person therefore prefers to work with specific information that can be processed subjectively. In business, SF styles frequently are found in personnel, sales, and public relations.

3. *The Intuitive–Thinking (NT) Style.* Sensing is not the only way to gather information according to the Jungian typology. Some people prefer to look beyond the obvious facts to less-apparent patterns, relationships, and possibilities that can be inferred from the facts. To the intuitive person, stopping at the senses is not interesting. Facts are viewed as redundant or too obvious, so intuition creates new information that is more subtle and abstract than that perceived by the senses. People who prefer intuition are often accused of not dealing with reality, because they frequently talk about the future or about complex concepts that have little observable content. The NT style combines

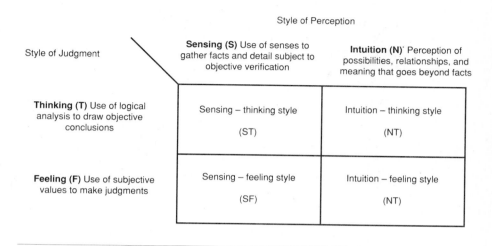

*The letter N is used for intuition because Jung's theory also includes introversion, for which the letter I is used.

Figure 6.10. The Jungian typology of cognitive decision styles. (From Robey, 1986. Reprinted with permission from Richard D. Irwin, Inc.)

intuition with the logical judgment of thinking and establishes a style very common among planners, entrepreneurs, and theoretical scientists.

4. *The Intuitive–Feeling (NF) Style.* When there is a preference for feeling, the result is a preference for abstractions dealt with subjectively. NF people are often strong humanists who approach work and life in an artistic fashion. Their intuition may lead them to neglect the well-being of specific individuals, but they easily become committed to general human causes. Improving the quality of work life and organizing labor groups for collective bargaining are causes in the workplace to which a person with NF style can easily become committed.

Because cognitive styles describe preferences and not abilities, a particular style does not prevent a person from performing a variety of organizational functions. Each style fits a specific type of information processing that goes on in an organization.

Since the implementation stages involved in technological change, such as diagnosis, initiation, and implementation, are different in their information-processing requirements, as noted earlier, then the variety of cognitive styles can be effective in managing change.

6.19 ECOLOGY, TECHNOLOGY, AND ENVIRONMENTAL MONITORING

Any information technology for managing change must consider the effects of technologies on the environment and also the design of expert systems for technology assessment.

Some of the issues involved in technology and ecology are reviewed briefly from the perspectives of Blackstone and Ophuls. Following this review, we will provide an outline of the steps in technology assessment by Coates.

6.19.1 Ecology and Technology: Blackstone and Ophuls

> It is well-known that new technology sometimes complicate our ethical values and decisions. Whether the invention is the wheel or a contraceptive pill, new technology always open up new possibilities for human relationships and for society. I am not suggesting that the answer to technology which has bad environmental effects is necessarily more technology. We tend too readily to assume that new technological developments will always solve man's problems. But this is simply not the case. One technological innovation often seems to breed a half-dozen additional ones which themselves create more environmental problems. We certainly do not solve pollution problems, for example, by changing from power plants fueled by coal to power plants fueled by nuclear energy if radioactive waste from the latter is greater than pollution from the former. Perhaps part of the answer to pollution problems is less technology. There is surely no real hope of returning to nature (whatever that means) or of stopping all technological and scientific development, as some advocate. Even if it could be done, this would be too extreme a move. The answer is not to stop technology, by to guide it toward proper ends, and to set up standards of antipollution to which all technological devices must conform. Technology has been and can be used to destroy and pollute an environment, but it can also be used to save and beautify it. What is called for is purposeful environmental engineering, and this engineering calls for a mass of information about our environment, about the needs of persons, and about basic norms and values which are acceptable to civilized men. It also calls for priorities on goals and for compromise where there are competing and conflicting values and objectives. Human rights and their fulfillment should constitute at least some of those basic norms, and technology can be used to implement those rights and the public welfare.
>
> —Blackstone (1983)

Blackstone argues that the right to a livable environment can be considered to be a human right:

> If the right to a livable environment were seen as a basic and inalienable human right, this could be a valuable tool (both inside and outside legalistic frameworks) for solving some of our environmental problems created by technological innovations. Traditionally we have not looked upon the right to a decent environment as a human right or as an inalienable right. Rather, inalienable or

natural rights have been conceived in somewhat different terms; equality, liberty, happiness, life, and property. If human rights are those rights which each human possesses in virtue of the fact that he is human and in virtue of the fact that those rights are essential in permitting him to live a human life (that is, in permitting him to fulfill his capacities as a rational and free being), then might not the right to a decent environment be properly categorized as such a human right? Given man's great and increasing ability to manipulate the environment and the devastating effect this is having, it is plain that new social institutions and new regulative agencies and procedures must be initiated on both national and international levels to make sure that the manipulation is in the public interest. It will be necessary to restrict or stop some practices and the freedom to engage in those practices. In the case of the abuse and wase of envirionmental resources, less individual freedom and fewer individual rights for the sake of greater public welfare and equality of rights seem justified. Both public welfare and equality of rights now require that natural resources not be used simply to the whim and caprice of individuals or simply for personal profit. Given the crisis of our environment created by technology, there must be certain fundamental changes in attitudes toward nature, man's use of nature, and man himself. Such changes in attitudes will have far-reaching implications for the institutions of private property and private enterprise and the values embedded in these institutions. There was a time in the recent past when property rights embodied the right to hold human beings in slavery. This has now been rejected, almost universally. It may be necessary that the notion of property rights be even further restricted to preclude the destruction and pollution of our environmental resources upon which the welfare and the very lives of all of us and of future generations depend. If a right or set of rights systematically violates the public welfare, this is the prima facie evidence that it ought not to exist.

In a system which emphasizes simply the interplay of market forces as a criterion, these factors (such as sights, smells and other aesthetic factors, justice, human rights—factors which are important to the well-being of humans) are not even considered. Since they have no direct monetary value, the market places no value whatsoever on them. Consequently, we must broaden our evaluative perspective to include the entire range of values which are essential not only to the welfare of man but also the welfare of other living things and to the environment which sustains all of life. And this must include a reassessment of rights.

Ophuls (1983) reinforces these arguments and argues that if

this inexorable process in destroying the environment is not controlled by prudent, and above all, timely political restraints on the behavior that causes it, then we must resign ourselves to ecological self-destruction. And the new political structures that seem required to cope [with the tragic consequences of technology] are going to violate our most cherished ideals, for they will be neither democratic nor libertarian. At worst, the new era could be an anti-utopia in which we are conditioned to behave according to the exigencies of ecological scarcity. The case for the coming ecological scarcity was most forcefully argued in the Club of Rome study "The Limits to Growth." That study says in essence that man lives on a finite planet containing limited resources and that we seem to be approaching some of these major limits with great speed—we are about to overtax the "carrying capacity" of the planet. The crisis of ecological scarcity poses basic value questions about man's place in nature and the meaning of human life. It is possible that we may learn from this challenge what Lao-tzu taught two and a half millenia ago.

Nature sustains itself through three precious principles, which one does well to embrace and follow.

These are gentleness, frugality and humility.

Ophuls (1983) summarizes the ethical perspective on ecology and technology:

> A very good life—in fact, an affluent, life by historic standards—can be lived without the profligate use of resources that characterizes our civilization. A sophisticated and ecologically sound technology, using solar power and other renewable resources, could bring us a life of simple sufficiency that would yet allow the full expression of the human potential. Having chosen such a life, rather than having had it forced on us, we might find it had its own richness.
>
> Such a choice may be impossible, however. The root of our problem lies deep. The real shortage with which we are afflicted is that of moral resources. Assuming that we wish to survive in dignity and not as ciphers in some ant-heap society, we are obliged to reassume our full moral responsibility. The earth is not just a banquet at which we are free to gorge. The ideal in Buddhism of compassion for all sentient beings, the concern for the harmony of man and nature so evident among American Indians, and the almost forgotten ideal of stewardship in Christianity point us in the direction of a true ethics of human survival—and it is toward such an ideal that the best among the young are groping. We must realize that there is no real scarcity in nature. It is our numbers and, above all, our wants that have outrun nature's bounty. We become rich precisely in proportion to the degree in which we eliminate violence, greed, and pride from our lives. As several thousands of years of history show, this is not something easily learned by humanity, and we seem no readier to choose the simple, virtuous life now than we have been in the past. Nevertheless, if we wish to avoid either a crash into the ecological ceiling or a tyrannical Leviathan, we must choose it. There is no other way to defeat the gathering forces of scarcity.
>
> For the first time in recorded history, science and technology have given man enough resources to ensure for every human being born upon this earth adequate material, intellectual and spiritual inputs to be able to live a full and rewarding life. If we can creatively utilize the wealth of the world and constructively canalize its resources, no one need go hungry, no child need be deprived of education, no person need be without adequate medical care, and the spectre of unemployment would be laid to rest for ever. But this will require not simply a readjustment of our existing perceptions, values, and priorities, it will need a paradigm shift, a break from existing orthodoxies and a bold thrust into a new dimension of awareness.

6.19.2 An Ode to the Earth: Chief Seattle (1988)

"The President in Washington sends word that he wishes to buy our land. But how can you buy or sell the sky? The land? The idea is strange to us. If we do not own the freshness of the air and the sparkle of the water, how can you buy them?

"Every part of this earth is sacred to my people. Every shining pine needle, every sandy shore, every mist in the dark woods, every meadow, every humming insect. All are holy in the memory and experience of my people.

"We know the sap which course through the trees as we know the blood that courses through our veins. We are part of the earth and it is part of us. The perfumed flowers are our sisters. The bear, the deer, the great eagle, these are our brothers. The rocky crests, the juices in the meadow, the body heat of the pony, and man, all belong to the same family.

"The shining water that moves in the streams and rivers is not just water, but the blood of our ancestors. If we sell you our land, you must remember that it is sacred. Each ghostly reflection in the clear waters of the lakes tells of events and memories in the life of my people. The water's murmur is the voice of my father's father.

"The rivers are our brothers. They quench our thirst. They carry our canoes and feed our children. So you must give to the rivers the kindness you would give any brother.

"If we sell you our land, remember that the air is precious to us, that the air shares its spirit with all the life it supports. The wind that gave our grandfather his first breath also receives his last sigh. The wind also gives our children the spirit of life. So if we sell you our land, you must keep it apart and sacred, as a place where man can go to taste the wind that is sweetened by the meadow flowers.

"Will you teach your children what we have taught our children? That the earth is our mother? What befalls the earth befalls all the sons of the earth.

"This we know: the earth does not belong to man, man belongs to the earth. All things are connected like the blood that unites us all. Man did not weave the web of life, he is merely a strand in it. Whatever he does to the web, he does to himself.

"One thing we know: our god is also your god. The earth is precious to him and to harm the earth is to heap contempt on its creator.

"Your destiny is a mystery to us. What will happen when the buffalo are all slaughtered? The wild horses tamed? What will happen when the secret corners of the forest are heavy with the scent of many men and the view of the ripe hills is blotted by talking wires? Where will the thicket be? Gone! Where will the eagle be? Gone! And what is it to say goodbye to the swift pony and the hunt? The end of living and the beginning of survival.

"When the last Red Man has vanished with his wilderness and his memory is only the shadow of a cloud moving across the prairie, will these shores and forests still be here? Will there be any of the spirit of my people left?

"We love this earth as a newborn loves its mother's heartbeat. So, if we sell you our land, love it as we have loved it. Care for it as we have cared for it. Hold in your mind the memory of the land as it is when you receive it. Preserve the land for all children and love it, as God loves us all.

"As we are part of the land, you too are part of the land. This earth is precious to us. It is also precious to you. One thing we know: there is only one God. No man, be he Red Man or White Man, can be apart. We *are* brothers after all."

6.20 TECHNOLOGY ASSESSMENT AND THE ENVIRONMENT

Any information technology used in the management of change must involve technology assessment. This can be done through expert systems, decision support systems, and a variety of data-base systems. The steps identified by Coates (1977) can be used in the design of such information systems.

Coates (1977) defines technology assessment as policy studies examining the full range of impact of the introduction of a new technology or the expansion of a present technology in new or different ways. One might look at it as an analysis of the total impact of a technology on society.

Technology assessment should be made with the knowledge that:

1. Decisions will be made, including in many cases the decision not to go ahead with a particular technology or project.
2. The organization, structure, and output of an assessment should be directed at informing decision makers about alternative actions and the probable consequences.
3. In approaching the problems of management of new technologies, one must bring into the analysis both certainty and uncertainty and establish a act of sound policies and actions based upon the interplay of not only what is known with some certitude but also what is not known.
4. The long-range indirect and unanticipated effects of a technology are often more significant than the immediate planned consequences.

Coates (1977) further observes that to be useful a technology assessment must go far beyond conventional engineering and cost studies to look at what else may happen in achieving an immediate goal to the total range of social costs.

The major steps involved in technology assessment are identified by Coates (1977) in Figure 6.11.

Coates cites the Environmental Policy Act of 1969 for the preparation of environment impact statements. Such statements must include (1) environmental impact of the proposed action, (2) any adverse environmental effects which cannot be avoided should the proposal be implemented, (3) alternatives to the proposed action, (4) the relationship the local short-term use of man's environment and the maintenance and enhancement of long-term productivity, and (5) any irreversible and irretrievable commitments of resources which would be involved if the proposed action were implemented.

Step 1 — Define the assessment task

- Discuss relevant issues and any major problems
- Establish scope (breadth and depth) of inquiry
- Develop project ground rules

Step 2 — Describe relevant technologies

- Describe major technology being assessed
- Describe other technologies supporting the major technology
- Describe technologies competitive to the major and supporting technologies

Step 3 — Develop state-of-society assumptions

- Identify and describe major nontechnological factors influencing the application of the relevant technologies

Step 4 — Identify impact areas

- Ascertain those societal characteristics that will be most influenced by the application of the assessed technology

Step 5 — Make preliminary impact analysis

- Trace and integrate the process by which the assessed technology makes its societal influence felt

Step 6 — Identify possible action options

- Develop and analyze various programs for obtaining maximum public advantage from the assessed technologies

Step 7 — Complete impact analysis

- Analyze the degree to which each action option would alter the specified societal impacts of the assessed technology discussed in Step 5

Figure 6.11. Seven major steps in technology assessment. (From Coates, 1977. Reprinted with permission from St. Martin's Press.)

REFERENCES

Anthony, R., *Planning and Control Systems: A Framework for Analysis,* Harvard University, Division of Research, Graduate School of Business Administration, Cambridge, Mass., 1965.

Barton, Leonard, and Yogan, J., 1985. "Marketing Advanced Manufacturing Processes" in D. Davis and Associates, *Managing Technological Innovation,* Jossey-Bass, San Francisco, 1986.

Bass, B., *Organizational Decision Making,* Irwin, Homewood, Illinois, 1983.

Blackstone, W., "Ethics and Ecology in Ethical Issues," in T. Donaldson and P. Werhane (Eds.), *Business: A Philosophical Approach,* Prentice Hall, Englewood Cliffs, N.J., 1983.

Business Week, Crisis in Middle Management, January (1983).

Cadwallader, M. L., "The Cybernetic Analysis of Change in Complex Social Organizations," *American Journal of Sociology,* September (1959).

Chief Seattle's Letter, cited in J. Campbell and B. Moyers, *The Power of Myth,* Doubleday, New York, 1988.

Child, J., 1981. *Organizations,* Harper & Row, New York, 1981.

Child, J., *Organizations: A Guide to Problems and Practice,* Harper & Row, 1984.

Coates, J., "Technology Assessment," in A. Teich (Ed.), *Technology and Man's Future,* St. Martin's Press, New York, 1977.

Daft, R., *Organizational Theory and Design,* West Publishing, St. Paul, Minn., 1989.

Deal, T., and A. Kennedy, *Corporate Cultures: The Rites and Rituals of Corporate Life,* Addison-Wesley, Reading, Mass., 1982.

De Bono, E., *New Think,* Avon Books, New York, 1971.

Donald, A., *Management Information and Systems,* 2nd ed., Pergamon Press, New York, 1979.

Dorsey, J., "A Communication Model for Administration," *Administrative Science Quarterly,* December (1957).

Hellriegel, D., and J. Slocum, *Management: A Contingency Approach,* Addison-Wesley, Reading, Mass., 1973.

Keen, G. and Scott Morton, M., *Decision Support Systems. An Organizational Perspective,* Addison-Wesley, Reading, Mass., 1978.

Hellriegel, D., and J. Slocum, *Management: A Contingency Approach,* Addison-Wesley, Reading, Mass., 1974.

Hussain, D., and K. Hussain, *Information Processing Systems for Management,* Irwin, Homewood, Ill., 1982.

Jung, C. G., *The Undiscovered Self,* Mentor Books, New York, 1957.

Kanter, J., *Management-Oriented Information Systems,* 2nd ed., Prentice Hall, Englewood Cliffs, N.J., 1977.

Kroeber, D., *Management Information Systems, A Handbook for Modern Managers,* Free Press and Collier Macmillan, New York, 1982.

Leavitt, H., "Applied Organizational Change in Industry: Structural Technological and Humanistic Approaches," in J. March (Ed.), *Handbook of Organizations,* Rand McNally, Chicago, 1965.

McDonough, A., Centralized Systems: Planning and Control. Wayne Thomson, PA, 1969.

Multinovich, J. S., and V. Vlohovich, "A Strategy for Successful MIS/DSS Implementation," *Journal of Systems Management*, (1984).

Murdick, R., and J. Munson, *MIS Concepts and Design*, Prentice Hall, Englewood Cliffs, N.J., 1986.

Ophuls, W., "The Scarcity Society in Ethical Issues," in T. Donaldson and P. Werhane (Eds.), *Business: A Philosophical Approach*, Prentice Hall, Englewood Cliffs, N.J., 1983.

Peters, T., and R. Waterman, *In Search of Excellence*, Harper & Row, New York, 1982.

Robey, D., *Designing Organizations*, Irwin, Homewood, Ill., 1986.

Simon, H., *The Shape of Automation for Man and Management*, Harper & Row, New York, 1965.

Thompson, J. D., *Organization in Action*, McGraw-Hill, New York, 1967.

Voich, D., Jr., H. Mottice, and W. Shrode, *Information Systems for Operations and Management*, South-Western Publishing, Dallas, 1975.

Zmud, R., *Information Systems in Organizations*, Scott, Foresman, Glenview, Ill., 1983

7
LEADERSHIP, STRATEGIC PLANNING, AND INNOVATION

7.1 ADVANCE ORGANIZER

We must learn how to make existing companies, large companies in particular, capable of innovation. A strategy is needed that will enable businesses first to identify the existing opportunities for innovation and then to give effective leadership for such innovation. It will no longer be sufficient to extend, broaden, modify, or attempt to adapt existing technologies. From now on, the need will be to innovate in the true sense of the word, to create truly new wealth-producing capacity, both technical and social (Drucker, 1985).

The adaptive corporation needs a new kind of leadership. It needs managers of adaptation equipped with a whole set of new, nonlinear skills. Leadership is the force behind every innovation. The efficacy of a leader is linked to his management of the strategic decision-making process, and, since such decision making is an instrument of change and innovation, we consider at length some of its elements.

Strategic planning at Northern Telecom from the perspective of J. Rankin (1988), vice president of Human Resources Management at Northern Telecom, provides a comprehensive view of the process. The strategic planning process, it is argued, should reflect the management style, values, basic assumptions, and practices of the company. The planning process at Northern Telecom accommodates the dynamics of change in every facet of the corporation's environment.

As they go through the strategic planning process, top managers should be able gradually to identify the consistent pattern to their culture, basic assumptions, beliefs, values, and so on. Making culture explicit is one way to facilitate more rapid strategic change and to assume flexibility at all levels of management. To match up corporate culture with business strategy, three strategies could be used: (1) ignore culture, (2) manage around the culture, and (3) change the culture to fit the strategy. The leader not only creates the rational and tangible aspects of organizations, such as technology and structure, but is also the creator of symbols, ideologies, language, beliefs, rituals, and myths. A dynamic analysis of organizational culture makes it clear that leadership is intertwined with culture formation, evolution, transformation, and destruction (Schein, 1985). The strategic management of corporate culture by the leaders is therefore critical. When a business is shifting its strategic direction, its culture may facilitate or inhibit such a strategic change.

The first ingredient in reinventing the corporation for the information age is a powerful strategic vision. Basically, the source of a vision is a leader. The systems approach forces him to look upon his organization as an information network. Adapting the organization to the information systems needs for effective strategic planning and the management of technological change is a major function of the leader. Another significant task of the leader who is reinventing the corporation is to select the right corporate structures. Since changes in corporate strategy also dictate changes in structure, the role of the leader in structural change is quite crucial to organizational effectiveness. Three uses of structure by the leader at each level of the organizational hierarchy are necessary in the management of technological change.

We conclude with the observation that a leader's style must reflect the ethical values of the corporate culture and the humanistic imperatives for effectively managing and designing tomorrow's organization.

Figure 7.1 summarizes these observations on leadership.

7.2 INTRODUCTION

The adaptive corporation needs a new kind of leadership. It needs "managers of adaptation" equipped with a whole set of new, nonlinear skills. Toffler (1985) further states that instead of constructing permanent edifices, today's adaptive executives may have to de-construct their companies to maximize maneuverability. They must be experts not in bureaucracy, but in the coordination of adhocracy. They must adjust quickly to immediate pressures, yet think in terms of long-range goals. And while in the past many managers could succeed by imitating another company's strategy or organizational model, today's leaders are forced to invent, not copy; there are no surefire strategies or models to copy. Above all, the adaptive executive today must be capable of radical action-willing to think beyond the thinkable; to reconceptualize products, procedures, programs, and purposes before crisis makes drastic change inescapable (Toffler, 1985).

There is perhaps no more pressing managerial problem than the sustained

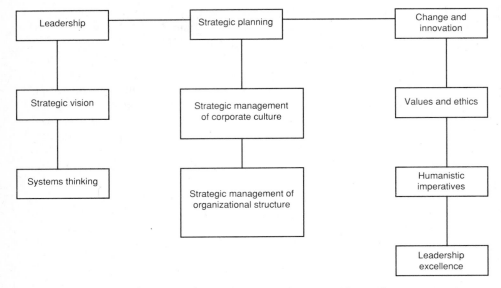

Figure 7.1. Leadership, strategic planning, and innovation.

management of innovation. It is the calculated outcome of strategic management and visionary leadership that provide the people, structures, values, and learning opportunities to make it an organizational way of life (Tushman and Nadler, 1986).

7.3 LEADERSHIP AND INNOVATION

Beyond making choices concerning strategy, structure, individuals, and the information organization, leaders also face the crucial, personal task of infusing their organizations with a set of values and a sense of enthusiasm that will support innovative behavior (Tushman and Nadler, 1986).

Several aspects of executive leadership behavior that can help or hinder innovation are identified by Tushman and Nadler (1986):

> 1. The executive team can develop and communicate a clear image of the organization's strategy and core values and the role of innovation in meeting the organization's strategy. The executive team must clearly and consistently articulate the importance of innovation and reinforce the necessary behavior.
>
> 2. The executive team can use formal and informal rewards to reinforce innovation. Innovative individuals and groups must receive recognition, attention, and support as well as formal rewards from the executive team to consistently reinforce behavior consistent with strategy and core values.
>
> 3. The senior executives cannot manage the organization alone. They must build executive teams with appropriate technical, social, and conceptual skills

to accomplish diverse tasks. The senior executive must also develop effective problem-solving skills in the top team. The team must be alert to external opportunities and threats, and possess the internal dynamics to deal with uncertainty.

4. Managing innovation requires visionary executive who provide clear direction for their organizations and infuse that direction with energy and value. Observation and research indicate that such executives frequently display three types of behavior. First, they work actively on envisioning or articulating a credible yet exciting vision of the future. Second, they personally work on energizing the organization by demonstrating their own excitement, optimism, and enthusiasm. Third, they put effort into enabling required behaviors by providing resources, rewarding desired behaviors, building supportive organizational structures and processes, and so on.

According to Gerstein (1987) corporations face a serious challenge is developing a strategic perspective at the top of their organizations. In considering future chief executives, the largest gap in candidates' current competency is in the area of information management and technology. These thoughts, emphasize two important themes: the centrality of strategic thinking, and the requirement that the CEO take a leadership role in shaping the company's technology strategy.

Leadership is a quality behind virtually every innovation. Tushman and Nadler (1986) observes that there are four qualities for leadership excellence. First, vision of what can be. Second, imagination, an essential ingredient for constructive risk taking. (Einstein, for example, considered imagination more important than knowledge.) Third, the ability to communicate—to move and motivate people, to inspire commitment for a project, an idea, or an ideal. Fourth, the willingness to persevere; it can be a long and tedious journey from a principle to a finished product.

The need for effective leadership in the management of the change process is important because of the dynamic nature of the organization. The organization operates in a dynamic environment which is constantly making changes necessary in the parts of the organization. Such changes in some parts make adaptive changes necessary in other parts. These adaptive changes inevitably modify the requirements for cooperative unity and therefore the character of that unity. The result, unless the organization disintegrates, is that the organization grows or evolves into a new state of dynamic equilibrium, a new form of the whole (Bakke, 1959).

It would be possible, of course, to conceive of the evolving form of the organization as a product of automatic adaptation to changes in its environment on the part of existing organizational resources and activity systems. We are asserting here that this dynamic adjustment (e.g., changes in resource allocations, bases of power and influence, technology, information flow, and the goal structure of the organization), initiated through systematic decision-making actions, is a self-conscious process and not a blind reaction to external forces. It is this set of situational contingencies and actions that creates the need for some synergic process activated by the leader (Bakke, 1959).

A number of organizational theorists comment on the role of the leader in the design and management of innovative organizations. Bennis (1966) has concluded that conventional approaches are "out of joint" with the recently emerging view of organizations as adaptive, problem-solving systems. He argued that the main challenge confronting today's organization is that of responding to changing conditions and adapting to external stress; and that there is a need for studies which reveal the processes by which the organization, through its members, searches for, adapts to, and solves its changing problems. Bennis further noted that the methodological rules by which an organization approaches its task and interacts with its environments are the critical determinants of effectiveness and, without the understanding of these dynamic processes, knowledge about organizational behavior is woefully inadequate (Bennis, 1966).

7.4 FUNCTIONS OF THE LEADER AND ADAPTIVE COPING

Some of the functions of the leader in the context of organizational change are described by Bennis (1966) and Schein (1970). The first is *reality testing,* or the capacity to test the reality of situations facing the organization—the ability of the organization to search out, accurately perceive, and correctly interpret the properties and characteristics of its environments. The second is *adaptability,* or the capacity to solve problems arising from changing environmental demands and to act with effective flexibility in response to these changing demands. The third is *integration,* or the maintenance of structure such that coordination is maintained and the various subunits do not work at cross-purposes (Schein, 1970).

An organization in a changing environment must have an adaptive-coping cycle, according to Schein (1970). The stages of the cycle are summarized as follows:

1. *Sensing:* the process of acquiring information about the external and internal environments by the organization.
2. *Communicating Information:* the process of transmitting information that is sensed to those parts of the organization that can act upon it.
3. *Decision-making:* the process of making decisions concerning actions to be taken as a result of sensed information.
4. *Stabilizing:* the process of taking actions to maintain internal stability and integration which might otherwise be disrupted as a consequence of actions taken to cope with changes in the organization's environment.
5. *Communicating Implementation:* the process of transmitting decision and decision-related orders and instructions to those parts of the organization that must implement them.

6. *Coping Actions:* the process of executing actions against an environment (external or internal) as a consequence of an organizational decision.
7. *Feedback:* the process of determining the results of a prior actions through further sensing of the external and internal environments.

7.5 LEADERSHIP, STRATEGIC PLANNING, AND CHANGE

A proper understanding of this strategy-making process will require a decision framework. We shall use the intelligence–design–choice framework developed by Simon (1957):

> *Intelligence activity* sets the stage for strategic decision by discovering a problem in need of solution or an opportunity available for development. In general, intelligence activity involves scanning the environment and collecting and analyzing information on various trends.
>
> *Design activity* begins once the area of action has been determined by intelligence activity. The two stages of design activity means of solving the problem or of exploiting the opportunity—and evaluation—determine the consequences of using these alternatives.
>
> *Choice activity* is concerned with choosing one of the alternatives that have been developed and evaluated. The "integration" of the various strategic decisions into a unified strategy is included in this category.

Although the intelligence, design, and choice activities are clearly delineated above, this is not always the case in practice. Nevertheless, the framework is a basically useful one for classifying strategy-making activity and for emphasizing the critical role of the leader in information processing and conversion in strategic planning.

7.6 LEADERSHIP AND STRATEGIC DECISION MAKING

The essence of management for all organizations, whether they are profit or nonprofit, is decision making. Management is the processing of information and conversion of information into decision outcomes. The effectiveness of a leader is linked to his management of the decision-making process. This calls for a number of complex skills which are all relevant for an effective leader. Peters and Waterman (1982), in their book *In Search of Excellence,* refer to this vaguely as a "bias for action." Some of the strategic skills for the leader in the decision-making process include (1) monitoring the decision environment, (2) reducing uncertainty in cause–effect analysis of the problem, (3) specifying decision objectives, (4) identifying decision criteria, (5) developing a viable strategy for managing goal conflicts and value conflicts,

(6) cultivating a climate for risk taking for innovative decision making, (7) clarifying a path of change produced by decision outcomes, (8) negotiating constraints that impact the decision-making process, (9) initiating structure, (10) developing management infrastructure for the decision, (11) designing control mechanisms to monitor the decision outcomes, (12) developing criteria to evaluate the effectiveness of the decisions, (13) enhancing task motivation to implement the decision (14) designing contingency plans for the decision outcomes, and (15) developing a bias for action and an organizational culture for rewarding excellence.

As an intervening element between the environment and the organization, organizational strategy does a number of things. First, it identifies the part(s) of the environment to be dealt with. It then prescribes a general course of action to be followed in attaining a goal or objective while dealing with that subenvironment. The strategy becomes an instrument of change. The process of this change is illustrated in Figure 7.2.

Newman (1963) cites a number of organizational features that are likely to vary with a change in strategy: centralization versus decentralization, degree of division of labor, size of operating units, mechanisms for coordination, nature and location of staff, management information systems, and characteristics of key personnel.

Environmental conditions provide much of the impetus for strategic decision making. Although not the only motivator of organization change, they

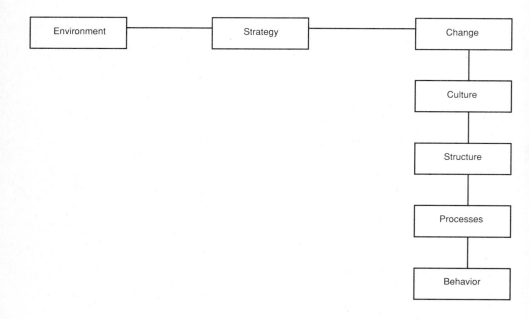

Figure 7.2. Strategy as an instrument of change.

are certainly among the most important. In order to survive, organizations must "fit" well with their environment; hence, any change in the environment will require a change in the organization. It is the function of strategy—an ongoing stream of decisions—and the leader to match or align organizational resources effectively with environmental opportunities and constraints. Because of their importance, strategic decisions must be closely linked with each other to form a consistent pattern for unifying and directing the organization. This is one of the primary functions of the leader.

A strategy is defined as a set of decision-making rules for the guidance of organizational behavior. Ansoff (1965) identifies three distinctive types of rules: (1) those for determining the relation between an organization and its external environment, generally referred to as the product market strategy; (2) those for establishing the internal relations and processes within the organization, frequently refered to as the organization concept; and (3) those by which the organization conducts its day-to-day business, called major operating policies. The second set of decision-making rules or organizational policies will be our principal focus. Some authors take issue with definitions that treat strategy as a plan or an explicit set of guidelines developed in advance of the activities they seek to direct. Mintzberg (1978), for instance, argues that this view limits our focus to abstract, normative aspects of the phenomenon. He suggests that we study strategy as *a pattern in a stream of significant decisions*.

Organizational policies are defined by Katz and Kahn (1973) as abstractions or generalizations about organizational behavior, at a level which involves the structure of the organization. This definition is in contrast to the notion that policies are behavior itself, or that they are official statements regardless of their relation to organizational structure and behavior. As abstractions about organizational behavior, policy statements may be either prospective or retrospective. This definition is further developed by Katz and Kahn (1973). Policies are, in their view, prospective generalizations about what organizational behavior shall be, at a level which involves changes in organizational structure. Such prospective statements of policy comprise a category of decisions: those decisions within an organization that affect the structure of the organization. Strategic planning is thus an aspect of organizational change—the decision aspect. Policy-making is also the decision aspect of that level of leadership that involves the alteration, origination, or elimination of organizational structure.

According to Davis (1986), scholars and managers interested in understanding the adoption and implementation of advanced manufacturing technologies and innovations must consider the influence of strategy in different areas of the firm, in addition to the values and attitudes of managers and decision makers, other characteristics of the organization's structure, and events in the organization's environment. Strategy integrates these different influences. Strategy is the process whereby the organization maneuvers in its environment. Strategy influences and is influenced by the adoption and implementation of innovations. Strategy, Davis (1986) continues, may be the most important element for making organization receptive to change and innovations.

Strategy forces the organization's decision makers to consider opportunities and threats and to structure the organization to optimize adaptive potential. Davis (1986) further argues that four types of strategy are important in the adoption and implementation of advanced manufacturing technologies: those of technology, manufacturing, marketing, and human resources.

7.6.1 Technology Strategy

We will briefly discuss one type of strategy, namely, technology strategy, as described by Davis (1986). Technology strategy emphasizes the importance of new technology to the firm and its contribution to overall business strategy. Mardique and Patch (1982) as cited by Davis state that firms must make technological choices in at least the following six areas: (1) selection of which technologies to invest in, determination of how these technologies should be embodied in existing products, which performance parameters should dominate, and how these new technologies should be evaluated; (2) level of competence required to become proficient in understanding and using the new technology, and proximity to the technological state of the art; (3) extent to which external or internal sources should be relied on as sources of new technology; (4) level of R&D investment considered optimal; (5) competitive timing of the firm; and (6) type of R&D organization and policies.

Firms wishing to be receptive to new ideas and practices such as advanced manufacturing technologies must think strategically about technology and develop long-range plans for its adoption. Because technological choices are also business choices, firms must search for technological threats and opportunities (Davis, 1986).

7.7 STRATEGIC CHANGE AND INNOVATION AND THEIR APPLICATION AT NORTHERN TELECOM

Strategic change requires a basic rethinking of the beliefs by which a company defines and carries on its business. Some of this rethinking is generated by changes in the external environment. Top managers are confronted with a wide variety of challenges to the traditional strategies of their companies. Some of these changes must be internalized within the culture of the organization and reflected in corporate strategies for change and innovation. Lorsch (1986) idenfies some areas in flux:

- Regulatory changes have revolutionized the competitive game in industries as diverse as airlines, trucking, banking, insurance, and telecommunications.

- Advancing electronic technology has spurred competition in the computer industry and also has made the computer a key competitive weapon for financial services, retailing, and telecommunications companies.
- The cost of technological and product development has caused even the giants of the aircraft industry to undertake joint ventures with foreign partners, formerly their competitors.
- Automotive companies facing changing consumer tastes and intense foreign competition are restructuring product lines, improving quality and productivity, and adopting foreign sources and joint ventures.
- Similarly, managers in a wide range of mature industries from steel to consumer durables are re-examining and changing traditional manufacturing and distribution policies to remain competitive in the world market.

There is considerable evidence that such pressures will accelerate in the years ahead, given the rapid rate of technological change, the increasing interdependence of global markets, and continuing regulatory change. This means that many top managers will increasingly find that they are confronted with major questions of how to position their companies in a new business and how to fundamentally change their firm's strategy in their existing business.

Northern Telecom's leadership in developing, evolving, and marketing advanced digital telecommunications systems has been the single most significant factor in its growth since the corporation's announcement of the "Digital World" in 1976. Its leadership in digital telecommunications technology will continue to be the basis for its growth throughout this decade.

At year end 1984, Northern Telecom had sold or had on order fully digital DMS (digital multiplex systems) switching and transmission equipment and PBXs (private branch exchanges) to serve the equivalent of 26.2 million lines in 67 countries. No company in the world has matched this success and experience in developing and supplying *fully* digital telecommunications systems.

In November 1982, Northern Telecom announced its OPEN World program to address the need for improving the management of information. The corporation committed to the application of its expertise in digital communications, and the underlying key technologies of semiconductor integrated circuits and software, to evolve the DMS and SL families and to introduce new products and services for the OPEN (Open Protocol Enhanced Networks) World.

During 1984, Northern Telecom introduced many new capabilities for the DMS and SL families toward fulfilling commitments made for the OPEN World program. The corporation and its research and development subsidiary, Bell-Northern Research, made considerable progress in development programs for the introduction, beginning in 1985, of major new products and services.

In February 1985, Northern Telecom hosted two important conferences for executives from organizations across North America and other countries.

The corporation launched the Meridian line of products comprising the Meridian SL-1 and SL-100 integrated services networks, including a local area network (LAN) capability, called LANSTAR, which will make it possible to transmit voice, data, graphics, and images within an organization over conventional telephone wires at the rate of 2.56 million bits of information per second. Previously installed SL PBXs can be upgraded to become Meridian SL-1 or SL-100 systems.

The corporation described in detail its plans to evolve and enrich the DMS family so as to offer through the public telecommunications network equivalent capabilities and services as those offered in private networks served by the Meridian SL-1 and SL-100 systems. Organizations can then choose the most attractive approach for them: their own private networks; using the public network capabilities provided by telecommunications companies; or a combination of both.

Northern Telecom also introduced the Meridian DV-1 Data Voice System, a data or integrated data and voice processing and communications system for organizations, departments, or branches of five to 100 people; families of digital integrated voice and data terminals; and a broad array of information services.

The company described its concept of a Dynamic Network Architecture for public telecommunications networks. A Dynamic Network Architecture will increase the flexibility and capabilities of public communications networks to handle the new opportunities developing as a result of such trends as the growth in computing in homes, factories, offices, and other institutions. Implementation of a Dynamic Network Architecture will involve the further evolution of the DMS switching and transmission systems; new applications for fiber-optics systems; and the development of other products and software.

7.8 STRATEGIC PLANNING AND ITS APPLICATION AT NORTHERN TELECOM

The following observations were made by J. Rankin (1988), vice president of Human Resources Management at Northern Telecom:

> The first rule of establishing a corporate planning process is that it should reflect the management style, values, basic assumptions and practices of the company it serves. If it exists in isolation, with its own set of rules and procedures, then it has limited utility.
>
> The planning process at Northern Telecom must accommodate the dynamics of change in every facet of the corporation's environment. This includes the rapidly evolving technological capability of the corporation's products, the massive changes evolving in its manufacturing processes, and the constantly changing political and regulatory scene.

7.8 STRATEGIC PLANNING AND ITS APPLICATION AT NORTHERN TELECOM

Northern Telecom operates in one of the world's most volatile business climates. Historically, telecommunications has existed as a regulated monopoly. In Canada and the United States, this resulted in regional monopolies—privately owned and operated, but highly regulated by local and federal governments. Progressive deregulation in the U.S. and Canada over the last 20 years has created new markets and new competition for home and business telephone systems, and in long-distance service.

In 1984, the divestiture of the Bell operating companies by American Telephone and Telegraph, AT&T, broadened the opportunities to sell to the telephone operating companies themselves, especially in such areas as customer premises equipment, which include private branch exchanges and business telephone systems.

In Northern Telecom's current decentralized business structure, many products are manufactured to support market needs, in more than one plant, often in more than one country. In the same way, marketing is planned and implemented by geographical area.

As the corporation has expanded internationally, it has developed a line management structure based on a system of operating subsidiaries, each with a geographical area of responsibility. Superimposed on this is a product-based strategic management structure, which insures that product strategies reflect global needs and opportunities, as well as those of individual countries.

By the early 1980s, we had defined our products into two principal lines: Integrated Office Systems, or IOS, and Integrated Network Systems, or INS. This changed the way in which our product line planning was done, and the senior strategic planning responsibility for each of these two main product lines was delegated to two senior executives called product line primes (PLPs).

These individuals, in addition to senior geographic responsibilities for sales and earnings of all Northern Telecom products, have the specific responsibility for the global direction, development, and success of the IOS and INS product lines. They play key roles in the overall Northern Telecom planning process.

The formal planning process is an annual, cyclical process. But the rapidly changing nature of our business also requires much faster reactions than are possible within a formal, annual process. And so, there also exists a routine management process that goes on continuously, involving day-to-day communications between key players. It introduces the deviations—the mid-term changes—that prevent the planning process from losing touch with the market place and new ideas.

The annualized process provides a baseline, and communicates crucial planning information to all the other players.

The formal planning process can be broken down into a series of inter-related processes: technology planning; market-planning; strategic-product-line planning; and operating planning.*

---

* *Author's Note:* Refer to Figure 7.3 for a flowchart of the strategic planning process.

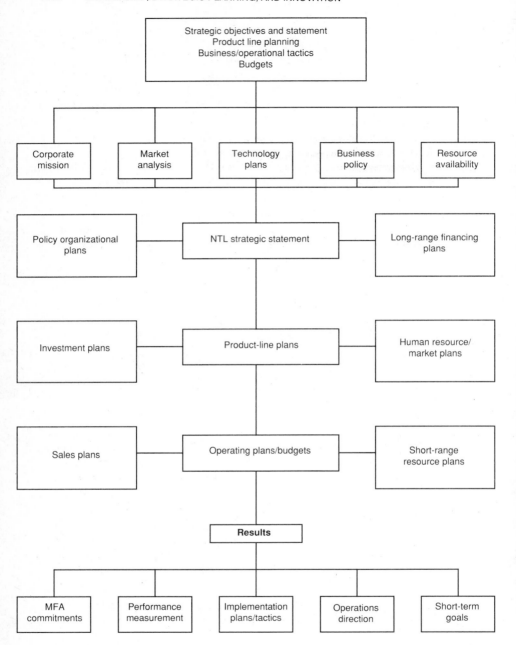

Figure 7.3. Overall planning process at Northern Telecom.

The technology-planning process and the market-planning process both feed into the strategic product-line-planning process. In turn, the strategic-product-line plan drives the operating plan. And it is the operating plan that drives our budgets. Ultimately, of course, it is in terms of the budget that we measure our overall short-term performance.

By comparison, the real value of the strategic product-line-planning process lies in the act of preparing the plans themselves. We'll now take a closer look at that process. The people who control product evolution are in the divisions. They, in effect, are the people who must do the strategic planning. They need some sort of framework to guide them, but it's absolutely critical that they do their own planning, rather than let a corporate group do it for them.

We, therefore, do no formal product planning at corporate headquarters. There are lots of strategies developed corporately, but that's quite distinct from the detailed strategic product-line-planning process.

While the real end-product of the strategic product planning process is the process itself, an important by-product of that process are the documented plans. These are loose-leaf binders that are produced by the individual product divisions. One book for each product prime. Even though there may be more than one division involved with a given product, the product prime is the person held responsible for drafting the plan for his product. And he has to take account of the global view for that product.

The plan focuses on technology and markets. By markets, I include both competition and customers. In effect, the strategic product plan takes a worldwide look at product strategies, investments, and results. Financial detail is not the most important part of the strategic product plan. This is another element that differentiates the strategic product plan from the operating plan, where the financial side is prepared in substantial detail.

It is, of course, important that product management understand the financial implications of the strategies and developments that they propose. If the implementation of his plan is going to require a huge increase in capital investment, people, and resources, the product prime should obviously be aware of that, and be able to estimate it properly.

The strategic product plans benefit the divisions by helping them structure and articulate what they want to do, and by promoting the discussion and debate between senior management and the divisions that must occur before the plans become enshrined in an operating plan, and subsequently, in the budget.

Finally, the whole process culminates in a meeting with the senior executive group, called the executive council, once a year in July. That's useful because it promotes the interchange of ideas between the IOS and INS product line people, and the members of the executive council. The meeting also establishes a framework for the total R&D budget for the corporation for the next three years, which is fed into the operating plan. These preliminary allocations are refined during the budgeting process that follows over the next five months. That, in effect, is what the strategic-product-line plan is about: setting the scene for deploying R&D, capital, and other resources in pursuit of global objectives. The detailed product plans are used by a large number of people across the corporation who have functional responsibilities.

For example, human resources people may want to get a better feel for what the issues and plans are within the divisions. They can do so by examining the product plans. Our subsidiary, Northern Telecom Electronics, itself a "product line" under a PLP, is responsible for manufacturing proprietary electronic components, such as microchips. They may want a better understanding of what some of the main product divisions consider the key technologies to be. They can get a feeling for that from the product plans, as well as influence others through their own product plan.

The operating plan, by comparison, outlines our goals, sets targets, and provides the broad guidelines for reaching those objectives. Our operating plan is a three-year rolling plan, the first year of which is the budget. When I said before that the operating plan focuses on management reporting and control, that means that the budget of the corporation, which is the first year of the operating plan, is used month-by-month, quarter-by-quarter, to compare performance against what the divisions have committed to achieving over the year. That process tells how much you are deviating from your budget.

Major global objectives are set at the beginning of each year for earnings per share and sales revenues. Those major objectives are set by the Chief Executive Officer. They are at the heart of the operating plan, which ultimately goes to the Board of Directors. Information and proposals from the product line plan feed into and drive the operating plan. Budgets that are proposed in the product line plan are really provided in the operating plan.

To recap, the data used in the operating plan are part of the management system, intended not just as a forecast, but to influence managers in very specific ways to take actions that will lead to desired results. The product line plan, on the other hand, really consists of proposals, some of which the corporation may decide not to proceed with.

This dual planning process allows for flexibility. It encourages people to propose strategies and products even though they might only have a 50/50 chance of being adopted. Thus, product planners are encouraged to be innovative, and to take more risks in their projections than would normally be found in a budget.

The operating plan process generally has much give-and-take between the bottom and top elements. For example, the projects and programs that are proposed bottom-up might be found to be inadequate as far as the amount of profit and/or sales growth they are going to generate. A senior group executive, for example, then feeds back, during the formulation of the plan, the information that he needs more earnings per share. A debate usually ensues where the line managers try to preserve targets that they are certain they can reach, while the group executive might get them to stretch.

Finally, it's worthwhile to recap the relationship between the strategic-product-line plan and the operating plan, and summarize how the two processes grew out of Northern Telecom's global expansion. The key point here is that once the corporation decided to operate geographically, with subsidiaries with manufacturing and marketing capability within a given geographical area, an overlying strategic product planning matrix became necessary. This keeps the products evolving in an orderly manner, and ensures that R&D investments cater to global requirements, as well as to the specific requirements of a single country.

Our planning process has become global because Northern Telecom has become global. It is formal and flexible because telecommunications requires huge investments in capital and R&D, while operating within a fast-changing, high-technology environment. Just as our marketing, manufacturing, and R&D must move nearer to the customer, so must our planning. As the corporation continues to change over the years, our planning process will have to keep pace.

From our perspective on the strategic planning process at Northern Telecom, we must not assume that it is simply an application of strategic information systems, integrated or distributed data bases, heuristics for problem solving, or technical review of operating plans and budgets. The underlying assumptions, beliefs, and values of the organization are also articulated in this process. These invisible dimensions of culture pervade the whole process of strategic planning, such as beliefs and assumptions about capital market expectations, product market competition, internal organization–management structure, and about the role of technological change. The strategic vision is constructed from these beliefs and assumptions. Also, a part of this strategic change involves acquisitions, diversification, new product development, test markets, and design of a management infrastructure for change.

As they go through this process, top managers should be able gradually to identify the consistent pattern to their culture and how beliefs are related to one another. Making culture explicit is one way to facilitate more rapid strategic change and to assure flexibility at all levels of management.

Another way to assure flexibility at the top is to encourage flexibility of thinking at subordinate levels of management (flexibility down the line). This can assure that succeeding generations of top management will be less blindly rigid in their adherence to cultural traditions. In addition, it can provide a source of new perspective to the present top managers as new ideas emerge from subordinate levels (Lorsch, 1986).

Two different routes can be used to encourage flexibility among middle managers. One is to stimulate new ideas within the organization. An in-company education program for middle managers, with outside experts as instructors, is one frequently employed means. Another is to encourage systematic rotation of managers among functions and businesses. In this way, their perspectives can be broadened. However, such inside activities always have the inherent limitation that new ideas are learned with company colleagues in the context of the company's culture (Lorsch, 1986).

7.9 LEADERSHIP AND ORGANIZATIONAL CULTURE: SCHEIN'S PERSPECTIVE

To match corporate culture with business strategy, three strategies can be used: (1) ignore the culture, (2) manage around the culture, and (3) change the culture to fit the strategy.

> The process of shaping culture is seen as a primary management role. The leader not only creates the rational and tangible aspect of organizations, such as technology and structure, but is also the creator of symbols, ideologies, language, beliefs, rituals and myths (Schein, 1985).

Many academics and practitioners believe culture can be positively influenced through consistent, thoughtful managerial action.

According to Schein (1985), a dynamic analysis of organizational culture makes it clear that leadership is intertwined with culture formation, evolution, transformation, and destruction. Culture is created in the first instance by the actions of leaders; culture is embedded and strengthened by leaders. When cultures become dysfunctional, leadership is needed to help the group unlearn some of its cultural assumptions and learn new assumptions. The unique and essential function of leadership is the manipulation of culture.

7.9.1 Leadership in Culture Creation

In a growing organization, the leadership externalizes its own assumptions and embeds them gradually and consistently in the mission, goals, structures, and working procedures of the group. Whether we call these basic assumptions the guiding beliefs, the theories in use, the basic principles, or the guiding visions on which a founder operates, there is no doubt that they become principal elements of the emerging culture of the organization.

At this stage, Schein (1985) argues that the leader needs both vision and the ability to articulate it and enforce it. A consistent set of assumptions is only forged by clear and consistent messages as the group encounters and survives its own crises. The culture creation leader therefore needs persistence and patience as well.

As groups and organizations develop, certain key emotional issues arise, these having to do with dependence on the leader, with peer relationships, and with how to work effectively. At each of these stages of group development, leadership is needed to help the group identify the issues and deal with them. During these stages leaders often absorb and contain the anxiety that is unleashed when things do not work as they should. The leader provides temporary stability and emotional reassurance while uncertainty and ambiguity are being reduced.

7.9.2 Leadership at Organizational Mid-life

The culture at this stage influences the strategy, structure, procedures, and ways in which the group members will relate to each other. Culture becomes a powerful influence on members' perceiving, thinking, and feeling; these predispositions, along with situational factors, will influence the members' behavior. Because it seems an important anxiety-reducing function, culture will be clung to even if it becomes dysfunctional in relationship to environmental opportunities and constraints (Schein, 1985).

Mid-life organizations show two basically different patterns, however. Some, under the influence of one or more generations of leaders, have developed a highly integrated culture even though they have become large and diversified; others have allowed growth and diversification in cultural assumptions as well and, therefore, can be described as culturally diverse with respect to their business, functional, and geographical units. How leaders manage culture at this stage of organizational evolution depends on which pattern they perceive and which pattern they decide is best for the future (Schein, 1985).

Leaders at this stage need, above all, the insight to know how to help the organization evolve to whatever will make it most effective in the future. In some instances, this may mean increasing cultural diversity, that is, allowing some of the uniformity that may have been built up in the growth stage to erode; in others, it may mean pulling together a cultural diverse set of organizational units and attempting to impose new common assumptions on them. In either case, what the leader most needs is *insight into the ways in which culture can aid or hinder the fulfillment of the organization's mission* and the *intervention skills to make desired changes happen* (Schein, 1985).

In the mature organization—if it has developed a strong unifying culture—culture now defines what is to be thought of as "leadership," what is heroic or sinful behavior, and how authority and power are to be allocated and managed. Thus, what leadership has created now either blindly perpetuates itself or creates new definitions of leadership, which may not even include the kinds of entrepreneurial assumptions that started the organization in the first place.

What leaders must do at this point in the organization's history depends on the degree to which the culture of the organization has, in fact, enabled the group to adapt to its environmental realities. If the culture has no facilitated adaptation, the organization either will not survive or will find a way to change its culture. If it is to change its culture, it must be led by someone who can, in effect, break the tyranny of the old culture (Schein, 1985).

A leader capable of such managed culture change can come from inside the organization if he has acquired objectivity and insight into elements of the culture. However, the formally designated senior managers of a given organization may not be willing or able to provide culture change leadership.

Leadership in this form may then have to come from other boundary spanners in the organization or from outsiders. It may even come from a number of people in the organization, so that it makes sense to talk of turnaround teams or multiple leadership. If the formally designated senior managers do not have the insight, skill, or vision to change the culture in whatever ways might be needed, then the culture will change under the impact of other forms of leadership, or it will atrophy and the group will disappear as a social entity. In any event, no change will occur without leadership from somewhere (Schein, 1985).

Schein (1985) summarizes some major implications for leadership and culture:

1. If leadership is critical both to the formation of culture and to culture change, we should examine more carefully the development of this form of leader, the *culture manager*.
2. If leadership is culture management, do we develop in our leaders the emotional strength, depth of vision, and capacity for self-insight and objectivity that are necessary for culture to be managed?
3. Do our management development processes nurture "creative individualists," emotionally strong boundary spanners with high objectivity and tolerance of deviant points of view.
4. If the environment in which organizations have to function is changing ever more rapidly, can we specify what kind of organizational culture will be most adaptive, or is the real implication that the more rapidly things change, the more dependent we are on leaders to manage the changes? If so, are we developing enough leaders so that they will be available when they are needed?

If there is to be a single final conclusion drawn from this intellectual journey through parts of organization theory, social psychology, and anthropology, it is that leadership and culture management are so central to understanding organizations and making them effective that we cannot afford to be complacent about either one.

In the overall context of implementing strategy, the element of culture, although very important, is no doubt the most nebulous and difficult to grasp. It is clear, however, that most organizations develop unique cultures, and that a specific culture has a direct impact on how well a particular strategy can be implemented. Recognizing the existence of culture and understanding its implications are necessary steps when contemplating the implementation of strategic change.

Corporate cultures have a powerful influence on managers' behavior. When a business is shifting its strategic direction, its culture may facilitate or inhibit such a strategic shift. Decision-making behavior, problem-solving style, and the management of communications, conflict, change, and creativity are parts of corporate behavior. The behavioral repertoire and patterns within the subcultures of a complex organization will predispose an organization to a variable rate of change. Each subsystem has its own unique time and goal orientations, means–ends chain, value systems, norms, rituals and symbols, ethical premises, strategic options, learning style, information-processing systems, and behavioral repertoire, hence the differences in the culture change of each subsystem. Because of these elements, some subcultures may be more oriented toward consolidative behavior, and others may be more oriented toward innovative behavior; hence, there is a variable rate of change among the organizational subsystems.

Without exception, the dominance and coherence of culture proved to be an essential quality of excellent companies. Moreover, the stronger the culture and the more it was directed to the marketplace, the less the need there was for policy manuals, organizational charts, or detailed procedures and rules (Peters and Waterman, 1982).

7.10 LEADERSHIP, STRATEGIC VISION, AND CHANGE

The first ingredient in reinventing the corporation for the information age is a powerful vision—a whole new sense of where the company is going and how to get there. Usually, the source of a vision is a leader, a person who possesses a unique combination of skills, that is, the mental power to create a vision and practical ability to bring it about. According to Naisbitt and Aburdene (1985), belief in vision is a radically new precept in business philosophy. It comes out of intuitive knowing; it says that logic is not everything; that it is not all in the numbers. The idea is simply that by envisioning the future you want, you can more easily achieve your goal. This vision is explicit in strategy formulation and in articulating the core values of the organization as noted earlier.

The leader who would create a strategic vision sufficiently compelling to motivate associates to superior performances must draw on the intuitive mind. Intellectual strategies alone will not motivate people. Only a company with a real mission or sense of purpose that comes out of an intuitive or spiritual dimension will capture people's hearts (Naisbitt and Aburdene, 1985).

Ohmae (1984) develops this notion of strategic vision by linking it to three major types of thinking (Figure 7.4) with the following comments:

> In strategic thinking, one first seeks a clear understanding of the particular character of each element of the situation and then makes the fullest possible use of human brainpower to restructure the elements into the most advantageous way. Phenomena and events in the real world do not always fit a linear model. Hence the most reliable means of dissecting a situation into its constituent parts and reassembling them in the desired pattern is not a step by step methodology such as systems analysis. Rather, it is that ultimate non linear thinking tool, the human brain. True strategic thinking thus contrasts sharply with the conventional mechanical systems approach based on linear thinking.

But it also contrasts with the approach that stakes everything on intuition, reaching conclusions without any real breakdown or analysis.

No matter how difficult or unprecedented the problem, a breakthrough to the best possible solution can come only from a combination of rational analysis based on the real nature of thinkings, and imaginative reintegration of all the different items into a new pattern. This is always the most effective approach to devising strategies for dealing successfully with challenges and opportunities.

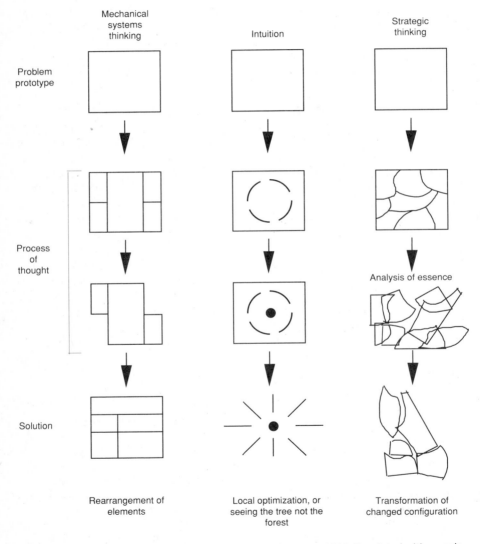

Figure 7.4. Three kinds of thinking processes. (From Ohmae, 1984. Reprinted with permission of McGraw-Hill, Inc.)

7.11 THE TRANSFORMATIONAL LEADER: TICHY'S AND DEVANNA'S PERSPECTIVES

Transformational leadership is about change, innovation, and entrepreneurship. This brand of leadership is a behavioral process capable of being learned and managed. It's a leadership process that is systematic, consisting of a purposeful and organized search for changes, systematic analysis, and the capacity to move resources from areas of lesser to greater productivity. This strategic transformation of organizations is not something that occurs through the idiosyncratic behavior of charismatic geniuses. It is a discipline with a set of predictable steps. Transforming an organization to make it strategically competitive is a complex task (Tichy and Devanna 1986).

Transformational dramas develop around three themes, as shown in Figure 7.5:

Act I. Recognizing the Need for Revitalization. The first act of the drama centers on the challenges the leader encounters when he/she attempts to alert the organization to growing threats from the environment.

Act II. Creating a New Vision. The second act of the drama involves the leader's struggle to focus the organization's attention on a vision of the future that is exciting and positive.

Act III. Institutionalizing Change. In the third and final act of the drama the leader seeks to institutionalize the transformation so that it will survive his/her tenure in a given position.

7.11.1 The Organization During Act I

Trigger Events. The need for change is triggered by environmental pressures. But not all organizations respond to signals from the environment indicating change. The external trigger event must be perceived and responded to be leaders in the first phase of the transformation process. Examples of failure to respond to trigger events litter the economic landscape in the form of bankruptcies, such as W.T. Grant and Penn Central, and near misses, such as International Harvester (recently renamed Navistar), Chrysler, and Continental Illinois. The failure of the American auto industry to adapt to changing consumer demand for small cars crippled Chrysler and hurt Ford and General Motors. The failure of the American steel industry to keep pace with technological innovation may well have sounded a death knell for American dominance in steel.

Felt Need for Change. Once organizational leaders accept the fact that their business environment is changing, key decision makers in the organization must be made to feel dissatisfaction with the status quo. The felt need for

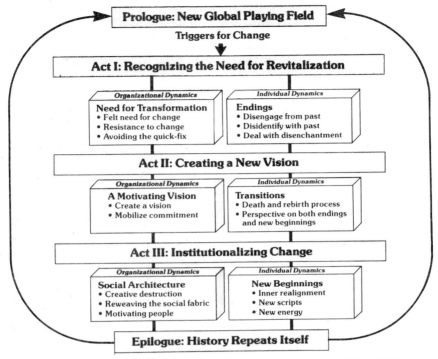

Figure 7.5. Tichy and Devanna's Transformational Leader, (Wiley, 1986)

change provides the impetus for transition, but this process does not always go smoothly. A key to whether resistant forces deter the organization from making the needed adjustments to environmental shifts is the quality of the leadership that is brought to bear.

7.11.2 The Organization During Act II

Creating a Vision. The leaders involved in organizational transformation need to create a vision that a critical mass of employees will accept as a desirable change for the organization. Each leader must develop a vision and communicate it in a way that is congruent with the leader's philosophy and style. At General Motors the vision emerged from several years of detailed committee work and staff analysis, while at Chrysler, Lee Iacocca relied on his intuitive and directive leadership philosophy and style. Both General Motors and Chrysler ended up with a new vision because transformational leaders shaped a new organization mission proactively. The long-term challenge to organizational revitalization is less "how" the visions are created and more the extent to which the visions correctly respond to environmental pressures and create transitions within the organization.

Mobilizing Commitment. The organization, or at least some critical mass within the organization, accepts the new mission and vision and makes it happen. It is in this stage of the transformational process that leaders must tap into a deeper sense of meaning for their followers.

7.11.3 The Organization During Act III

Institutionalizing Change. Revitalization is just empty talk until the new vision becomes reality. The new way of thinking becomes day-to-day practice. New realities, actions, and practices must be shared so that changes become institutionalized. At a deeper level this requires shaping and reinforcing a new culture that fits with the revitalized organization. How people are selected for jobs, appraised and rewarded on their performance, and developed for future responsibility are of overwhelming importance.

What happens at the organizational level is by itself not sufficient to create and implement change. Major transitions unleash powerful conflicting forces in people, and individual psychodynamics of change must be understood and managed. Change invokes simultaneous personal feelings of fear and hope, anxiety and relief, pressure and stimulation, threats to self-esteem and challenges to master new situations. The task of transformational leaders is to recognize these mixed feelings, act to help people move from negative to positive emotions, and mobilize the energy needed for individual renewal.

7.11.4 The Individual During Act I

William Bridges, in *Transitions: Making Sense of Life's Changes,* outlines a three-phase process of individual change involving endings, transition states, and new beginnings. During each of these phases a set of psychological tasks can be identified. Individuals must work through these if they are to complete the change process successfully.

Endings. All individual transitions start with endings. Employees who cling to old ways of doing things will be unable to adjust to new demands. They must follow a process that includes disengaging from the past; disidentification with its demands; disenchantment with its implications and disorientation as they learn new behaviors.

Transformational leaders provide people with support by helping replace past glories with future opportunities. This will happen only if they are able to acknowledge individual resistance that is derived from a sense of loss in the transition. Leaders should encourage employees to accept the failures without feeling as if they had failed. It does not help to treat transitions as if the past did not exist. The past will hold the key to understanding what went wrong, as well as what worked, and can frequently provide a useful map to the future.

7.11.5 The Individual During Act II

Neutral Zones. Employees need the time to work through their feelings of being disconnected with the past and not yet emotionally committed to the future. This phase causes the most trouble in action-oriented Western cultures, for it tends to be viewed as nonproductive. Yet the difference between success and failure in organizational transformations can occur during this stage. Passing successfully through the neutral zone requires taking the time and though to gain perspective on both the ending—what went wrong, why it needs changing, and on what must be overcome to make a new beginning. It is during this phase that the skills of a transformational leader are really put to the test. A timid bureaucrat who revels in the "good old days" will not provide the needed support to help individuals traverse the neutral zone. A strong dictatorial leader may also fail, by forcing a new beginning before people have worked through their feelings and emotions.

7.11.6 The Individual During Act III

New Beginnings. Once a stage of psychological readiness to deal with a new order of things is reached, employees must be prepared for the frustration that accompanies failure as they replace thoroughly mastered routines with a new act. Adequate rehearsal time will be needed before everyone learns their new lines and masters their new roles so that the play can become again a seamless whole rather than a set of unintegrated scenes.

How Can Leaders Effectively Transform Their Organizations? To lead effectively in turbulent times we must return to basic questions about the nature and purpose of the organization in question. A reexamination of its technical systems undoubtedly will result in new mission and strategy. It should also result in a major revamping of the financial, marketing, production, and human resource systems as well so that they will drive the organization toward the new goals. This technical realignment will force a consideration of the political allocation system in most organizations. In many, new criteria will determine who gets ahead, how they get rewarded, and who has the power to make decisions. And, finally, they must look at the values and beliefs they have helped to install in their members and ask if this culture supports the change the organization wishes to make. Because of the dynamic nature of organizations, these systems rarely reach a state of equilibrium where they do not require some attention, but the major initial wrenching that is necessary will provide a major challenge to transformational leaders.

The technical, political, and cultural (TPC) framework provides the contextual map to guide this effort and becomes an important conceptual tool for organizing the actions and decisions of transformational leaders.

Technical Design Problems. All organizations face technical design problems. People, money, and technical resources must be arranged to help minimize environmental threats and maximize environmental opportunities. Strategic planning, goal setting, organizational design, and the design of management systems are all tools sued to solve technical problems in organizations. The quest is for a solution that provides profitability.

Political Allocation Problems. Similarly, all organizations face the problem of allocating power and resources. Questions of who is to be involved in the decision-making process and how rewards and benefits are to be allocated are essentially problems of political allocation. Unlike the technical arena, in which there are formalized tools, such as strategic planning and organization design, the political arena frequently suffers from a lack of formalized, systematic analysis. This is unfortunate, since the greatest leadership challenges tend to lie in the strategic political arena. It is here that the tension between organizational goals and objectives and individual aspirations most often comes into play.

Cultural Value Problems. No organization can write a set of procedures so complete that they specify people's behavior in all situations. Consequently, organizations are held together in part by normative glue. If we consider the excellent organizations discussed by Peters and Waterman, we find that a common thread is their ability to articulate their values to their employees. This helps to inform decisions at all levels of the organization. Whereas it may not be as critical for blue-collar workers to share the value system of the top management team as it is for the top management team to share a common set of beliefs and values, it should be recognized that conflicting values and beliefs within the organization lead to confrontation rather than cooperation.

An important caveat in this discussion is to distinguish between a difference of opinion and a value conflict. All organizations benefit from open and honest debate on important issues and it is vital that they recognize the difference between healthy deviance and destructive heresy on the part of their members.

The Strategic Rope. The TPC issues can be seen as three intertwined strands of a rope. The rope metaphor underscores several points. First, from a distance, individual strands are not distinguishable. Similarly, a casual observer cannot distinguish the TPC systems in organizations. Second, close examination of the rope will reveal that each strand is made up of many substrands, just as close examination of organizations will reveal many TPC systems. Finally, the strength of the rope depends not only on the strength of the strands that make it up but also on their connection. A rope can unravel; an organization begins to come apart when its systems work at cross-purposes.

While the entrepreneurial founder of an organization weaves the rope from scratch, the transformational leader must unravel the old rope and reweave it. The task of the entrepreneur is difficult and the failure rate among new ventures bears testimony to that fact, but it is an essentially different problem from that faced by the transformational leader. People and organizations find this situation difficult, and human nature throws its weight on the side of resistance to change or failure to revitalize.

7.12 LEADERSHIP, STRUCTURE, AND CHANGE

A major task of the leader who is reinventing the corporation is to select the right corporate structures. The critical question, then, is which structures will realize the vision (Naisbitt and Aburdene, 1985). Since changes in corporate strategy also dictate changes in structure, the role of the leader in structural change is quite critical to organizational efficacy. Since technological change invariably involves changes in the characteristics of the organizational structure, as noted earlier, structural change is therefore imperative in the effective management technological change. The three uses of structure by the leader at each level of the organizational hierarchy are also essential in the management of technological change. At the strategic apex, they are particularly critical for technological change because the change of structure is radical; Toffler (1985) considers this radical change imperative for the information age because of turbulence in the environments of most organizations. At the other levels, the change in structure is more incremental and may be adequate for the implemenation and routinization stages of the change process.

The research focus on leadership, structure, and changes is evaluated by Hage and Aiken (1970), who argue that there has been inadequate research on the relative importance of leadership style as opposed to structural properties in understanding change dynamics. It may well be that everything we have said in this book about structural change is simply a function of variations in leadership style—the charismatic leader in the highly changing organization and the tradition or "caretaker" leader in nonchanging organizations (Hage and Aiken, 1970).

The influence of leadership on organizational structure is noted by Katz and Kahn (1973), who observe that three basic types of leadership behavior occur in organizational settings: (1) the introduction of structural change, or *policy formulation*; (2) the interpolation of structure, that is, piecing out the incompleteness of existing formal structure, or *improvisation*; and (3) the use of structure formally provided to keep the organization in motion and in effective operation, or *administration*. Every instance of leadership involves the use, interpolation, or origination of organizational structure to influence others. When people are influenced to engage in organizationally relevant behavior, leadership has occurred. When no such attempt at influence is made, there has been no leadership. Katz and Kahn (1973) further observe that the distribution of leadership acts is by no means random. Some positions (offices) are defined largely in terms of expectations involving such influential acts. The presidency of a company is an example, and so is the office of first-level supervisor. The exercise of leadership by persons occupying such positions is facilitated by the organizational resources (rewards, punishments) made available to them, and above all by the power of legitimacy, the implicit contract which each member makes to accept influence in prescribed matters from designated "leaders." The distribution of information in organizations follows a similar pattern, so that positions designated formally as offices of leadership receive information that increases the expertise of their occupants relative to those they are expected to lead.

Furthermore, Katz and Kahn (1973) note that there is a relationship between the three patterns of leadership and the hierarchical levels of positions in the organization. Except in democratically constituted systems, only the top echelons of line and staff officers are really in a position to introduce changes in structure. The piecing out of structure is found most often in the intermediate levels of the organization. And the lowest supervisory level has open to it mainly the exercise of leadership by the skillful use of existing structure. In other words, the degree of freedom to supplement existing structure is not as great as the lowest officer level as it is at the second and third levels of command. And the freedom to originate, eliminate, and change organizational structure is not as extensive at the intermediate levels as it is at the top levels. It is true, of course, that the top level can exercise all three patterns of leadership, so that as we ascend the hierarchy we find more types of action available to the officer groups.

According to Katz and Kahn (1973), the exercise of these three patterns of organizational leadership also calls for different cognitive styles, different degrees and types of knowledge, and different affective characteristics. Hence, the leadership skills appropriate to one level of the organization may be irrelevant or even dysfunctional at another level. The consistent and equitable employment of devices characteristic of the good administrator at the lowest level may be of little use to the policymaker at the top. The cognitive and effective requirements of three patterns of organizational leadership are illustrated in Figure 7.6.

In terms of the three basic functions of leadership—the use, interpolation, and origination of structure—the emphasis of effective leadership is, according to Katz and Kahn (1973), away from the unvarying use of existing structure and toward the origination of structure. Such leadership manifests the ability to change in response to external demands for change. The subordination of structure means that leaders assert freedom from the requirements of existing structure; they propose to use structure for the achievement of organizational goals, rather than to be used by it.

Katz and Kahn (1973) assert that one mark of effective leadership is the ability to subordinate structure when situational requirements are clear and popular support is adequate. The subordination of structure, in other words,

Types of Leadership Process	Appropriate Organizational Level	Abilities and Skills	
		Cognitive	Affective
Origination: change, creation, and elimination of structure	Top echelons	System perspective	Charisma
Interpolation: supplementing and piecing out of structure	Intermediate levels pivotal roles	Subsystem perspective: two-way orientation	Integration of primary and secondary relations: human skills
Administration: use of existing structure	Lower levels	Technical knowledge and understanding of system and rules	Concern with equity in use of rewards and sanctions

Figure 7.6. Leadership patterns, their loci in the organization, and their skill requirements. (Adapted from Katz and Kahn, 1973).

is the behavior which flows reasonably from an adequate attainment of system perspective. The inability to alter and originate structure is perhaps an inconspicuous disease of leadership. In the long run, the disease is absolutely and certainly fatal to organizations (Katz and Kahn, 1973).

As with the other aspects of effective leadership, the importance of the ability to subordinate structure varies with the echelon of leadership. It is of particular import at the apex of the organization. It varies also with the stability of the organizational environment and with the stage of the organization in its total life cycle. During periods of rapid environmental and technological change or rapid development of the organization itself, the ability to modify the original structure is of the greatest importance. During times of environmental stability and organizational maturity, the requirements for structural subordination will be minimal.

Finally, it is argued that the critical task of the intermediate levels of management is to piece out the organizational structure, or guide subordinates to do so, in ways that optimize organizational functioning. On the cognitive side, this involves some degree of internal system perspective, specifically technical know-how about tasks of the relevant subsystems and knowledge of those subsystems with immediately adjacent subsystems. In terms of affective orientation, the basic requirement is the ability to integrate primary and secondary relationships. This type of orientation has been associated with human relations skills (Katz and Kahn, 1973).

Another way of looking at leadership and structure is through the contingency perspective provided by Lawrence (1985). The leader focusing on information complexity and resource scarcity can determine what mode of strategic decision making will be dominant. Since both dimensions of the organizational domain, information complexity and resource scarcity, are issues in managing technological change, and because technological change involves a variety of structural configurations that a leader must adopt, this perspective by Lawrence is particularly relevant here (Figure 7.7).

The description of the model is based on Lawrence's observations. In area 1, with high information complexity and low resource scarcity, there is an overabundance of information to be handled, but little press for conserving resources. The expected form is adhocracy, with its loosely coupled organizational units. Affluent universities and hospitals are the usual examples of organizations in this niche. Innovation can be expected, but not efficiency. It is in this context that the theory would hypothesize the dominance of the garbage-can model, with occasions for decision making used as receptacles for assorted problems, solutions, and participations. There is an abundance of differentiation but little pressure for integration, so all the various points of view and ideas are aired, but whatever resolution emerges is usually more arbitrary than rational.

At the other extreme, in area 9, organizations must find ways to survive with few resources and a very stable information domain. It is under these conditions that the dominant use of the action-generator mode could best contribute to organizational survival. This is where machine bureaucracies are to be expected. These organizations rely heavily on detailed behavior programs and action generators in order to minimize costs and thereby cope with the scarcity of resources. This can work as long as nothing of significance changes in the environment. These organizations have no lack of resources at hand for responding to environmental change. Their learning capacity will have atrophied from lack of use and lack of resources. They are the organizations that are cited as places in which strategic decision making is virtually nonexistent.

Another relevant example is area 3, which combines a high scarcity of resources with high information complexity. These conditions lead to high mortality rates for organizations. Lots of variation exists, but no slack resources. Firms are small and simple, with no capacity for research and development work. They are pressed to rely primarily on incremental, trial-and-error ways of coping, without much time for reflection. If they hit on a successful strategic method, they can build up slack resources and move into a more benign domain.

By contrast, in area 7, with the lack of environmental challenge implicit in both abundant resources and a quiescent information domain, organizations can be expected to turn inward and make decisions on internal criteria. Power

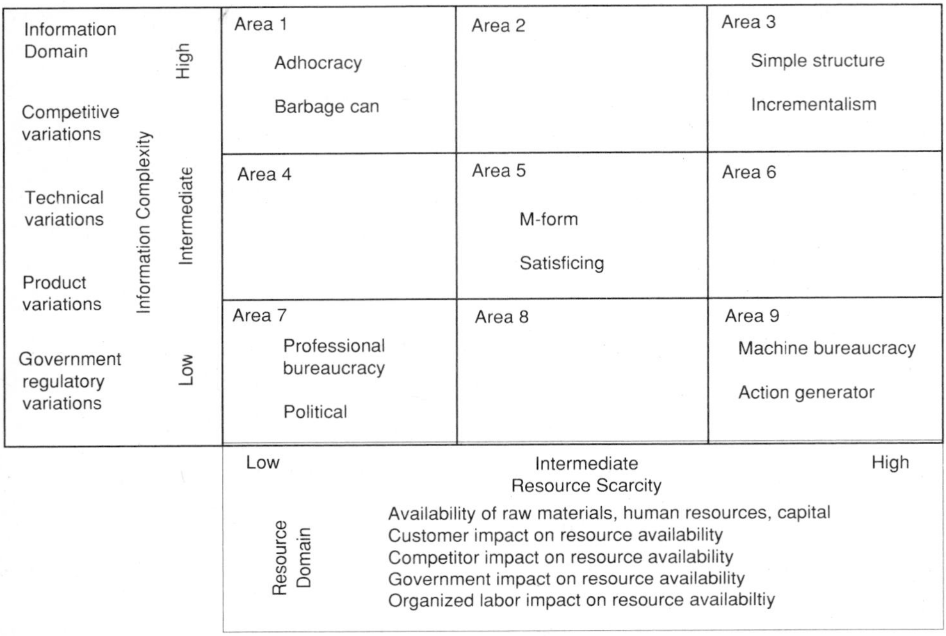

264 LEADERSHIP, STRATEGIC PLANNING, AND INNOVATION

Information Domain		Area 1 Adhocracy Barbage can	Area 2	Area 3 Simple structure Incrementalism
Competitive variations Technical variations Product variations	High Intermediate Low (Information Complexity)	Area 4	Area 5 M-form Satisficing	Area 6
Government regulatory variations		Area 7 Professional bureaucracy Political	Area 8	Area 9 Machine bureaucracy Action generator

```
            Low              Intermediate            High
                           Resource Scarcity
Resource    Availability of raw materials, human resources, capital
Domain      Customer impact on resource availability
            Competitor impact on resource availability
            Government impact on resource availability
            Organized labor impact on resource availabiltiy
```

Figure 7.7. Neo-contingency analytical framework. (From Lawrence and Dyer, 1985. Reprinted with permission from Jossey-Bass.)

distribution among members and groups will be an overriding issue, and power will tend to flow to those with ascribed status factors and dominating personalities rather than to those who can meet environmental challenges. So strategic decision making will be predominantly political in nature. It follows that organizations that stay in this niche, without real competition, will tend to grow older, larger, and more bureaucractic.

Area 5 is of special interest in our neo-contingency model. Firms in this category are facing vigorous competition for both resources and ideas, but not overwhelming pressures in either regard. This classification fosters organizations that are especially attractive from a societal standpoint, since they are more apt to be both innovative and efficient. It is here that we expect to find the M-form (multidivisional) organization that can combine the strengths of decentralized, divisionalized decision making with some joint decision processes that can create synergies across divisions. It is here that we expect firms to use more elements of normative decision models, but, in descriptive terms, our theory would suggest that "satisficing" would be the dominant mode.

This application of the neo-contingency model builds on the differences among strategic decision-making modes, suggesting how these differences can be explained by incorporating them in a more complete analysis of organizations and their environments. It provides an example of how organization theory can contribute to a broader understanding of the strategic decision-making process.

Naisbitt and Aburdene (1985) observe that the old bureaucratic layers are giving way to the more natural arrangements of the new information society. It is as if all of the boxes in the organizational chart were thrown into the air and programmed to fall into a new set of patterns that best facilitates communications networks, hubs, lattices, circles, and wheels. New structures, no matter how brilliantly designed, cannot be applied top down. Companies must experiment with them, and each new model must be tailored to the company and its specific needs. The most innovative companies will adopt a variety of structural configurations. It is the role of the leader in designing the structure to translate his strategic vision into reality and to cope with the demands of technological change.

Some corporations are not adaptive to change, flexibility, and growth. As we enter the information age, they have not changed their structures to meet its challenge and complexity. They have retained their highly centralized and bureaucratic structures rather than experiment with matrix structures, adhocracy, and other decentralized forms, such as framework or modular functions. The time has come to restructure the corporation. A more innovative structure with flexible integrating mechanisms is needed. Structure can compromise performance, especially when it is designed by an administrative elite that converts it into a network of power, influence, and status. In such a case, organizational politics is more critical than organizational excellence. The effective leader must design a new structure or restructure the organization for the information age or it will become a corporate dinosaur. The design of structure is crucial, since it impacts on basic organizational processes such as communications, decision making, and control as well as on the behavioral domain and the ability of the system to cope with change from the environmental domain. If the corporation is visualized as an information-processing system, then new concepts in information theory and cybernetics will have to be used in the design of the organizational structure. Such classical concepts as span of control, unity of command, scalar chain, hierarchy of authority, centralization, bureaucracy, and standardization have marginal utility in organizational design. As information-processing systems, organizations must be designed using concepts such as feedback; information subsystems; decision units; differentiation of information units; integration of information units; change, complexity, and uncertainty of information in the environmental domain; information distributed data bases; information networks; information connectivity; and so on. Information processing becomes the rationale for organizational design rather than power and authority.

7.13 LEADERSHIP, HUMANISTIC VALUES, AND EFFECTIVENESS

Many organizations today, particularly those at the leading edge of technology, are faced with ferment and flux. In increasing instances, the bureaucratic model—with its emphasis on relatively rigid structure, well-defined functional specialization, direction and control exercised through a formal hierarchy of authority, fixed systems of rights, duties, and procedures, and relative impersonality of human relationships—is responding inadequately to the demands placed on it from the outside and from within the organization. Bennis (1966) further states that there is increasing need for experimentation, learning from experience, flexibility and adaptability, and growth. There is a need for greater inventiveness and creativity, and a need for collaboration among individuals and groups. Greater job mobility and the effective use of temporary systems seem essential. An environment must be created in which people will be more fully utilized and challenged and in which they can grow as human beings.

In *Changing Organizations,* Bennis (1966) has pointed out

> "the bureaucratic form of organization is becoming less and less effective, that it is hopelessly out of joint with contemporary realities, and that new shapes, patterns, and models are emerging that promise drastic changes in the conduct of the corporation and in managerial practices in general. At least one of the new models, the one with which our recent experience is most closely connected, is organic and systems-oriented. We feel that, for the present at least, this model is one that can suggest highly useful responses to the newer demands facing organizations."

Organizations are questioning and moving away from the bureaucratic model, in part because man is asserting his individuality and his centrality, in part because of growing dissatisfaction with the personally constraining impact of bureaucracies. In this flux, organizations and man must find a way with each other. In our view, this way will be found through changing values—values that can hopefully serve the needs for effectiveness and survival of organizations and the needs for individuality and growth of emergent man. Those concerned with organization theory and with organizational development have, in our judgment, a most important role to play in this quest.

Growing evidence strongly suggests that humanistic values not only resonate with an increasing number of people in today's world, but also are highly consistent with the effective functioning of organizations built on trust.

7.14 LEADERSHIP, VALUES, AND ETHICS

A leader's style must reflect the values of the organizational culture in the decision-making process. The more ethical the values in the decision-making process, the more dynamic the organizational culture in promoting trust, integrity, and harmony.

Power research tells us more about a manner in which a leader influences the organization than about the power source and values behind his influence. Power and moral leadership can be a magnificant combination for excellence and harmony. Therefore, an effective leader must have some moral and ethical values in managing the enterprise. Some guidelines for the use of power that are ethical and non-Machiavellian are (1) building and exercising expert power; (2) promoting an image of expertise; (3) maintaining credibility; (4) acting confident and decisive, especially in a crisis situation; (5) keeping informed through professional development-information as a strategic power resource; (6) demonstrating a bias for action; (7) recognizing the priorities and power of the organization's subsystems; (8) avoiding winning through intimidation and preventing erosion of self-esteem of members by institutionalizing a dynamic of caring; (8) providing a rationale for decisions and insisting on compliance; (9) monitoring to verify actions; (10) being credible as a source of reward; (11) using moral incentives as a reward; and (12) being ethical in all decisions, for actions reveal motives, motives design habits, habits decide character, character is power, not only knowledge, and character determines leadership effectiveness. It is from integrity in the leader's behavior that trust emerges. The dynamics of caring can serve as a motivator if a leader is not too obsessed with power in influencing behavior when managing the decision-making process. The leader's style expresses the harmony in his goals, desires, beliefs, self-concept, and atitudes. His values, reflected in all of these elements that are cognitive determinants of his behavior, must be integrated, consistent, and in unity. The style reflects the values of the leader in the goal-setting process, in the decision-making process, and in the motivational strategy he adopts. What we need, according to Sai Baba (1985), is not more information, but transformation to develop character.

Dag Hammarskjöld (1987) wrote in his notebook, subsequently published as *Markings,* on the issue of leadership:

> Remember that your position does not give you the right to command, it only lays upon you the obligations of so living your life that others may receive your orders without being humiliated.

This is a noble vision of the use of power in leadership by a person himself in a leadership position, as former Secretary General of the United Nations.

7.15 VALUES, SELF-ESTEEM, AND LEADERSHIP EXCELLENCE

There is a significant relationship between an individual's values, self-worth, and effectiveness. Self-esteem is a key to effective leadership. To be able to influence others, the leader must be sure of his abilities. If secure within himself, he will have high but realistic expectations of what can be achieved. What leaders believe about themselves guides what they believe about others. At the heart of a secure identity lies a certainty about one's primary values. If a leader is clear on what those values are and is convinced that they are right for him, that is, congruent with his self-image, he is more likely to be able to influence others toward those goals, or values (Manning, 1981).

Values provide the leader with standards or criteria for taking action, for justifying one's own and others' actions, and for comparing oneself with others. An understanding of one's values contributes significantly to an understanding of self. Over time, a person develops a set of rules that becomes his value system for making choices. An important aspect of one's value system is the ordering of one's values according to their relative importance. The relative ordering of values is what distinguishes one person's value system from another's. Values also influence the perception of situations and problems, the process of choice, interpersonal relationships, and the elements (symbols, artifacts, rituals) in the culture of an organization. They also set limits to ethical behavior and may constitute the core of ethical imperatives. Finally, values about how people should relate to each other, exercise power, define what is good, and so forth serve to reduce uncertainty and anxiety (Schein, 1985).

7.16 VALUES AND DECISION MAKING

In every type of organization—business, educational, religious, government, medical, and voluntary—leaders engage in decision making. Decision making is the essence of management. It is therefore in the decision-making process that a leader's values, as well as his moral vision and imagination, are tested.

Values provide the leader with a set of guidelines to steer him through the entire decision-making process. Value judgments taken from his personal values and overlaid by the values of the organization occur in the formulation of objectives, the search for relevant alternatives, the ranking and evaluation of alternatives, the moment of choice, and the point of implementation.

Values may be thought of as the guidance system a manager uses when confronted about choices among alternatives. A value can also be viewed as an explicit or implicit conception of what an individual or group, selecting from among available alternatives, regards as desirable ends and means to these ends. The role of values and ethics in the decision-making process is rather explicit (Harrison, 1975):

1. In the setting of organizational objectives, it is necessary to make value judgments regarding the selection of opportunities and necessary improvements within time and resource constraints.
2. In the development of a range of relevant alternatives, it is essential to make value judgments about the various possibilities that have emerged from the search activity.

3. At the time of choice itself, the values of the decision maker as well as ethical considerations of the moment are significant factors in the process.
4. The timing and means for implementing the choice necessarily require value judgments as well as an awareness of ethical interests.
5. Even in the follow-up and control stage of the decision-making process, value judgments are unavoidable in taking corrective action to ensure that the implemented choice has a result compatible with the original objective.

Values and ethics pervade the entire process of choice. Also, they are an integral part of the leader's belief system as they are reflected in his behavior in arriving at a choice and putting it into effect. Decisions are complex sets of value premises which involve choice among these value premises.

7.16.1 Values and Character

> Character is the concretization of the leader's moral and ethical values. Character, therefore, is a strategic source of power.

"One must struggle to create happiness by developing one's character—the badge that designates one as a person" (Herzberg, 1982). The need for status, power, and security must be integrated with the need for personal growth and character. Values are the foundations of one's character. Moral and ethical values produce integrity, truth, and love that create splendor in one's character. Character is the concretization of morality (Sai Baba, 1985). Character is a source of power. It is more powerful than information, expertise, and knowledge because it relates fundamentally to one's values, belief systems, and ideal self-concept. It is the spiritual core of one's being and the synthesis of all behavior and action. In the information age what we need is not more information or disinformation, but transformation of our values, goals, and character. Information can then be used ethically to make decisions regarding life choices. Disinformation will be then perceived as a strategy of deceit that violates the basic principle of truth.

Herzberg (1982) suggests six levels of growth which can be used to assess a leader's character development:

1. *Knowing More, That Is, Learning.* Is learning conducive to growth, change, feedback to monitor's changes in one's values, attitudes, and desires? Is learning only related to routine job procedures and work methods?
2. *Understanding More.* The assembling of knowledge must be accompanied by integration of this knowledge. Understanding provides

feedback, clarification of values and priorities, and the priorities assigned to one's goals.
3. *Creativity.* This is the leap of insight that makes it possible to fall in love with the product of one's learning. It is the use of a variety of cognitive styles to process information.
4. *Effectiveness in Ambiguity.* The ability to make good decisions in gray areas. Tolerance for ambiguity must be high for creative decision making.
5. *Individuation.* This brings happiness in the form of unique abilities. It is the exceptional values that set one apart.
6. *Real Growth.* The ability to look into two identity mirrors to differentiate personality adjustment and status from character accomplishments and self-respect. Do you behav ethically toward others or do you rely on illusory growth for satisfaction?

Character growth is not necessarily the growth of one's ego, which can compromise one's values and ideals. The power of the ego in its play for recognition, status, power, and aggrandizement can make one subordinate one's values and ideals in the pursuit of these goals. The incentives associated with these goals and the desire for power can make one a prisoner of one's ego. Character is the mechanism for liberation from one's ego, its demands and expectations. Character provides the moral values to monitor the power play of the ego. Character as the concretization of morality is the most valid and reliable measure of one's spiritual growth. With integrity as the central value in the leader's character, his goals, self-concept, and attitudes can be evaluated against this central value. Similarly, the absence of such blemishes on one's character as envy, lust, conceit, greed, and malice also indicates the growth of truth and ingegrity as central values in one's character. The blemishes in one's character are potential stressors on the leader's performance. They provide some measure of discrepancy between one's actual versus ideal self-concept. Furthermore, they are associated with negative emotions and attitudes that disrupt the order and rhythm of the mind. The central values in one's character must be consistent with one's perceived self-image, or value conflict will emerge. Conflicts between instrumental and ultimate values may also produce stress in the leader's decision-making strategies. It is in the character of the leader that the dialectics of trust, love, truth, duty, compassion, and integrity is found—the imperatives for ethical decision making.

7.16.2 Values and Vedantic Ethics

> The values of the Vedantic ethical system focus on character as the concretization of morality. Character is power, a compound of five values—truth, love, right action, peace, and nonviolence—which must be reflected in one's decision-making strategies.

Vedanta is a Hindu philosophical system of thought. Ethics and morality are the basis of spiritual life according to Vedantic philosophy. Without ethics, philosophy is purely speculation and abstraction. Morality is the basis of all things and truth is the substance of morality. Spirituality devoid of moral values and principles is futile. Morality is religion in practice; religion is morality in principle. Religion gives us data upon which an ethical science can be built (Sivananda, 1974). Morality cannot be changeable, transient, or optional. Morality must be anchored in the love of God. It is the motive that counts in the performance of an action. Right and wrong are to be determined not by the objective consequences, but by the nature of the subjective intent of the agent. If the inner motive is pure, all the subsequent consequences are pure and good. Virtues survive only when kept in perpetual practice. All ethics have as their aim the realization of the self, the evaluation of consciousness, and the refinement of the senses. The basic values are truth (Sathya), right action (Dharma), love (Prema), Peace (Santhi), and Nonviolence (Ahimsa). These five values include in a balanced way profound moral insights. They are derived from the universal order which upholds societal harmony (Shah, 1985). These five basic values and the ideals that flow from them lead to the realization of our true possibilities as human beings. Morality is designed to shape one's character. Character influences the unfolding of consciousness. It is character, the concretization of morality, which determines the heights to which one depends on truth, the core value in Vedantic ethics. However, our society is shielded from truth, and from realizing the importance of truth, by frantic competitive activity around us, the desperate seeking of greater and greater profit, the definition of reality by advertising agencies charged not with dispensing the truth, but with dispensing profit creating images of reality. The result is moral confusion and moral ambivalence (Bauslaugh, 1986).

> Consider, then, the problem of honesty as a major value in the exercise of authority. Without question it is easier to achieve and to manipulate power if one is willing to be discreetly dishonest. Machiavelli argued that dishonesty is not only useful but necessary, that no good can be achieved without power, that power cannot be achieved without deceit, and that deceit is therefore necessary for achievement of good. This is a powerful argument, and one that is particularly attractive to those in power, for it excuses their dishonesty. Various versions of this argument not only persist today, but they form the basic moral framework in which businessmen, administrators and politicians conduct the affairs of our society. That framework is different from strict Machiavellianism, however, in that it allows for a peculiar dichotomy between moral theory and practice. In theory everyone believes in honesty—in practice honesty is irrelevant; we have it both ways with the most critical of human values.
>
> This approach to morality has a disastrous impact upon the collective moral consciousness of our society. Authority is exercised by so many people, at so many levels, in such intense competition with others, that dishonesty is used not to bring peace and prosperity but to gain personal, political or corporate advan-

tage. It is used in the competition for power: for personal gain and access to the perquisites of power and information—hence the strategy of disinformation. Deceit is a pervasive integral component of power manipulation.

—Bauslaugh (1986)

The transformational leader for the information age is one who has a strong sense of ethics, moral vision, and imagination. This vision is strategic and it is based on his character. Without character, a leader has only knowledge, skills, and techniques to guide him. The ethical crisis in modern society in all fields—law, medicine, business, religion, and government—is indicative of the absence of character in our leaders.

REFERENCES

Ansoff, I., *Corporate Strategy*, Penguin, Harmondsworth, England, 1965.

Bakke, E., "Concept of Social Organization," in Maison Haire (Ed.), *Modern Organization Theory*, Wiley, New York, 1959.

Bass, B., *Organizational Decision Making*, Irwin, Homewood, Ill., 1983.

Bauslaugh, G., "Authority and Moral Leadership." *Humanist in Canada*, Ottawa, 1986.

Bennis, W., *Changing Organizations*, McGraw-Hill, New York, 1966.

Chandler, A., *Strategy and Structure*, MIT Press, Cambridge, Mass., 1962.

Child, J., "Organization Structure, Environment and Performance: The Role of Strategic Choice," *Sociology*, **6,** (1982).

Daft, R., *Organizational Theory and Design*, West Publishing, St. Paul, Minn., 1986.

Davis, D., "Integrating Technological Manufacturing, Marketing, and Human Resource Strategies," in Donald D. Davis and Associates (Eds.), *Managing Technological Innovation*, Jossey-Bass, San Francisco, 1986.

Drucker, P. *Innovation and Entrepreneurship*. Harper & Row, New York, 1985.

Etzioni, A., *The Active Society: A Theory of Societal and Political Processes*, Collier MacMillan, London, 1968.

Gerstein, M. S., *The Technology Connection*, Addison-Wesley, Reading, Mass., 1987.

Hage, J., and M. Aiken, *Social Change in Complex Organizations*, Random House, New York, 1970.

Hammarskjöld, Dag, *Markings*, Ballantine Books, New York, 1987.

Harrison, F., *The Managerial Decision Making Process*, Houghton Mifflin, Boston, 1975.

Herzberg, F., "The Lonely Struggle to Develop Character," *Industry Week*, New York, Jan, 1982.

Katz, D., and R. Khan, *The Social Psychology of Organizations*, Wiley, New York, 1966.

Katz, D., and R. Khan, *The Social Psychology of Organizations*, Wiley-Interscience, New York, 1973.

Kelly, J., *Organizational Behavior: Its Data, First Principles and Applications,* Irwin, Homewood, Ill., 1980.

Lawrence, P., "On Defence of Planning as a Rational Approach to Change," in M. Pennings and Associates (Eds.), *Organizational Strategy Change,* Jossey-Bass, San Francisco, 1985.

Lorsch, J., "Managing Culture: The Invisible Barrier to Strategic Change," *California Management Review,* (1986).

Manning, F., *Managerial Dilemmas and Executive Growth,* Reston Publishing, Reston, Va., 1981.

March, J., and H. Simon, *Organizations,* Wiley, New York, 1958.

Mintzberg, H., "Patterns of Strategy Formation," *Management Science,* **10,** (1978).

Naisbitt, J., and P. Aburdene, *Reinventing the Corporation,* Warner Books, New York, 1985.

Newman, W. H., *Administrative Action,* 2nd ed., Prentice Hall, Englewood Cliffs, N.J., 1963.

Ohmae, K., *The Mind of the Strategist: "Business Planning for Competitive Advantage,"* McGraw-Hill, New York, 1984.

Peters, T., and R. Waterman, *In Search of Excellence,* New York, Harper & Row, 1982.

Rankin, J., *Strategic Planning at Northern Telecom,* Speech presented in Vancouver, Canada, Toronto, 1988.

Robey, D., *Designing Organizations: A Macro-Perspective,* 2nd ed., Irwin, Homewood, Ill., 1986.

Sai Baba, Sri Sathya, *Education and Tranformation,* Sanathana Sarathi, Prasanthi, Nilyam, India, Anantapur District, 1985.

Schein, E., *Organizational Psychology,* 2nd ed., Prentice Hall, Englewood Cliffs, N.J., 1970.

Schein, E., *Organizational Culture and Leadership,* Jossey-Bass, San Francisco, 1985.

Shah, S., *Education in Human Values, Medicine for a Modern Epidemic,* Sai Chandana, Prasanthi Nilayam, Puttarparthi, India 1985.

Simon, H., *Administrative Behavior,* 2nd ed., Free Press, New York, 1957.

Sivananda, S., *Bliss Divine,* Divine Light Publications, Madras, India, 1974.

Stonich, P., *Implementing Strategy: Making Strategy Happen,* Bellinger, Cambridge, Mass., 1982.

Tannenbaum, A., *Control and Organization,* McGraw-Hill, New York, 1968.

Tannenbaum, A., and S. Davis, "Values, Man and Organization," in W. French, C. Bell, Jr., and R. Zawacki (Eds.), *Organizational Development: Theory, Practice, and Research,* Business Publications, Dallas, 1979.

Tichy, N., and M. Devanna, *The Transformational Leader,* Wiley, New York, 1986.

Toffler, A., *The Adaptive Corporation,* Bantam, New York, 1985.

Tushman, M., and D. Nadler, "Organizing for Innovation," *California Management Review,* **28**(3), (1986).

8
DESIGNS FOR CHANGE

8.1 ADVANCE ORGANIZER

Managers of large enterprises have urgent sociohumanistic responsibilities to create self-actualizing organizations that will emphasize human values over technological imperatives in their design.

The design of organizations is open to options made possible by technology. Firms are not locked into one conceptual scheme, but are free to create structures that fit with their mode of operation and corporate philosophy. Computer and telecommunications aids can be tailored to make a given organizational scheme work (Lund and Hansen, 1986).

We consider a variety of conceptual schemes for organizational design. The need for structural change is indicated by reference to a number of structural problems such as adaptability, role ambiguity, and integration (Duncan, 1976); and environment, diversification, growth, and technology (Child, 1981, 1984). A new organizational paradigm (Trist, 1978) is therefore mandated by this evidence of structural problems and the need for structural change. Some new templates for today's organizations are advanced (Drucker, 1974). Attributes of the innovative organization, as developed by Peters and Waterman (1982) and Thompson (1973), are discussed. The characteristics of organic-mechanistic systems noted by Davis (1986) are appropriate for technological change. Another conceptual scheme for organizational design encompasses temporary problem-solving systems (Bennis, 1973). The framework structure and modular units of the adaptive corporation (Toffler, 1985) and collateral organizations (Zand, 1979) also emerge as another alternative design for change.

The matrix design with its potential for rapid change and high information-processing capacity is considered at length. Adhocracy as an innovative organizational design also is a possibility to be considered in the management of technological change. We conclude the chapter with a consistent call for more humanistic organizational designs (Argyris, 1973) and a framework for such a design (Davis, 1985).

Figure 8.1 outlines the subjects of discussion.

8.2 THE NEED FOR A NEW CORPORATE DESIGN

> The new corporate design for the information society will be based on the joint optimization of information imperatives of the new technologies and the humanistic imperatives (intuition, imagination, creativity, conditions for self-actualization) of the new corporate person.

There is a growing recognition that yesterday's hierarchical structures do not work in the new information society. Yet the question is, If not bureaucracy, what? Most of us do not know how to organize ourselves any other way. The old bureaucratic layers are giving way to more natural arrangements of the new information society. Naisbitt and Aburdene (1985) further observe that it is as if all the boxes in the organization chart were thrown into the air and programmed to fall into a new set of patterns that best facilitate communication—networks, hubs, lattices, circles, and wheels.

Davis (1986) reinforces these observations. He states that the adoption of advanced manufacturing technologies is facilitated by certain configurations of management attitudes, values, and organizational structure, practices, context, and environment. But precise delineation of the most appropriate structural configuration is difficult because of gaps in our knowledge about organization and technology. New forms of computerized automation only partly fit existing knowledge about organizational behavior and theory. Design of organizations is open to options made possible by technology. Firms are not locked into one conceptual scheme, but are free to create structures that fit with their mode of operation and corporate philosophy, according to Lund and Hansen (1986). Organizations that are not receptive to new technologies and that do not modify themselves to absorb these successfully fail to adapt to changing environments. Such firms will emerge as corporate dinosuars, states Toffler (1985). The problem for organizations is twofold. First, they must become more receptive to the adoption of advanced technologies. Second, managers must ensure that the performance of these technologies is not hindered by the surrounding social system of the adopting organization.

Another design is suggested by Toffler (1985) for the information age.

276 DESIGNS FOR CHANGE

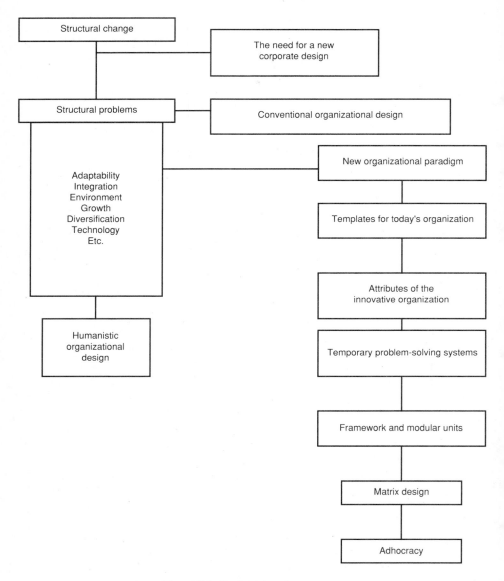

Figure 8.1 Designs for change.

Instead of rigid conventional departments, the firm is divided into a highly flexible structure composed of "framework" and "modules." Instead of being treated as an isolated unit, it is pictured as occupying a position at the center (and as part of) a shifting constellation of related companies, organizations, and agencies. This is a framework of an adaptive organization.

The framework is the thin coordinative wiring that strings together a set of temporary modular units. The constellation consists of the company and the interdependent or semiautonomous outside organizations on which it relies. Constellation implies a movement of certain functions both downward and outward. This cannot take place, however, unless there is a corresponding strengthening of the system's integrative machinery. The adhocratic units or modules of the corporation must be coordinated. Effective restructuring requires not an indiscriminate "loosening of the reins of central control," but, on the contrary, a selective, controlled movement of power upward as well. Certain functions can only be carried out centrally. In organizational terms, these may be referred to as "framework functions." They are concerned primarily with the definition, coordination, and maintenance of standards, and the provision of specialized services or resources to companies and organizations that are part of the constellation as a whole. Toffler (1985)

Child (1984) observes that when environments become more volatile, with shortened planning-time horizons, a premium is placed on adaptability. This means that it is necessary for managers to secure more up-to-date intelligence about events and to evaluate and coordinate the information more rapidly to decide how to respond to changing conditions. One path of reorganization which appears to follow from these environment changes leads to a more complex and specialized structure. This structure is described by Child.

A complex and differentiated structure is more difficult to coordinate and it becomes necessary for the organization to generate an increased capacity for integrating information across different specialized functions and for transmitting the end result effectively to the point of management decisions. It is not feasible to rely on programs, plans, and referral of exceptions for top-level decisions as the means of coping with eventualities that arise. The organization must be designed with flexibility to cope with greater uncertainty. Organization structures have to be reexamined to see how they facilitate effective communications laterally between units and functions and whether full advantage is being taken of aids such as planning staff and computerized information systems which improve the way the vertical flow of information operates. Improved structural mechanisms for both lateral and vertical integration are necessary. Some regrouping of activities, in order to reduce the complexity of communications and to strengthen links between those functions which need to work together in a particular area, may be a possible design strategy (Child, 1981).

Naisbitt and Aburdene (1985) discuss a variety of organizational forms as companies re-invent structures to realize the new corporate vision. The old bureaucratic layers are giving way to the more natural arrangements of the new information society. It is as if all of the boxes in the organization chart were thrown into the air to fall into a new set of patterns that best facilitate communication—networks, hubs, lattices, circles, and wheels. The organizational model is based on a variety of communications networks and distributed data-processing (DDP) configurations.

Burton (1988) notes that the alternative view of organization is quite different. The domain of concern is much broader. A new information system is not simply a technical change which enhances organizational efficiency. It changes information—who knows what when. It is not a neutral technological innovation. It modifies the power and possibilities for opportunism which a number of individuals hold. It creates a new basis for coalitions. Agendas will be different. Some individuals have goals and incentives enhanced with new information; others do not see it in their own self-interest. This political perspective of organizational design recognizes the power rationale in organizational design and the crucial role of power and information.

Computer technology can change what it means to manage and to act in an organization. In fact, such a change is happening and is going to happen regardless of what the designers think they are doing. When we accept the fact that computer technology will radically change management and the nature of office work, we can move toward designing that change as an improvement in organizational life (Flores et al., 1988).

Most IS (information systems) organizations consist of an operations department and an applications development department. In addition to these two components, there are a variety of add-ons, possibly even the new information center. Central processing, distributed processing, micros, office systems, communication, etc., are likely to be crammed within the same basic structure that was designed 20 years ago to accommodate the management of a central processing oriented technology. Adjustments are needed to accommodate the management of a growing inventory of technology (Owen, 1987).

Child (1984) also observes that change has become so pervasive that companies have established specialized organizational planning departments with the task of constantly monitoring the requirements for structural changes and of designing appropriate schemes to accomplish this. Weick and Daft (1984) suggest that organizations learn and adapt very slowly, reinforcing observations on the need to institutionalize innovation. Organizations pay obsessive attention to habitual cues, long after their practical value has lost all meaning. Important strategic assumptions are buried deep in the minutiae of management systems and other habitual routines. Weick and Daft (1984) suppose that inflexibility stems from the mechanical pictures of organizations in the classical mode. In excellent companies, Peters and Waterman (1982) point to different attributes of communications systems that foster innovation that are different from those of the classical mode. Zand (1979) recognized this need to supplement the classical mode two decades ago and suggested the collateral organization as a mechanism for flexibility to manage tasks with characteristics of complex, nonroutinized work generally characteristic of technological innovation. Duncan (1979) also argues that one of the major functions of organizational design is to structure the organization so as to build in innovativeness; he suggests the ambidextrous organization, a dual structure for two stages of innovations. Bennis (1973) earlier perceived the need for temporary problem-solving systems linked together by coordinating and task-evaluating specialists in an organic flux—this is the organizational form, he argues, that will gradually replace bureaucracy as the dominant organizational model. This need for new corporate design has also been emphasized consistently by Mintzberg (1979), who notes that a hybrid structure—a divisionalized adhocracy—could emerge for a complex and dynamic environment.

It is evident from the foregoing observations that an organizational perspective is necessary to understand the factors related to the successful use of advance manufacturing technologies. Unlike numerically controlled equipment and other simpler forms of automation, computer-controlled technologies, such as robotics, computer-aided design/computer-aided manufacturing (CAD/CAM), computer-aided engineering (CAE), flexible manufacturing systems (FMS), and computer-integrated manufacturing (CIM), are so complex that they stretch the ability of managerial and organizational systems to absorb them. These new technologies, Davis (1986) further observes, are often embedded in organizations designed and managed in a fashion more suitable for older, simpler production methods. Necessary, then, is a rethinking of the purpose of manufacturing organizations and how best to organize them.

While emphasis has been placed on the need for the new corporate paradigm to be designed by a different set of principles for the effective implementation of new technologies, we must recognize the need for the adjustment and adaptation of individuals in such a system. Agyris, Bennis, Trist, Davis Likert, and other industrial humanists have consistently argued for a humanistic perspective in organizational design—the need for individuals to develop their value potentialities within the organization. Argyris (1959), in his classic study of personality in the organization, states that human personality is always attempting to actualize its unique organization of parts resulting from a continuous, emotionally laden, ego-involving process of growth. Unique differences of individuals cannot be ignored by technology or task specialization. Self-actualization cannot be compromised by task specialization. If the principles of formal organization are used as ideally defined, then the employees will tend to work in an environment leading to psychological failure.

The imperatives of humanism conflict with the imperatives of technology and have the potential to produce a high incidence of technostress (Benson, 1985). The humanistic perspectives argue for the actualization of the self and the development of the value potentialities of the individual. This development is facilitated by freedom and autonomy in his work setting. The worker brings to his job situation certain hopes, expectations, and values—creative, experiential, social, and ethical values. Emotion and perception are important elements of the individual complex. The "inner world" of the worker is viewed as just as important as external reality in determining productivity. The worker obtains his sense of identity through interpersonal dynamics and is more responsive to the norms of the work group. The worker is also responsive within a corporate culture that meets his needs for esteem, self-actualization, will to meaning, and character growth as a source of ethical excellence (Sankar, 1991). The task of deriving a set of design principles to blend the humanistic and technological imperatives in the new corporate paradigm for the information society is quite a complex task for any organizational architect.

8.2.1 The Individual in the New Corporate Design

A new corporate design based on the systems and cybernetic principles can be as dysfunctional as the bureaucratic machine if a culture audit is not undertaken concurrently to blend the technical imperatives of the new technologies with the humanistic imperatives of the new corporate person. Our view of design is consciously oriented toward improving the quality and effectiveness of organizational life, not just providing computer support for current practices. All innovative technology leads to new practices which cause social and organizational changes whether anticipated or not (Flores et al., 1988).

The information-processing perspective in the previous sections captures many significant aspects of coordinating the activities of people in organizations, but it leaves out some of the most important factors about why people are there in the first place, how hard they work, and whether they find their activities satisfying or alienating.

According to Fromm (1968), if we are only concerned with input–output figures, a system may give the impression of efficiency. If we take into account what the given methods do to the human beings in the system, we may discover that they are bored, anxious, depressed, tense, and so on. The result would be a twofold one: (1) their imagination would be hobbled by their psychic pathology, they would be uncreative, their thinking would be routinized and bureaucratic, and hence there would be inertia in the system; and (2) they would suffer from tension, stress, and alienation, all of which will reduce their creative potential.

Argyris (1973) not only comments on the conditions that inhibit individual growth and self-actualization, but also identifies some of the major requirements of tomorrow's organization. To meet the challenges of complexity, change, and uncertainty, modern organizations need (1) much more creative planning, (2) the development of valid and useful knowledge, (3) increased concerted and cooperative action with internalized long-range commitment, and (4) increased understanding of criteria for effectiveness that meet the challenges of complexity. These requirements depend on (1) continuous and open access between individuals and groups, (2) free reliable information where (3) interdependence is the foundation for individual and departmental cohesiveness and (4) trust, risk taking, and keeping each other is prevalent, so that (5) conflict is identified and managed in such a way that destructive win–lose stances are minimized and effective problem solving is maximized.

The culture of the new design will be based on joint optimization, a sociotechnical paradigm which is an ethical imperative. Ethical ground rules are the heart of organizational culture. Ethics is the fulcrum in culture for producing change (Pastin, 1986). A new design that neutralizes the basic information pathologies of both the classical and contingency design and that stress the information imperatives noted earlier has a greater chance of developing an entrepreneurial culture than the typical bureaucratic culture. Within such a culture, the ethos of the organization facilitates risk-taking behavior, the exercise of discretion and autonomy in decision making, and the prevalence of organizational "slack" for absorbing the consequences of errors in risk-taking behaviors. The value system of the entrepreneurial culture rewards the exercise of initiative, judgment, creativity, and ethics in decision making. This is the type of culture that organizations must design and nurture to facilitate the emergence of new technologies and their integration into all the domains of the organization (Sankar, 1988). Designing the new corporation must go hand in hand with a culture audit of its norms, dominant values, philosophy, rules of the game, climate, and patterns of assumptions, organizational symbols and communications rituals, artifacts, and symbols. Virtually every successful new model for re-inventing the corporation was based on the mutual interest of corporations and the people within them. One of the most fundamental shifts is movement away from the authoritarian hierarchy to the new lateral structures, lattices, networks, and small teams where people manage themselves (Naisbitt and Aburdene, 1985). The new corporate design will therefore create a new corporate culture that is more humane, value-based, and dynamic.

8.3 SYMPTOMS OF STRUCTURAL PROBLEMS IN CONVENTIONAL ORGANIZATIONAL DESIGN: DUNCAN'S PERSPECTIVE

To be effective, an organization must attain its goals and objectives, adapt to the environment, and be designed in such a way that its managers experience low role conflict and ambiguity. There are thus certain kinds of information the manager responsible for organizational design should be sensitive to in monitoring whether the appropriate structure is being used. A number of potential problem areas in the structural design have been identified by Duncan (1979) and are cited below.

8.3.1. Adaptability

Can the organization *anticipate* the need for change in its technology, structure, procedures, processes, work flow, job design, etc.? Can the organization *facilitate* the adoption of innovations or changes? Can the organization *initiate* changes in areas that are relevant for its effective performance? The adaptability of the organizational structure is one factor by which the effectiveness of the organization can be measured.

Organizational decision makers may not be able to anticipate problems before they occur. There may be a tendency in the organization to wait until problems occur and then react to them because the organization simply does not have enough information to develop contingency plans. Decision makers may make errors in trying to predict trends in their decision environment. Without proper coordination across divisions, the organization may lose control over the relationship between its internal functioning and its environment (Duncan, 1979).

The organization may not be able to get critical information to the right decision center for setting corporate policy in a particular area. Because of decentralized structure and lack of effective coordination through some form of lateral relations, effective decision making is therefore jeopardized. The organization, having identified a problem vis-à-vis its environment, may simply not be able to take corrective action quickly enough.

A poor capacity to innovate and be receptive to new ideas is often another sign of structural failure. Innovation of any kind encompasses a multistage process in which ideas are generated, selected for further development, and adopted for regular use. The type of structure an organization has can affect this process at any of its stages, but the same structure will not be appropriate for every stage. A well-designed structure can assist in the integration and control of the whole innovative process (Duncan, 1979).

8.3.2 Role Ambiguity

Symptoms of poor fit between structure and environment may also show at the level of the individual in terms of some increase in either role conflict or role ambiguity. It is important that the organization monitors the level of role conflict and ambiguity among its managers and the resulting stress they experience.

Individuals may be experiencing increased role conflict. This may occur when the organization is implementing a functional structure in a dynamic environment. The environment may be changing and the individuals may be required to make quick responses to this changing environment. Having to wait for new policy changes to come down the hierarchy may delay the organization in responding appropriately. In addition, decision makers at the top of the organization will suffer from role conflict when the environment is changing rapidly. In the functional organization, when new situations occur they are referred to higher levels of the organization for decision and action. The result is that top-level decision makers become overloaded and the organization's response to the environment slows down. Duncan (1979) observes that in a dynamic environment the functional organization constrains the decision-making adaptation process.

Individuals in the organization may also experience increased role ambiguity, that is, they may be unclear as to what is expected of them in their roles. Role ambiguity is likely to occur when the decentralized organization is implemented without some effective use of lateral relations. Individuals may feel they do not have the information for decision making. Divisional managers may not know the corporate staff's policy on various issues, and corporate staff may not be interacting adequately with the divisions (Duncan, 1979).

8.3.3. Integration

There are genuine differences of opinion between departments whose activities are interdependent. Open disagreement between parts of an organization presents one of the more easily identifiable problems in the areas of communication and integration. Problems can include duplication of effort between units that are not exchanging information adequately, and poor morale among staff who feel that they are not given enough information about changes planned for their areas of work. As the organization differentiates its functions, activities, and processes as a response to a changing environment, the need for integration becomes urgent. The information-processing requirement of the organization will necessitate the development of various integrative devices. If the information processing is deficient, it could be a signal for some reorganization, such as reducing management levels, regrouping activities, or introducing new coordinative mechanisms (Duncan, 1979).

8.3.4 Control

One of the essential requirements for the planning and control process to be effective is that people should have a good idea of what they are expected to achieve. A common problem of employees and managers is lack of a clear definition of what their authority or responsibilities are, or that they have not had the opportunity of establishing their work objectives in discussion with management. The classical approach to management calls for a clear statement of objectives, responsibilities, and authority in job descriptions, manuals of procedure, and the like. However, in a rapidly changing environment, job descriptions and other aspects of formalization are not appropriate. It will be more appropriate to keep the definitions of people's jobs open to adjustment to give workers more discretion or autonomy on the basis of mutually agreed objectives and performance criteria (Duncan, 1979).

8.4 REORGANIZATION: THE NEED FOR STRUCTURAL CHANGE: CHILD'S PERSPECTIVE

8.4.1 Environment

As conditions change in the external and internal environments, it becomes imperative for managers to examine the implications of the change for the design of their organizations' structures and jobs and to decide whether any reorganization is required. Change has become so pervasive that companies have established specialized organization–planning departments whose task is to constantly monitor the requirements for structural changes and design apporopriate schemes to accomplish this.

The following observations about the need for reorganization are based on the views of Child (1981). He notes that when environments become more volatile, with shortened planning-time horizons, a premium is placed on adaptability. This means that it is necessary for managers to secure more up-to-date intelligence about events and to evaluate and coordinate the information more rapidly to decide how to respond to changing conditions. One path of reorganization that appears to follow from these environmental changes leads to a more complex and specialized structure.

Child (1981) believes that a complex and differentiated structure is more difficult to coordinate and it becomes necessary for the organization to generate greater capacity for integrating information across different specialized functions and for transmitting the end result effectively to the point of management decision. It is not feasible to rely on programs, plans, and referral of exceptions for top-level decisions as the means of coping with eventualities that arise. The organization has to be designed with flexibility to cope with greater uncertainty. Organization structures have to be reexamined to see how they facilitate effective communication laterally between units and functions and whether full advantage is being taken of aids such as planning staff and computerized information systems, which improve the way the vertical flow of information operates. Improved structural mechanisms for both lateral and vertical integration are necessary. Some regrouping of activities, in order to reduce the complexity of communications and strengthen links between those functions that need to work together in a particular area, may be a possible design strategy (Child, 1981).

8.4.2 Diversification

Another factor in reorganization is diversification. In early stages of diversification into a new field of activity, only minor adjustments to the organization structure are made by management. A special coordinator may be appointed to look after the new operation in consultation with functional departments.

The question of at what point an organization moves away from a functional structure, or creates a new division out of an existing one, is a particularly difficult one. Indeed, there is always the choice of not moving over to a fully fledged divisional structure, but instead, grafting a divisional-type split of certain departments like marketing into a basically functional structure. This type of mixed structure will have to be complemented with clear policies on priorities (Child, 1981).

8.4.3 Growth

Growth is another source of reorganization. Pressures for reorganization arise when the top executive or the top team can no longer maintain effective day-to-day control, and when the growth of the organization calls for a greater specialization between, and broadening in the range of, management functions. Growth, Child (1981) observes, generally leads to extended hierarchies and problems of communication. If it occurs on the basis of diversification into many unrelated fields, it can generate serious problems of control and coordination. Continued expansion is a prime source of pressures for reorganization: "It is for this reason that size has emerged in so many studies as the major predictor of the type of structure adopted by organizations in practice" (Child, 1981).

8.4.4 Technology

Recent and more advanced computer applications have involved a more fundamental change in which functions can be closely integrated through sharing common computer files as part of specially designed systems that are intended to govern the flow of work across departments (Child, 1981). The impact of integrated information systems on management structures and jobs can be considerable and may lead to the elimination of established boundaries, both horizontally between departments and vertically between middle and top management. Horizontally, the thrust tends to be away from a functional division of responsibilities and toward matrix and product forms, since these accord with the logic of the major work flows which the stages of information processing must parallel (Child, 1981).

In the case of vertical boundaries, the computer permits relevant data to be updated continuously and for this information to be passed to top management for review. These data can be easily combined with other relevant facts to provide a much broader analysis than before and provide the capacity for senior management itself to make the decision. In such situations, the middle manager could disappear from the scene as a decision maker. A similar possibility emerges with advanced automation of production or work flow processes, whereby a whole plant becomes integrated into the scope of a single control system. This allows for all relevant information to pass rapidly up to senior management for decision making. Such advanced technology can lead to quite substantial structural changes, particularly a decrease in the number of middle management levels and a centralization of much operational decision making.

Out of the whole gamut of technological changes, argues Child (1981), computers and automation introduce the most far-reaching pressures for reorganization. Other developments, such as a move from small-batch to mass production, are also relevant, since they occasion changes in appropriate systems of production control and in the employment of specialists. Most technological advances tend to enhance the numbers and importance of certain specialist groups. The integration of specialists into traditional line hierarchies is generally difficult. Consequently, organizations have experimented with team structures that detach specialists from functional departments and allocate them to teams concerned with the management and servicing of a particular work flow (Child, 1981).

Organizations that are not receptive to new technologies and do not modify themselves to absorb these successfully fail to adapt to changing environments. Such firms will become corporate dinosaurs (Toffler, 1985). The problem is twofold for organizations: first, they must become more receptive to the adoption of advanced technologies; second, managers must ensure that the performance of these technologies is not hindered by the surrounding social system of the adopting organization.

Trist (1981) makes the following observations regarding the pace of technological change and its implications:

> People and the organizations they build are having to learn to cope with a quality in the environment that has become increasingly salient in recent years. This quality, which may usefully be referred to as turbulence, arises from the growing interdependence and complexity and the resulting higher level of uncertainty that exist in the world today. Among other factors, this turbulence is being induced by the accelerating rate of technological change. To put it succinctly, we have lost the stable state, and in order to regain it and reduce environmental turbulence, we can less and less depend on our inherited organizational values and forms. We have to seek new ones. Improving the quality of working life (QWL) is an essential part of the search for alternative organizational patterns.

8.5 A NEW PARADIGM FOR ORGANIZATIONAL DESIGN: TRIST'S DESIGN

Table 8.1 sets out the key features of the new organizational paradigm, which can potentially lead to a high QWL for all members of the enterprise. They contrast strongly with those of the old organizational paradigm, set out on the left, and which has been instrumental in constraining most employees to a low QWL. This section is based on the observations of Trist regarding the need for a new organizational paradigm. Our traditional organizations follow the technological imperative, which regards the human as simply an extension of the machine and therefore an expendable spare part. By contrast, the emergent paradigm is founded on the principle of joint optimization, which regards the human being as complementary to the machine and values his unique capabilities for appreciative and evaluative judgment. He is a resource to be developed for his own sake rather than to be degraded and cast aside.

Traditional organizations are also characterized by maximum work-breakdown, which leads to circumscribed job descriptions and single skills—the narrower the better. Workers in such roles are often unable to manage the uncertainty or the variance that characterizes their immediate environment and they therefore require strict external controls. Layer upon layer of supervision comes into existence, supported by a wide variety of specialist staffs and formal procedures. A tall pyramidic organization results, which is autocratically managed throughout, even if the paternalism is benign. By contrast, the new paradigm is based on an optimum task grouping, which encourages multiple, broad skills. Workers in such a role system become capable of a much higher degree of internal control, having flexible group resources to meet a greater degree of environmental variance. This leads to a flat organization characterized by as much lateral as vertical communication. Moreover, a participative management style emerges, with the various levels mutually articulated rather than arranged in a simple hierarchy (Trist, 1981).

TABLE 8.1 Old Versus New Paradigm

Old Paradigm	New Paradigm
The technological imperative	Joint optimization
Humans as extensions of the machine, expendable spare parts	Humans as complementary to machines, resources to be developed
Maximum task breakdown, single narrow skills	Optimum task grouping and multiple, broad skills
External controls (supervisors, specialist staffs, procedures)	Internal controls (self-regulating subsystems)
Tall organization chart, autocratic style	Flat organization chart, participative style
Competition, gamesmanship	Collaboration, collegiality
Organization's purposes only	Members' and society's purposes also
Alienation	Commitment
Low risk taking	Innovation

Source: E. Trist, *The Socio Technical Perspective in Perspectives on Organizational Design,* A. H. Van deVen and W. Joyce (Eds.), Wiley, New York, 1981.

As Trist (1981) has stated, traditional organizations serve only their own ends. They are, and indeed are supposed to be, selfish. However, the new paradigm imposes the additional task on them of aligning their own purposes with those of the wider society as well as with those of their members. By so doing they become both "environmentalized" and "humanized," and thus become more truly purposeful, rather than merely remaining the impersonal and mindless forces that are increasing environmental turbulence.

A change in all these regards from the old paradigm to the new brings into being conditions that allow commitment to grow and alienation to decrease. Equally important is the replacement of a climate of low risk-taking with one of innovation. This implies high trust and openness in relations. All these qualities are mandatory if we are to transform traditional technocratic bureaucracies into a continuous adaptive learning system. And this is the central task. This transformation is imperative for survival in a fast-changing environment. It involves nothing less than the working out of a new organizational philosophy which is described by Trist.

I use the term philosophy advisedly to indicate that far more is involved than methods or techniques. These of course have their place, but a philosophy involves questions of basic values and assumptions. Those of the new paradigm are radically different from those of the old. The old is based on technocratic and bureaucratic principles, the new on socioecological and paraticipative principles. Each sub-system has a wide repertoire of response capability. It can thus better meet uncertainty and contain turbulence. This is one of the most important features of self-regulating systems of autonomous work groups and of open mutually articulated levels.

8.6 NEW TEMPLATES FOR TODAY'S ORGANIZATIONS: DRUCKER'S VIEWS

According to Drucker (1974), organizational structures are becoming increasingly short-lived and unstable. The classical organization structures of the 1920s and 1930s, which still serve as textbook examples, stood for decades without needing more than an occasional touching up. American Telephone and Telegraph, General Motors, Du Pont, Unilever, and Sears Roebuck maintained their organizational concepts, structures, and basic components through several management generations and major changes in the size and scope of the business. Today, however, a company no sooner finishes a major job of reorganizing itself than it starts all over again.

Drucker (1974) claims that changes in the objective task of the organization have generated new design principles that do not fit traditional organizational concepts. He refers to three new design principles, *team organization, simulated decentralization,* and *system structure,* and describes each system as follows:

> In team organization, a group—usually a fairly small one—is set up for a specific task rather than for a specific skill or stage in the work process. In the past 20 years we have learned that whereas team design was traditionally considered applicable only to short-lived, transitory, exceptional task-force assignments, it is equally applicable to some permanent needs, especially to the top-management and innovating tasks.
>
> In an organization that is both too big to remain functionally organized and too integrated to be genuinely decentralized, simulated decentralization is often the organization answer. It sets up one function, one stage in the process, or one segment as if it were a distinct business with genuine profit and loss responsibility; it treats accounting figures, transfer prices, and overhead allocations as if they were realities of the markeplace. For all its difficulties and frictions, simulated decentralization is probably the fastest growing organization design around these days. It is the only one that fits, albeit poorly, the materials, computer, chemical, and pharmaceutical companies, as well as the big banks; it is also the only design principle suited for the large university, hospital, or government agency.
>
> Finally, in systems structure, team organization and simulated decentralization are combined. The prototype for this design principle was NASA's space program, in which a large number of autonomous units—large government bodies, individual research scientists, profit-seeking businesses, and large universities—worked together, organized and informed by the needs of the situation rather than by logic, and held together by a common goal and joint top management. The large transnational company, which is a mix of many cultures, governments, businesses, and markets, is the present embodiment of an organization based on the systems concept.
>
> Organizational design should simultaneously structure and integrate three different kinds of work: (1) the operating task, which is responsible for producing the results of today's business; (2) the innovative tasks, which creates the

company's tomorrow; and (3) the top-management task, which directs, gives vision, and sets the course for the business of both today and tomorrow. No one organizational design is adequate for all three kinds of work; every business will need to use several design principles side by side.

Drucker states that each organizational structure must satisfy the need for
- Clarity, as opposed to simplicity. A modern office building is exceedingly simple in design, but is it very easy to get lost in one; it is not clear.
- Economy of effort to maintain control and minimize friction.
- Direction of vision toward the result rather than the effort.
- Understanding by each individual of his own task as well as that of the organization as a whole.
- Decision making that focuses on the right issues, is action oriented, and is carried out at the lowest possible level of management.
- Stability, as opposed to rigidity, to survive turmoil, and adaptability to learn from it.
- Perpetuation and self-renewal, which require that an organization be able to product tomorrow's leaders from within, helping each person to develop continuously; the structure must also be open to new ideas.

All of these criteria for organizational design can be incorporated into the structure of the innovative organization. The organization that initiates change and the organization that manages technological change will be subject to some of these design criteria.

8.7 SOME ATTRIBUTES OF THE INNOVATIVE ORGANIZATION: PETERS AND WATERMAN

Weick and Taft (1984) suggest that organizations learn and adapt very slowly. They pay obsessive attention to habitual internal cues, long after their practical value has lost all meaning. Important strategic business assumptions (e.g., control versus risk-taking bias) are buried deep in the minutiae of management systems and other habitual routines whose origins have long been obscured by time. Weick supposes that inflexibility stems from the mechanical pictures of organizations.

Peters and Waterman (1982) observe that innovative companies not only are unusually good at producing commercially viable new widgets, but are especially adroit at continually responding to change of any sort in their environments. As the needs of their customers change, the skills of their competitors improve, the mood of the public shifts, the forces of international trade realign, and the government regulations are revised, these companies track, revamp, adjust, transform, and adapt. In short, as a whole *culture,* they innovate.

According to Peters and Waterman (1982),

> That concept of innovation seemed to us to define the task of the truly excellent manager or management team. The companies that seemed to us to have

achieved that kind of innovative performance were the ones we labelled excellent companies. In excellent companies there are five attributes of communication systems that seem to foster innovation:

1. Communication systems are informal. At 3M there are endless meetings, through few are scheduled. Most are characterized by people casually gathering together—from different disciplines—to talk about problems. At Digital the chief executive meets regularly with an engineering committee of about 20 engineers from all levels of the organization. He sets the agenda and periodically disbands and reconstitutes the committee to maintain a fresh flow of ideas. He sees his role as that of a catalyst or devil's advocate. Proponents of successful ideas work primarily through the informal rather than the formal organization. A championing system at the heart of the organization means a de facto informal culture.

2. Communication intensity is extraordinary. Intel executives call this process "decision making by peers"—an open, confrontation-oriented management style in which people go after issues bluntly, straightforwardly. Senior managers at Exxon and Citibank display this intensity also. They make a presentation and then the screaming and shouting begin. The questions are unabashed; the flow is free; everyone is involved. Nobody hesitates to cut off the chairman, the president, a board member.

3. Communication is given physical supports. Physical trappings such as blackboards help spur the intense informal communication that underpins regular innovation. Intel's new buildings in Silicon Valley were designed to have an excess of little conference rooms. Management wants people to each lunch there, do problem solving there. The rooms are filled with blackboards. MIT's Allen has been studying physical configurations for years. His results from research and engineering settings are striking. If people are more than 10 meters apart, the probability of communication at least once a week is only about 8 or 9 percent (versus 25 percent at 5 meters).

4. Forcing devices. There is still another aspect of the communication system that spawns innovation. IBM's "Fellow" program is the classic. There are 45 fellows, heralded as "dreamers, heretics, gadflies, mavericks and geniuses." A Fellow is given virtually a free rein for five years. His role is quite simple: to shake up the system. Other forcing devices are Technology Centers at Datapoint, for they are places where people from disparate disciplines are to get together in name of innovation.

5. The intense, informal communication system acts as a remarkably tight control system even as it spawns rather than constrains innovation. We believe that controls throughout the excellent companies are the truly tight ones. You can't spend much time at one of these companies without lots of people checking up informally to see how things are going.

8.8 ORGANIC-MECHANISTIC SYSTEMS AND INNOVATION

Davis (1986) also reinforces some of above conclusions in terms of the characteristics of the innovative organization. He observes that technical progress has consistently been related to organizational form. Firms having an organic form tend to be technologically innovative and receptive to change. This organic form is characterized by informal definition of jobs; lateral, networklike communication patterns; consultative rather than authoritative communication; diffusion of knowledge seeking throughout the firm; commitment to a technological ethos of managerial progress that is highly valued; and importance and prestige attached to extraorganizational affiliations and activities (Burns and Stalker, 1961).

Mechanistic forms of organization, Davis (1986) continues, are less innovative and more resistant to change than organic forms. Mechanistic forms are characterized by rigid breakdown of roles and jobs into functional specialization, precise definition of duties, responsibilities, and power; hierarchical control; authority and communication; reliance on vertical interactin; insistence on loyalty and obedience to superiors and greater authority; and importance attached to internal than to extraorganizational knowledge and activities (Burns and Stalker, 1961).

There relationships have received empirical support in many different contexts. Decentralized and participative forms of decision making, low reliance on written rules and regulations to govern work behavior, concentrations of professionals of different types, and complexity in general are the organizational features most frequently related to technological innovation and innovation adoption (Davis, 1986).

To prepare the organization for new technology, certain changes are necessary, according to Davis (1986): those in information flow and control, power and authority, supervision, and performance appraisal. Formerly, central management tasks such as deciding among priorities, allocating resources, and scheduling are easily subsumed into software. Expert systems, natural-language processing, and other forms of artificial intelligence may further decentralize knowledge and decision making. The organic form is more receptive than the mechanistic form to these changes.

8.9 TEMPORARY PROBLEM-SOLVING SYSTEMS AND INNOVATION: BENNIS' PERSPECTIVE

According to Bennis (1973), the social structure of organizations of the future will have some unique characteristics. The key word will be temporary; there will be adaptive, rapidly changing temporary systems. These will be task forces organized around problems to be solved. These groups will be arranged on organic rather than mechanical models; they will evolve in response to a problem rather than to programmed role expectations. The "executive" thus becomes a coordinator or "linking pin" between various task forces. People will be differentiated not vertically, according to rank and role, but flexibility and functionally according to skill and professional training. Adaptive problem solving, temporary systems of diverse specialists, linked together by coordinating and task-evaluating specialists in an organic flux—this is the organizational form that will gradually replace bureaucracy (Bennis, 1973).

The need for temporary adaptive problem-solving systems is critical because of the pathologies of the bureaucratic organization noted by Bennis (1973):

1. Bureaucracy does not adequately allow for personal growth and the development of mature personalities.
2. It develops conformity and "group think."
3. It does not take into account the informal organization and emergent and unanticipated problems.
4. Its systems of control and authority are hopelessly outdated.
5. It has no adequate juridical process.
6. Communications (and innovative ideas) are thwarted or distorted due to hierarchical divisions.
7. The full human resources are not being utilized due to mistrust, fear of reprisals, etc.
8. It cannot assimilate the influx of new technology or scientists entering the organization.
9. It modifies personality structure so that people become and reflect the dull, gray, conditioned "organization man."

This type of structure cannot cope with worker values of the future for involvement, participation, and autonomy in their work, as well as job requirements that call for more responsibility and discretion. The tasks of the firm, Bennis (1973) observes, will be more technical, complicated, and unprogrammed. Essentially, they will call for the collaboration of specialists in project or team form organization.

8.10 THE INNOVATIVE ORGANIZATION: THOMPSON'S PERSPECTIVE

According to Thompson (1973), the innovative organization will be characterized by structural looseness generally, with less emphasis on narrow, nonduplicating, and nonoverlapping definitions of duties and responsibilities. Job descriptions will be of the professional type rather than the duties type. Communications will be freer and legitimate in all directions. Assignment and resource decisions will be more decentralized than is customary. The innovative organization will not be as highly stratified as existing ones. This is implied in the freedom of communication, but the decline in the importance of extrinsic professional rewards and the increasing growth of interest in professional esteem would bring this about.

In the innovative organization, Thompson (1973) continues, deparmentalization must be arranged so as to keep parochialism to a minimum. Some overlappping and duplication and some vagueness about jursidictions make a good deal of communication necessary. People have to define and redefine their responsibilities continually, case after case.

Thompson (1973) also refers to the need for different structural configurations for initiation of change and the implementation of change. If formal structures could be sufficiently loosened, it might be possible for organizations and units to restructure themselves continually in light of the problem at hand. Thus, for generating ideas, planning, and problem solving, the organization or unit would "unstructure" itself into a freely communicating body of equals. When it is time for implementation, requiring a higher degree of coordination of action (as opposed to stimulation of novel or correct ideas), the organization could then restructure itself into the more usual hierarchical form, tightening up its lines somewhat.

Should it prove impossible for organizations to become flexible enough to allow restructuring themselves in the light of the problem at hand, it would be preferable to retain a loose structure in the interest of generating new ideas and suffer from some fumbling in the attempt to coordinate action for the purpose of carrying them out.

Integrative departmentalization, combined with freedom of communication, interunit projects, and lessened subunit chauvinism, will create extra departmental professional ties and interests, resulting in an increase in the diversity and richness of inputs and in their diffusion, thereby stimulating creativity.

8.11 THE ADAPTIVE CORPORATION: TOFFLER

Toffler (1985) refers to organizations that do not initiate drastic change or rethink their goals and restructure themselves as organizational dinosaurs. For many of these firms, the years between 1955 and 1970 were of almost uninterrupted, straight-line growth in an equilibrial environment. Since then, this straight-line strategy has become a blueprint for corporate disaster. The reason, he continues, is simple: instead of being routine and predictable, the corporate environment has grown increasingly unstable, accelerative, and revolutionary. Managers in this context must cope with nonlinear forces, that is, situations in which small inputs can trigger vast results and vice versa.

There was a time when a company's tables of organization (formal and informal) stayed put for long periods. By the time a company had successfully made the transition from the original "one-man, one-rule" structure of its founder to a many-layered hierarchy, it had also, most likely, bolted into place a permanent departmental framework. Line and staff were clearly delineated. Suborganizations reporting to the top of the hierarchy provided fixed corporate services (e.g., legal, financial, and personnel). Once this iron framework was installed, the company might shrink or grow, according to its fortunes, but the basic elements of its structure usually held firm. Reorganizations were few and far between (Toffler, 1985).

However, to survive today's onrushing changes, Toffler argues that we must be prepared to reconsider the very models on which our obsolete organizations are based. Three of the most common problems currently facing companies are (1) organizational mismatch; (2) overreliance on top-down hiearchy, and (3) just plain flab (Toffler, 1985). We will review the first two here:

1. *Mismatch.* Existing organizational structures in most companies are designed to produce a few basic kinds of decisions repetitively. Under the traditional bureaucratic system, for every problem in the environment, there is a matching component of the organization, marketing, manufacturing, finance, etc. Today, an increasing number of problems arise that cannot be neatly matched with any one component of the organization. The result is a growing number of mismatches between the organizational structure existing at any given moment and the requirements at that moment. The wrong kinds of problems go to the wrong departments for solution; problems are misconceived, bent out of shape to fit preexisting organizational lines; or the departmental lines themselves are continually gerrymandered in a futile search for the "perfect" permanent organization. At the level of subunits, the acceleration of change means that the corporation faces an even more rapid flow of "one-time" opportunities and problems. A "one-time" or temporary problem, however, requires a "one-time" or temporary organization to resolve it. The result, Toffler (1985) continues, is a necessary proliferation of modular, temporary, or self-destruct units—task forces, problem-solving teams, ad hoc committees, and other groups assembled for a special and temporary purpose. This shift from permanent to adhocratic firms is in fact a vast fundamental adaptation by society to the imperatives of high-speed social change (Toffler, 1985).

2. *Hierarchy.* Sharply vertical hierarchies, with orders flowing smoothly down a chain of command, have long been regarded as extremely efficient, and this form of control is characteristic of industrial-era organizations. This control system, according to Toffler (1985), depends on two factors: heavy and accurate feedback from the field and relative homogeneity in the types of decisions required. Today, the strict vertical hierarchy is losing its efficiency because the two fundamental conditions for its success are disappearing. Decision makers are increasingly confronting more and more varied types of decisions, and feedback from the field is increasingly inadequate. In consequence, effective decisions today must be taken at lower and lower levels within the organization.

8.11.1 A Framework and Modular Units

Toffler (1985) suggests another design for the information age. Instead of rigid conventional departments, the firm is divided into a highly flexible structure composed of a "framework" and "modules." Instead of being treated as an isolated unit, the structure is pictured as occupying a position at the center (and as part of) a shifting constellation of related companies, organizations, and agencies. This is the design of an adaptive organization.

The framework is the thin coordinative wiring that strings together a set of temporary modular units. The constellation consists of the company and the interdependent or semiautonomous outside organizations on which it relies. Constellation implies both a movement of certain functions downward and a movement of other functions outward. This cannot take place, however, unless there is a corresponding strengthening of the system's integrative machinery. The adhocratic units or modules of the corporation must be coordinated. Effective restructuring requires not an indiscriminate "loosening of the reins of central control," but, on the contrary, a selective, controlled movement of power upward as well. Certain functions can only be carried out centrally. In organizational terms, these may be referred to as "framework functions." They are concerned primarily with coordination, the definition and maintenance of standards, and the provision of specialized services or resources to companies and organizations that are part of the constellation as a whole.

8.12 COLLATERAL ORGANIZATION AND INNOVATION: ZAND'S PERSPECTIVES

Research into the relation between the structure of a problem and the effectiveness of different organizations, according to Zand (1979), suggested that a work group benefits from using more than one model of organization. Authority of production-centered organizations work best with "well-structured problems"; knowledge/problem-centered organizations work best with "ill-structured" problems (Figure 8.2). Zand refers to the concept of a secondary mode of working as a collateral organization. Hence, a collateral organization is a supplemental organization coexisting with the usual, formal organization; it is a mechanism for flexibility to manage tasks with characteristics of complex, nonroutinized mental work, which also generally characterize technological innovation, according to Zand (1979).

There are no experimental data on whether authority/production groups, given ill-structured problems, shift to a knowledge/problem mode. Observation of authority/production organizations suggests that when they are confronted with an ill-structured problem, such as entering a volatile market undergoing rapid technical change, managers do not shift to another mode but try to redefine the problem, forcing it to fit the existing hierarchy and division of labor. Burns and Stalker (1961) found that companies unable to shift to a knowledge/problem mode were unsuccessful in the new environment.

A collateral organization is distinguishable from and linked to the formal organization as follows (Zand, 1979):

1. The purpose of the collateral organization is to identify and solve problems not solved by the formal (primary) organization.

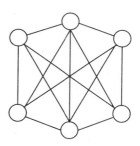

Elements	Authority/Production	Knowledge/Problem
1. Levels of authority	Many	Few
2. Division of labor	High	Low
3. Links to others in the organization	Few	Many
4. Source of influence and power	Position in the hierarchy	Ability to identify and solve problems
5. Use of rules and procedures	High	Low
6. Primary purpose	Maximize output	Analyze or invent knowledge to solve problems

Figure 8.2 Types of organizations. (From Zand, 1979. Reprinted with permission of Business Publications.)

2. A collateral organization creatively complements the formal organization. It allows new combinations of people, new channels of communication, and new ways of seeing old ideas.
3. A collateral organization operates in parallel or in tandem with the formal organization. Both the collateral and the formal organizations are available; a manager chooses one or the other, depending on the problem. A collateral organization does not displace the formal organization.
4. A collateral organization consists of the same people who work in the formal organization. There are no new people.
5. The outputs of the collateral organization are inputs to the formal organization. The ultimate value of a collateral organization depends on successfully linking it to the formal organization, so its outputs are used.
6. A collateral organization operates with norms (i.e., expectations of how

people will behave) that are different from those of the formal organization. The different norms facilitate new ideas and new approaches to obstacles.

A collateral organization has the following characteristics (Zand, 1979):

1. All channels are open and connected. Managers and specialists are free to communicate without being restricted to formal channels in the hierarchy.
2. There is a rapid and complete exchange of relevant information.
3. Norms encourage careful questioning and analysis of goals, assumptions, methods, alternatives, and criteria for evaluation.
4. A manager can approach and enlist others in the organization to help solve a problem, without being restricted to his formal subordinates.

8.13 MATRIX ORGANIZATION: A DESIGN FOR CHANGE; GALBRAITH

PRINCIPLE FOR MANAGING CHANGE

Effective implementation of an innovation is facilitated by the analysis of diagnostic data about (1) the perceived attributes of the innovation (degree of complexity, specialization, and uncertainty); and (2) the perception of managers of its utility in optimizing organizational performance.

An emerging organization design, termed *matrix organization,* is spreading through industry and parts of the public sector. Although the exact definition of the term is not well established, it typically means a balanced compromise between functional and product organization. Timely response to change requires information and communication channels that efficiently get the information to the right people at the right time. Matrix structures encourage constant interaction between product unit or product and functional department managers for information processing and problem solving. Information is channeled vertically and horizontally as people exchange technical knowledge. The result is quicker response to competitive conditions. Matrix organization also makes it possible for top management to delegate ongoing decision making, thus providing more time for long-range planning (Galbraith, 1971).

A matrix is an organization that represents a compromise between functional and product departmentalization and employs a multiple authority structure. In the matrix organization, individuals from various departments

TABLE 8.2 Differences Between Functional, Matrix, and Product Organization Forms

Parameters	Functional	Matrix	Product
1. Complexity of information	Low	Medium	High
2. Variety of information	Low	High	Medium
3. Uncertainty of information	Low	High	Medium
4. Diffentiation of information subsystems	Low	High	Medium
5. Changes in environment	Low	High	Medium
6. Product changes	Low	Moderate	High
7. Technology	Standard	Complicated	New
8. Product portfolios	Small	Diverse	Several

are assigned to one or more project teams to work together for the duration of the assignment. In the functional organization, each department has its own specialists (e.g., accounting, marketing, production, personnel). In the product organization, all of the different specialists needed to produce a given product are in the same unit.

The matrix is designed to exploit the strengths of both functional and product structures yet avoid the weaknesses of either.

The strengths of the functional structure are in its ability to facilitate a high level of specialization, centralized control, and efficiency. Its weaknesses are in the displacement of goals of the organization because of its heavy focus on the speciality (e.g., finance versus marketing versus production, etc.), poor coordination among specialized units, and inadequate information flow between these units (Galbraith, 1971).

The strengths of the product structure are in its relative autonomy, coordination of activities around a product line, and its adaptability to changes in product characteristics. Also, the evaluation of performance is based on the product. The weaknesses of the product structure are reflected in the drift of the unit or subsystem toward greater autonomy, the need for more control mechanisms to monitor its performance, and the duplication of effort in having functional areas such as marketing, finance, personnel, and so on for each product line.

The matrix structure combines the advantages of both functional and product structures.

A useful way to think about the differences between functional, product, and matrix organization forms is presented in Table 8.2. It indicates that organization designs may range from purely functional to purely product organizations. In the functional organization, heads of various functions make all decisions that affect their departments. Their authority is based on their position in the organization's hierarchy and on policies, rules, regulations, and job descriptions. On the other hand, a product organization places the decision-making authority with each product manager. These managers make decisions in a semiautonomous manner with respect to the other product managers. The matrix form uses advantages of both the product and functional forms (Galbraith, 1971).

8.13.1 Essential Characteristics of Matrix Organizations

1. Some managers report to two bosses rather than to a single boss. The unity-of-command principle is broken. A dual chain of command is a distinctive characteristic.
2. Firms tend to adopt matrix forms when it is absolutely essential that they be responsible to two sectors: markets and changing technologies.
3. A matrix organization is more than matrix structure. It must be reinforced by matrix systems, such as dual control and performance evaluation systems, a manager whose leadership style is informal, and a management system that fosters open discussion of problems.
4. Most matrix organizations assign dual command responsibilities to functional departments and to product departments.
5. There is the normal vertical hierarchy within functional departments, which is "overlaid" by a form of lateral influence. The matrix is unique in that it legitimates lateral channels of influence.
6. Every matrix organization contains three unique and critical roles: the top manager who heads up and balances the dual chain of command, the matrix bosses who share subordinates, and the two-boss employees who report to different bosses (Galbraith, 1971).

8.13.2 Conditions for the Use of the Matrix Design

The factors that determine choice of a matrix design are diversity of the product line, the rate of change of the product line, interdependencies among subunits, and level of technology. These are described by Galbraith (1971) and are discussed below.

Condition 1

Product Lines. The greater the diversity between product lines and the greater the rate of change of products in the line, the greater the pressure to move toward product structures. When product lines become diverse, it becomes difficult for general managers and functional managers to maintain knowledge in all areas; the amount of information they must handle exceeds their capacity to absorb it. Similarly, the faster the rate of new product introduction, the more unfamiliar the tasks being performed.

Managers are therefore less able to make precise estimates concerning resource allocations, schedules, and priorities. During the process of new product introduction, these same decisions are made repeatedly. The decisions concern trade-offs between engineering, manufacturing, and marketing. This means there must be greater product influence in the decision process.

Condition 2

Interdependence. The functional division of labor in organizations creates interdependencies among the specialized subunits. That is, a problem of action in one unit has a direct impact on the goal accomplishment of the other units. Organizations usually devise mechanisms that uncouple the subunits, such as in-process inventory and order backlogs. The degree to which inventories and backlogs develop is a function of how tight the schedule is. If there is a little slack in the schedule, then the functional departments can resolve their own problems. However, if rapid response to market changes is a basis of competition, then schedules are squeezed and activities run in parallel rather than in series. This means that problems in one unit directly affect other units. The effect is a greater number of joint decisions involving engineering, manufacturing, and production. A greater need for product influence in these decisions arises due to the tight schedule.

Although the tightness of the schedule is the most obvious source of interdependence, tight couplings can arise from reliability requirements and other design specifications. If the specifications require a more precise fit and operation of parts, then the groups designing and manufacturing the parts must also "fit and operate" more closely. This requires more coordination in the form of communication and decision making (Galbraith, 1971).

Condition 3

Level of Technology. If tight schedules and new products were the only forces operating, every organization would be organized around product lines. The level of technology or degree to which new technology is being used is a counteracting force. The use of new technologies requires expertise in market research and in the technical specialties involved in engineering, production engineering, and manufacturing.

However, if the expertise is critical to competitive effectiveness, the organization must acquire it internally. If the organization is to make effective use of the expertise, the functional form of organization is superior.

8.13.3 Advantages of the Matrix Structure

Knight (1976) summarizes a number of advantages of the matrix structure, among them flexibility, technical excellence, balancing of conflicting objectives, and motivation.

> **Flexibility.** Matrix organizations are asserted to have a capacity to respond quickly and adaptively to a changing and uncertain environment. This is partly due to additional lateral communication channels. The information filtration that is characteristic of the organizational hierarchy is limited in the matrix structure. Because of frequent contact between members of different depart-

ments, information permeats the organization and more quickly reaches those people who need to take account of it. Decisions can be made more rapidly, and changes in the givens of a situation can quickly be translated into changes of plan and acted on (Knight, 1976).

Technical Excellence. Matrix structures facilitate high-quality and innovative solutions to complex technical problems. The use of resources with diverse viewpoints and perspectives of different specialties and departments in cross-functional, interdisciplinary project teams add to the quest for technical excellence. The matrix facilitates the efficient allocation of specialists (Knight, 1976).

Balance. The need to balance conflicting objectives lies at the origin of the matrix organization. It contains a set of built-in checks and balances on time, cost, and performance, and is an effective means for balancing customers' requirements for project completion and cost control with the organization's need for economic efficiency and development of technical capability for the future (Knight, 1976).

Motivation. Project teams tend to operate in a more participative and "democratic" manner than functional hierarchies. The special knowledge and responsibility for given aspects of the work ensure participation of all members of the team with no attention to status or rank. This results in people at lower levels in the organization finding themselves with a greater say in significant decisions and is likely to enhance their commitment to the task. Knight (1976) lists three ways in which working in a matrix structure can help to develop individual employees. First, it enlarges their experience and broadens their outlook, by putting them in situations in which they have to weigh a wider range of considerations and issues than those arising within their own specialty. Next to the job rotation practiced by some companies, it is the best way of exposing people to the different facets of an operation and diverse criteria that must be satisfied. Second, the matrix increases the responsibility of individuals representing their specialty within a mixed team, and involves them in challenges and decisions that are normally met at a higher level in a departmental hierarchy. Finally, it exposes them to a wider arena and gives persons of high potential a chance to demonstrate their capabilities and make a name for themselves (Knight, 1976).

Change and Innovation. Daft (1989) observes that the matrix structure is best when environmental uncertainty is high and when goals reflect a dual requirement, such as for both product and function. The dual-authority structure encourages communication and coordination to cope with rapid environmental change. The matrix is also good for nonroutine technologies that have interdependencies both within and across functions. The matrix is an organic structure that facilitates adaptation to unexpected problems. Luthans (1985) also reinforces this observation and argues that this type of organization design will undoubtedly become more widely used in the future to meet the demands of increased differentiation and the ever-increasing complexity and uncertainty of the external environment (Knight, 1976).

8.13.4 Problems of the Matrix Organization

Conflict. Robbins (1983) argues that since the matrix dispenses with the unity concept, ambiguity is significantly increased. Confusion and role ambiguity create uncertainty and conflict—two elements for political behavior in an organization. There is a possibility of power struggles in the matrix. Because the rules, procedures, and regulations are stipulated in a bureaucracy, the incidence of, or potential for, power plays is limited. When the rules are vague and ambiguous, however, a high incidence of power struggles between functional and project managers results. This environment can produce stress. The dual chain of command introduces role conflict and role ambiguity.

Knight (1976) argues that the matrix expresses a set of preexisting conflicts between organization needs and environmental pressures. It structures these conflicts and internalizes them within the organization, but does not remove them. The main institutionalized form of conflict within the matrix is that between functional managers and project managers who are competing for the control of the same set of resources.

Knight (1976) lists a number of factors that increase the potential for conflict, based on his study of matrix management:

- Diversity of disciplinary expertise
- Low power of project manager
- Poor understanding of project objectives by team
- Role ambiguity of team members
- Lack of agreement on superordinate goals
- Perception by functional staff that their roles are being usurped
- Low perception of interdependence
- Managerial level (more conflict at higher levels)

The management of conflict is a principal role of the project manager. For the effective performance of the matrix, the sources of power available to the project manager to manage conflict must be diagnosed.

Groupitis. Robey (1986) argues that since the matrix depends on team functioning, there may be an overdependence on group decision making even where it is unnecessary. This slows down projects and creates conflict with functional managers. In such a case, project leaders must take independent action.

Anarchy. Because of the dual chain of command in the matrix and the role ambiguity, some managers may perceive the matrix as anarchy, or disorganization. This may be interpreted as a license to break rules and ignore policies in the interest of being organic, flexible, and modern. The matrix is not a free-form structure (Robey, 1986).

Managing the Decision Context. Davis et al. (1977) identify another problem area in the matrix organization. The second necessary condition for a matrix organization to be effective is that a very high volume of information be processed and focused for use in making key decisions. If the organization is to cope with such an information-processing load, the top leader must be only one among several key decision makers—he must delegate. However, he cannot delegate to other decision makers the job of setting the stage; he must himself manage the decision context.

Managerial Style. From researching nine large organizations utilizing a matrix organizational structure, Argyris (1973) noted that the theory of matrix organization and team approach made sense according to executives, but found that it was problematic in practice.

People still seemed to polarize issues, resisted exploring ideas thoroughly, mistrusted each other's behavior, focused on trying to protect one's own function, overemphasized simplified criteria of success (e.g., figures on sales), worked too much on day-to-day operations and short-term planning, engaged in routine decisions rather than focus more on long-range risky decisions, and emphasized survival more than the integration of effort into a truly accepted decision.

Argyris (1973) concluded that the problem was with managerial style, not the matrix per se. The behavior styles needed for the effective use of the matrix organization are different from the styles managers have developed within the traditional pyramidal organization. In addition, the group dynamics that are effective in the pyramidal structure are different from those effective in the matrix organization.

8.14 ADHOCRACY AND CHANGE: MINTZBERG'S PERSPECTIVES

An *adhocracy* has been described as a rapidly changing, adaptive, temporary system organized around problems to be solved by groups of relative strangers with diverse professional skills. Robbins (1981) describes the key characteristics of adhocracies in terms of the three structural dimensions cited above.

Adhocracies have a high degree of horizontal differentiation based on formal training. Since adhocracies are staffed predominantly by professionals with a high level of expertise, horizontal differentiation is great. Vertical differentiation is low. The number of levels in the organizational hierarchy in a bureaucracy is absent because of the need for flexibility. Many levels in the hierarchy, in addition to contributing to information filtration and limited feedback, also restrict the capacity of the organization to adapt. Additionally, professionals operate with some degree of autonomy and discretion and their own standards for control and supervision, according to Robbins (1981).

Emphasis on rules, regulations, and procedural specifications is limited in an adhocracy. The need for flexibility limits the degree of formalization. Rules and regulations are effective where standardization of behavior is stressed. Because of the ad hoc nature of the problem-solving posture in this type of structure, formalization becomes superfluous. Rules that govern experts are diffuse and abstract because uniformity and standardization are not valued properties of the structure.

Decision making in adhocracies is decentralized. The adhocracy depends on decentralized "teams" of professionals for decision making. Employees of the organization are assigned to temporary work units on an "as needed" basis. They are part of that team until the unit's objectives are achieved. As long as they are part of the team, employees participate actively in its key decisions. Democratic decision processes are prevalent in contrast to bureaucracy's hierarchical decision making. That is, influence in an adhocracy is based on the power of expertise rather than on the authority of position (Robbins, 1981).

Robbins concludes that the adhocracy takes on the form of a very adaptable structure. Multidisciplinary teams are formed around specific objectives. The departmental stability found in a bureaucracy is replaced by constantly changing units. Teams are formed regularly within the organization to pursue specific objectives. Individuals within the adhocracy find their responsiblities in a continual flux, adjusting as they move from assignment to assingment. Responsibilities are not delineated clearly as they are in the bureaucracy. The adhocracy represents a viable alternative when it is important that the organization be adaptable and creative, when individual specialists from diverse disciplines are required to collaborate to achieve a common goal, and when tasks are technical, nonprogrammed, and complex. In contrast to bureaucracy, adhocracy is clearly an inefficient structure. It lacks the precision and expediency of the bureaucracy, but compensates by its effectiveness in terms of flexibility and innovation.

8.14.1 Conditions of Adhocracy

Mintzberg (1979) identifies a number of conditions of the adhocracy: the basic environment, disparate forces in the environment, and frequent product change, among others.

Basic Environment. The adhocracy is clearly positioned in an environment that is both dynamic and complex. A dynamic environment calls for an organic structure, and a complex one calls for a decentralized structure. An adhocracy is the only configuration that is both organic and relatively decentralized. Innovative work, being unpredictable, is associated with a dynamic environment; and the fact that the innovation must be sophisticated means that it is complex (Mintzberg, 1979).

Disparate Forces in the Environment. The organization must create different work constellations to deal with different aspects of its environment and then integrate all their efforts. In such a case, the adhocracy is feasible and a logical structural alternative to the divisionalized form generally adopted by multinational firms. These firms group their major divisions by either region or product line. But recent changes in their environments have resulted in a near balance of the pressures to adopt each of these two bases of grouping. The matrix structure allows the establishment of regional and product divisions at the same level of the hierarchy. A product manager in a given region could report to both an all-product regional division manager and an all-region (worldwide) product division manager.

Mintzberg (1979) further notes that a hybrid structure—a divisionalized adhocracy—could emerge. Its markets are diversified, like all organizations that use the divisionalized form, but parts of its environment are more complex and dynamic. Those multinational firms with interdependencies among their different product lines, and facing increasing complexity as well as dynamism in their environment, will opt for the divisionalized adhocracy hybrid.

Frequent Product Change. A number of organizations are drawn toward adhocracy because of the dynamic conditions that result from very frequent product change. The extreme case is the unit producer—the manufacturing firm that custom-makes each of its products to order, as in the case of an engineering company that produces prototypes. Because each customer order constitutes a new project, the organization is encouraged to structure itself as an operating adhocracy. Again, some manufacturers of consumer goods operate in markets so competitive that they must change their products almost continuously. Here, dynamic conditions, when coupled with some complexity, drive the structure toward the adhocracy form. Product competition requires serious innovation and more complex decision making, often based on sophisticated research and development activity. Adhocracy becomes the preferred configuration in this situation, concludes Mintzberg (1979).

8.15 ORGANIZATIONAL DESIGN AND HUMANISM: ARGYRIS' PERSPECTIVE

Students of management theory also ask what happens when a form of organization does not correspond to human nature. They turn to psychology for answers and are told that employee alienation is caused by having a system foreign to their nature imposed on them. Argyris (1959), in particular, defends this conclusion in his book *Personality and Organization* (Scott et al., 1981).

Alienation is a very important subject in management, according to Scott et al. (1981), since it is responsible for reducing organizational consensus and heightening organizational conflict. Alienation is a serious organizational disease. It has been examined in management circles by virtually every behavioral scientist with the belief that whatever might be done to cure the disease ought to be given due consideration. Thus, sociology, psychology, and social psychology, to name the most influential of these sciences in management, have taken their turn at diagnosing organizational malaise.

The formal organization in which technological change is managed consists of a number of conditions, processes, and properties that reinforce the alienation of employees.

The most basic property of formal organizations is its logical foundation or essential rationality. It is the planner's conception of how the intended consequences of the organization may best be achieved. The underlying assumption made by creators of the formal organization is that within respectable tolerances man will behave rationally, that is, as the formal plan requires him to behave. Organizations are formed with particular objectives in mind, and their structures mirror these objectives, according to Argyris (1973):

Designing Specialized and Fractionized Work: (Argyris 1973)

Low	Designing production rates and controlling speed of work	High
Low	Giving orders	High
Low	Evaluating performance	High
Low	Rewarding and punishing	High
Low	Perpetuating membership	High

An inspection of the literature suggests that an overwhelming number of organizations manifested organizational structures that looked like pyramids (of different shapes). The logical purpose of these pyramidal structures was (1) to centralize information and power (at the upper levels of the structure) and (2) to specialize work. The logic explicitly stated that management should be in control over the key organizational activities. Thus, management should be high on the six organizational activities outlined above.

These continua together represented a logic which was difficult for designers of organizations to ignore, if the organizations were to have pyramidal structures. The model assumed that the more an organization approximated the right end of the continua, the more it reproduced the ideal of formal organization. The model, however, said nothing a priori about where any given empirical organization would be along these continua.

8.15.1 Problems of Individual Adjustment

Task Specialization. However, there is the argument that human nature resists formal organizational principles. The principle on which other organization principles are based is *task* (work) *specialization*. The outcome of task specialization is organizational efficiency. Inherent in this assumption are other suppositions. The first is that the human personality will behave more efficiently as the task that it is to perform becomes specialized. The second is that there is one best way to define the job so that it can be performed at greater speed. The third is the assumption that any individual differences in the human personality may be ignored by transferring more skill and thought to machines.

Argyris (1973) further observes that a number of difficulties arise concerning these assumptions when the properties of human personality are considered. First, the human personality is always attempting to actualize its unique organization of parts resulting from a continuous, emotionally laden, ego-involving process of growth. Unique differences of individuals cannot be ignored by technology or task specialization. Self-actualization cannot be compromised by task specialization, which requires the individual to use only a few limited abilities. Moreover, as specialization increases, the less complex motor abilities are used more frequently. These, research suggests, tend to be of lesser psychological importance to the individual. Therefore, the principle violates two basic givens of the healthy adult human personality. It inhibits self-actualization and provides expression for a few, shallow, superficial abilities that do not provide the "endless challenge" desired by the healthy personality. Task specialization requires a healthy adult to behave in a less mature manner, but it also requires that he feel good about it. Because of the specialized nature of the worker's job, few psychological rewards are possible.

Chain of Command. Following the logic of specialization, organizational planners create a new function, that is, leadership, the primary responsibility of which is to control, direct, and coordinate the interrelationships of the parts and to make sure that each part performs its objective adequately. The planner thus makes the assumption that administrative and organizational efficiency is increased by arranging the parts in a determinate hierarchy of authority.

If the parts being considered are individuals, then they must be motivated to accept direction, control, and coordination of their behavior. The leader is therefore assigned formal power to hire, discharge, reward, and penalize the individuals in order to mold their behavior in the pattern of the organization's objectives.

The impact of such a state of affairs is to make the individuals dependent on, passive, and subordinate to the leader. As a result, the individuals have little control over their working environment. At the same time, their time perspective is shortened because they do not control the information necessary to predict their futures. These requirements of formal organization act to inhibit four of the growth trends of the personality, because to be passive, subordinate, and to have little control and a short time perspective exemplify in adults the dimensions of immaturity, not adulthood.

Consequently, the management helps create a psychological set that leads workers to feel that the basic causes of dissatisfaction are built into industrial life, that the rewards they receive are wages for dissatisfaction, and that if satisfaction is to be gained they must seek it outside the organization.

There are three assumptions, Argyris (1973) argues, inherent in the above solution that violate the basic givens of human personality. One is that a whole human being can split his personality so that he will feel satisfied in knowing that the wages for his dissatisfaction will buy him satisfaction outside the plant. Another is that the employee is primarily interested in maximizing his economic gain. The third is that the employee is best rewarded as an individual producer. The work group to which he belongs is not viewed as a relevant factor.

There are some basic incongruences between the growth trends of a healthy personality and the requirements of formal organization. If the principles of formal organization are used as ideally defined, the employees will tend to work in an environment where they are (1) provided minimal control over their workaday world; (2) expected to be passive, dependent, and subordinate; (3) induced to perfect and value the frequent use of a few superficial abilities; and (4) expected to produce under conditions leading to psychological failure. As the jobs become more mechanized (i.e., take on assembly-line characteristics), this incongruence increases.

In summary, Argyris (1973) expresses his findings in terms of a number of propositions. There is a lack of congruency between the needs of healthy individuals and the demands of the formal organization. Employees will tend to experience frustration because their self-actualization will be blocked. They will tend to experience failure because they will not be permitted to define their own goals in relation to their central needs, and paths to these goals. They will experience short-time perspective because they have no control over the clarity and stability of their future. The nature of the formal principles of organization causes subordinates, at any given level, to experience competition, rivalry, intersubordinate hostility, and to develop a focus toward parts rather than whole.

8.16 HUMANISTIC ORGANIZATIONAL DESIGN

The classical design according to Davis (1985) placed a heavy emphasis on the excessive division of labor and an overdependence on rules, procedures, and hierarchy. The worker became isolated from other workers. The result was higher turnover and absenteeism. Quality declined and workers became alienated. Conflict arose as workers tried to improve their lot. Management's response to this situation was to tighten the controls, increase supervision, and organize more rapidly. These actions, calculated to improve the situation, only made it worse.

Because of the problems inherent in the classical design, management moved toward a more behavioral, participative, humanistic design (Figure 8.3). The objective was to make the job environment supportive rather than threatening to employees, a place that stimualted their drives and aspirations and helped them grow as whole persons. Work had to be psychologically as well as economically rewarding.

Classical design	Humanistic design
Closed system Job specialization Centralization Authority Tight hierarchy Technical emphasis Rigid procedures Command Vertical communication Negative environment Maintenance needs Tight control Autocratic approach	Open system Job enlargement Decentralization Consensus Loose project organization Human emphasis Flexible procedures Consultation Multidirectional communication Positive environment Motivational needs Management by objectives Democratic approach

Figure 8.3 Humanistic organizational design. (From Davis, 1985. Reprinted with permission from McGraw-Hill, Inc.)

The organic adaptive structure that incorporates most of the elements of the humanistic design should increase motivation, according to Bennis (1973), and thereby effectiveness, because it enhances satisfactions intrinsic to the task. There is harmony between the individual's needs for meaningful, satisfactory, and creative tasks and a flexible organizational structure. People will be more intellectually committed to their jobs and will require more involvement, participation, and autonomy in their work.

Coping with rapid change; living in temporary work systems; setting up meaningful relations and then breaking them; all augur social strains and psychological tensions, according to Bennis (1973). Learning how to live with ambiguity and to be self-directing will be the task of education and the goal of maturity.

Bennis (1973) concludes that in these new organizations, participants will be called on to use their minds more than at any other time in history. Fantasy, imagination, and creativity will be legitimate in ways that today seem strange. Social structures will no longer be instruments of psychic repression, but will increasingly promote play and freedom on behalf of curiosity and thought.

REFERENCES

Argyris, C., *Personality and Organization,* Harper & Row, New York, 1959

Argyris, C., "Today's Problems with Tomorrow's Organization," in J. Yun and W. Strom (Eds.), *Tomorrow's Organization,* Scott, Foresman, Glenview, Ill., 1973.

Bennis, W., "Beyond Bureaucracy: Will Organizational Man Fit the New Organization?" in J. Yun and W. Storm (Eds.), *Tomorrow's Organization*, Scott, Foresman, Glenview, Ill., 1973.

Benson, G., "Mindlessness as Next to Mechanicalness," *Training and Development Journal* (1985).

Burns, T and A. Stalker, *The Management of Innovation*, Tavestock Publications Ltd., London, 1961.

Burton, R., "Information and Organizational Theory," Academy of Management Symposium, Annual Meeting, Chicago, 1988.

Child, J., *Organizations*, Harper & Row, New York, 1981.

Child, J., *Organizations: A Guide to Problems and Practice*, Harper & Row, London, 1984.

Daft, R., *Organizational Theory and Design*, West Publishing, St. Paul, Minn., 1989.

Davis, D., "Integrating Technological, Manufacturing, Marketing and Human Resources Strategies," in Donald Davis and Associates (Eds.), *Managing Technological Innovations*, Jossey-Bass, San Francisco, 1986.

Davis, K., "Trends in Organization Design," in K. Davis and R. News (Eds.), *Organizational Behavior*, McGraw-Hill, New York, 1985.

Davis, S., and P. Lawrence, *Matrix*, Addison-Wesley, Reading, Mass., 1977.

Dewar, R., and J. Hage, "Elite Values Versus Organizational Structure in Predicting Innovation," *Administrative Science Quarterly*, **18**, (1974).

Drucker, P., "New Templates for Today's Organization," *Harvard Business Review*, January–February (1974).

Duncan, R., "The Ambidextrous Organization: Designing Dual Structure for Innovation," in R. Kilmann, L. Pondy, and D. Slevin (Eds.), *The Management of Organizational Design: Strategies and Implication*, North Holland, Amsterdam, 1976.

Duncan, R., "What is the Right Organizational Structure:? Decision Tree Analysis Provides the Right Answer," *Organizational Dynamics*, **7**(3), Winter (1979).

Flores, F., M. Graves, B. Hatfield, and T. Winograd, "Computer Systems and Design of Organizational Interactions," *AMC Transactions on Office Information Systems*, **6**, 1988.

Fromm, E., *Revolution of Hope: Toward a Humanized Technology*, Harper & Row, New York, 1968.

Galbraith, J., "Matrix Organizational Design: How to Combine Functional and Product Forms," *Business Horizons*, February (1971).

Knight, K. "Matrix Organization: A Review," *Journal of Management Studies*, May (1976).

Lund, R. T., and J. A. Hansen, *Keeping America at Work: Strategies for Employing the New Technologies*, Wiley, New York, 1986.

Luthans, F., *Organizational Behavior*, McGraw-Hill, New York, 1985.

Mintzberg, H., *The Structuring of Organizations: A Synthesis of Research*, Prentice Hall, Englewood Cliffs, N.J., 1979.

Naisbitt, J. and P. Aburdene, *Re-inventing the Corporation*, Warner Books, New York, 1985.

Owen, D., "Information Systems Organization: Keeping Pace with the Pressures," in S. Madnick, *The Strategic Use of Information Technology,* Oxford University Press, New York, 1987.

Pastin, M., *The Hard Problems of Management: Gaining the Ethics Edge,* Jossey-Bass, California, 1986.

Perrow, C., "A Framework for the Comparative Analysis of Organizations," *American Sociological Review,* **32,** (1967).

Peters, T., and R. Waterman, *In Search of Excellence,* Harper & Row, New York, 1982.

Robbins, S., *Organization Theory: The Structure and Design of Organizations,* Prentice Hall, Englewood Cliffs, N.J., 1983.

Robey, D., *Designing Organizations,* Irwin, Homewood, Ill., 1986.

Sankar, Y., "Corporate Culture, Values, and Ethics," *International Journal of Value Based Management,* **I** (1988).

Sankar, Y., *Corporate Culture in Organizational Behavior,* Harcourt Brace Jovanovich, 1991.

Sapolsky, H., "Organizational Structure and Innovation," *Journal of Business,* **40,** (1967).

Scott, W., and T. Mitchell, *Organization Theory: A Structural and Behavioral Analysis,* Irwin, Homewood, Ill., 1972.

Scott, W., T. Mitchell, and P. Birnbaum, *Organizational Theory: A Structural and Behavioral Analysis,* Irwin, Homewood, Ill., 1981.

Thompson, V. "Bureaucracy and Innovation" in *Organizational and Managerial Innovation,* L. Rowe & W. Boise (Eds.), Goodyear, California, 1973.

Toffler, A., *The Adaptive Corporation,* Bantam, New York, 1985.

Trist, E., *The Socio Technical Perspective in Perspectives in Organizational Design,* A. H. Van de Ven and W. Joyce (Eds.), Wiley, New York, 1981.

Weick, K., *The Social Psychology of Organizing,* Addison-Wesley, Reading Mass., 1969.

Weick, K. and R. Daft, "Toward a Model of Organization as Interpretation Systems," *Academy of Management Review,* 1984.

Youker, R., "Organization Alternatives for Project Management," *Project Management Quarterly,* March (1977).

Zand, D., "Collateral Organization: A New Change Strategy," in W. French, Jr., C. Bell, and R. Zawacki (Eds.), *Organizational Development: Theory, Practice, and Research,* Business Publications, Dallas, 1979.

9
A MODEL FOR IMPLEMENTING CHANGE

9.1 ADVANCE ORGANIZER

The most obvious point about implementation is that we do not understand the dynamics of implementation. The characteristics of the adapting organization must be considered when implementing manufacturing technologies (Davis, 1986a). In terms of the scope of adoption and implementation research, hundreds of studies have revealed several characteristics of managers and organizational structure, process, and context that are related to the adoption of innovations. The implementation of an innovation involves the diffusion of that innovation, the effecting of organizational change (the use of the innovation), and the management of systems (innovation–organizational interface) (Schultz and Slevin, 1975).

Any function as complex as the management of change requires some kind of conceptual framework of thinking to guide it. We provide such a conceptual framework for implementing change. In general, the implementation model indicates that an organization's capacity to implement an innovation is contingent on (1) the characteristics of the innovation being adopted; (2) the functions of the managers at each stage of the change process (the evaluation, initiation, implementation and routinization stages); (3) the nature and character of the domains of the organization (the behavioral, structural, process, technical, and management systems domains); and (4) the type of strategies adopted by the manager to modify the characteristics of the innovation and/or organization. The implementation process is a complex contingent process.

The conceptual model for implementing and managing change focuses on a number of principles for managing change. A detailed analysis of each of the

9.1 ADVANCE ORGANIZER

four stages of the change process is given with specific operating principles for managing each stage. Next, the five domains of the organization affected in the implementation of an innovation are described. It is argued that the implementation of an innovation may create changes in the behavioral, structural, management systems, technical, and process domains of the adopting organization. The principles for managing each of these domains as they relate to the implementation process are also developed. Managerial strategies associated with these principles of change are also discussed. In addition, the politics of implementing change is considered, since an innovation may create changes in the sources of power within an organization.

The chapter concludes with implementation dynamics from a participative perspetive. Participative management as a strategy for implementing new technologies is recommended because it develops an adaptive learning capacity, it is ethically the correct procedure for planning changes that affect other people, and it is a way of confronting political issues involved in change. The participative aspects of management by objectives (MBO) are reviewed briefly and a model based on MBO is also developed.

The framework in Figure 9.1 highlights the themes of this chapter on implementation dynamics.

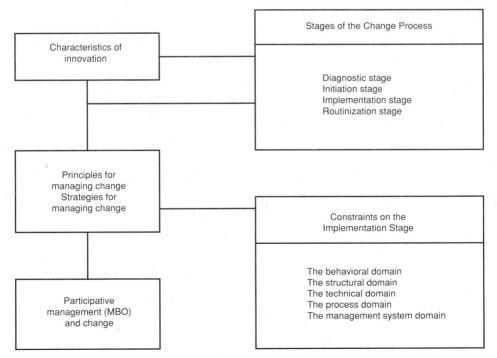

Figure 9.1. A model for implementing change.

9.2 INTRODUCTION

The characteristics of the adopting organization must be considered when implementing advanced manufacturing technologies. The organizational perspective requires that we focus on the firm's domains, such as behavior, structure, processes, management systems, strategy, corporate culture, job design modules, and so on.

An organizational perspective is necessary to understand the factors related to successful use of advanced manufacturing technologies and to achieve the level of performance they are capable of providing. Unlike numerically controlled equipment and other simpler forms of automation, computer-controlled technologies such as robotics, computer-aided design/computer aided manufacturing (CAD/CAM), computer-aided engineering (CAE), flexible manufacturing systems (FMS), and computer-integrated manufacturing (CIM) are so complex that they stretch the ability of managerial and organizational systems to absorb them (Davis, 1986b).

With reference to the scope of adoption and implementation research, Davis (1986b) further observes that hundreds of studies have revealed several characteristics of managers and organizational structure, processes, and context that are related to the adoption of innovations. Examples include the values and attitudes of top managers, the form of decision making, the extent of rules and regulations, the form of strategic planning, and so on. Organizations are likely to adopt innovations, according to Davis (1986b), when they have strategies that stress technological advancement; structures that allow decentralized decision making; limited formal rules and regulations; high proportions of managerial specialists who are professionally active; and managers who value new ideas and are receptive to change. Characteristics of the innovation itself are also important.

The problems associated with the implementation of new technologies are noted by Davis (1986a).

Adoption and implementation of advanced manufacturing technologies are thus facilitated by certain configurations of management attitudes and values, organizational structure and practices, and organizational context and environment. But precise delineation of the most appropriate configuration is difficult because of gaps in our knowledge about organizations and technology. New forms of computerized automation only partly fit existing knowledge about organizational behavior and theory.

An important principle that innovation research teaches us, according to Ettlie (1986), is that there is rarely a single sociotechnical solution to a problem. Organizations typically try to select technologies for their future that conform with certain internal values and experiences and that are consistent within limits with the technologies that will later become available. Some of these technologies constitute radical or discontinuous departures from the experience in the firm, the industry, or even the world.

Hetzner et al. (1986) observe that studies of the implementation process of new technologies must be considered as collections not only of hardware, but of knowledge embedded in both the machines themselves and in the software and organizational systems used to operate and control the machines. As such, the implementation of manufacturing technology is as much a problem of knowledge dissemination and utilization as of engineering.

9.3 CONCEPTIONS OF THE IMPLEMENTATION PROCESS

The fundamental nature of the implementation problem is noted by Schultz and Slevin (1975).

> "It is, in any interpretation, a problem involving human participants, social interactions, organizational structure, and the management of change—in short, a complex behavioral process."

Several models of the implementation process have been developed, most of which provide the means for testing specific hypotheses about implementation behavior.

The research findings on implementation set the stage for increased theoretical work on implementation and organizational change. They also support two propositions, according to Schultz and Slevin (1975): (1) theories of implementation should be behavioral, because the implementation process is behavioral; and (2) theories of implementation should be cast in a form that makes their implications testable.

Introducing technological change into an organization presents a different set of challenges to management than does the work of competent project administration. Frequently, however, the managers responsible for shepherding a technical innovation into routine use are much better equipped by education and experience to guide that innovation's development than to manage its implementation (Barton and Gogan, 1986).

Keen and Scott Morton (1978) argue that the most obvious point is that we do not understand the dynamics of implementation. This is really an extraordinary fact; implementation is the avowed purpose of a large number of highly skilled and experienced professionals in the computer field, but while many of them are certainly successful implementers, they seem unable to identify any general principles underlying their success.

Another dimension to effective implementation which compounds the problem is noted by Sproul and Hofmeister (1986). They note that previous work on implementation has typically emphasized either the objective characteristics of programs and organizations or the motivational characteristics of people. Although these variables should not be ignored, how people think about a new program is an important component of any implementation effort. Three cognitive processes, interpretation, attribution, and inference, contribute to people's mental representations of an innovation. These representations have behavioral consequences for the implementation process; they also represent the invisible dimensions of this complex process.

The model of the innovation–adoption process by Rogers (1976)—that is, knowledge of the innovation, formation of a favorable or unfavorable attitude toward the innovation, decision to adopt, implementation, and decision to retain or discard the innovation once implemented—describes the adoption of innovations such as advanced manufacturing technologies, according to Davis (1986a). One becomes aware of a new system such as computer-aided design (CAD), considers the pros and cons of the new system, and makes a purchase decision. Once purchased, the new system is implemented within the organization and its performance is evaluated through analysis of productivity, return on investment, and other outcomes. If the innovation is still viewed as appropriate, it is continued.

9.4 A MODEL OF THE IMPLEMENTATION PROCESS

Figure 9.2 shows a model of the implementation process. (Note that the various panels, A–E, will be referred to in the following text.) Any function as complex as the management of change requires some kind of conceptual framework of thinking to guide it. Some theoretical considerations are being and have been applied in making decisions about the management of change. What is lacking is a coherent and consistent conceptual framework for implementing change. In general, the implementation model indicates that an organization's capacity to implement an innovation is contingent on a number of factors. The critical determinants are (1) the characteristics of the innovation being adopted, (2) the functions of the manager at each stage of the change process, (3) the nature and character of the domains of the organization, and (4) the types of strategies adopted by the manager to modify the characteristics of the innovation and/or organization. These strategies will produce either incremental or innovative change, which will be reflected in changes in managerial/organizational behavior. Our position is that for an innovation to be effectively implemented in an organization, it must be compatible with or fit that organization. This fit between the organization and innovation profiles must occur at three levels: behavioral (attitudes, perception, motivation, etc.), structural (levels of complexity, formalization, centralization), and process (job design/work flow, information flow, pattern of power distribution, management style, etc.). If an innovation requires a major (innovative) change in behavioral or structural characteristics, or organizational processes, the possibility of successful implementation will be reduced. An innovation may vary in the effects it produces on managerial behavior, organizational structure, and processes. Some innovations may trigger changes predominantly in the behavioral sector, others in organizational processes (e.g., information flow or organizational procedure, regulation, operations manual), and still others in the structural sector. Because of

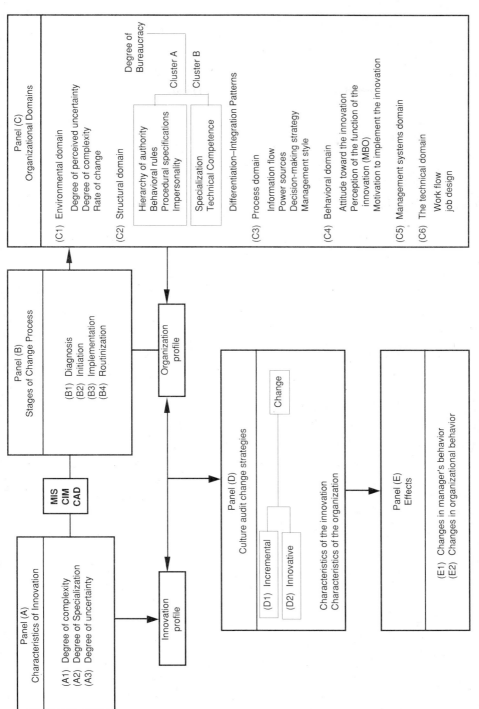

Figure 9.2. A model of implementation process: a systems framework.

the interdepencence among the three major components of the organization, the innovation must be monitored by the manager for the effects of change in these components.

The stages in implementing an (engineering, technical, or organizational) innovation are

1. Evaluating the distinctive characteristics of the innovation in terms of its degree of complexity, specialization, and uncertainty.
2. Mapping the functions of the manager in the various stages of the change process.
3. Evaluating the types of constraints (environmental, structural, process, and behavioral) that are likely to inhibit and/or facilitate the adoption of the innovation.
4. Designing of a set of strategies for changing the characteristics of the innovation and/or of the organizational components.
5. Monitoring the effects produced by the innovation on managerial and organizational behavior.

The implementation of an innovation is a complex contingent process. Effective implementation is contingent on the characteristics of the innovation, the characteristics of the organization, and the type of strategies adopted by the manager to negotiate the constraints (behavioral, environmental, structural, and process) that affect the innovation. Effective implementation clearly indicates that the management of change is a complex and critical function of the manager.

9.4.1 Stages of the Implementation Process (Panel A)

Stage 1: Identify the distinctive characteristics of the innovation.

Specifically, there is a positive relationship between the characteristics of the innovation (complexity–uncertainty), the level of motivation expressed for the innovation, and the probability of its effective adoption and diffusion. The functions and objectives of the innovation will partially determine its degree of complexity, specialization, and perceived uncertainty. These in turn will influence the type of resistance toward the innovation. It is tempting to classify innovations on the basis of their attributes and to explain innovation success (and failure) in terms of these attributes. For example, one research study showed that technical innovations, such as new products and automation, were more readily adopted than administrative innovations, such as organizational restructuring (Robey, 1986). Because administrative innovations often change the balance of power, those losing power may firmly resist changes during implementation. Technical innovations or changes may also alter power relationships, but the effect is usually less direct and sometimes not perceived until after the change is implemented.

Other attributes may also be important. The complexity or uncertainty of an innovation, and the specialized information processing required by it, will adversely affect its chances for adoption. Research on innovation attributes is hampered by the obvious fact that what is simple for one innovation may be complex for another. In other words, complexity and uncertainty are defined in terms of the organization's perception of the innovation, not its objective properties. This difficulty in determining the attributes of innovations and their impact on adoption has led most theorists to pay greater attention to correlates of innovation success that are more easily observed (Robey, 1986).

9.4.2 Some Dimensions of Technical Innovations (Panel A)

By far the most significant technological trends in manufacturing for the next decade or more are (1) the increased use of computers to control and integrate manufacturing operations and information and (2) the expanding capacity of electronic communication systems to handle enormous quantities of data over long distances at modest cost.

The merging of these two technologies yields a profusion of product and process innovations. Of the range of technical principles and devices that are being applied or are being readied for application, only the combination of computer and telecommunications technologies has the potential for massive technological changes in the coming decade (Lund and Hansen, 1986).

Slocum and Sims (1980) observe that technology can be analyzed in terms of three technology dimensions: work flow uncertainty, task uncertainty, and task interdependence. Work flow uncertainty refers to knowledge about when inputs will arrive at an individuals' station to be processed. Task uncertainty refers to the degree to which the individual employee lacks knowledge about how to accomplish the task. Task uncertainty is likely to be high where there is incomplete technical knowledge about how to produce desired outcomes. Task interdependence is the degree to which decision making and cooperation between two or more employees (groups) is necessary for them to perform their own jobs. The three types of interdependent task relations are pooled, sequential, and reciprocal; each has increasing degrees of interactions.

In the design of new jobs or the redesign of existing jobs, it is often necessary to consider and make changes in one or more of the three technological dimensions: task uncertainty, work flow uncertainty, and task interdependence (pooled, sequential, or reciprocal).

Another dimension of technology is noted by Hackman (1980), namely, discretion. When little discretion is required or permitted by the technology, work procedures are necessarily standardized and structured to a considerable extent. Jobs are usually segmented and routinized and contain little variety, autonomy, identity, and significance for workers. The dimensions of technology are summarized in Figure 9.3.

The relevance of these dimensions of technology to effective implementa-

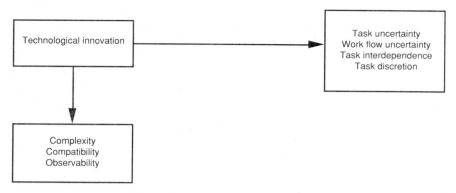

Figure 9.3. Dimensions of technology.

tion is noted by Ettlie (1986). Many implementation attempts fail or are marginally successful because the process design as it is ultimately installed and integrated into the work flow of the organization is incorrectly matched to the product needs of the user organization. This disparity, he continues, often results from an inflexibility of design, a premature rigidity of the process configuration, an incomplete understanding of the product characteristics that are crucial for process capabilities, or all three.

This segment on the dimensions of technologies are based on the observations of Davis (1986a). Perception of the characteristics of innovations such as advanced manufacturing technologies determines, in part, the likelihood of their adoption. Rogers, after reviewing hundreds of studies, has described the characteristics of innovations that contribute most to their adoption: relative advantage of the degree to which the new technology is perceived to be better than that which precedes it, compatibility (the degree to which the new technology is consistent with existing values, past experiences, and the needs of the potential adopting organization), complexity (the degree to which the new technology is relatively difficult to understand and use), triability (the degree to which the new technology may be experiemented with on a trial basis), and observability (the degree to which the results of the new technology are observable to others).

Davis (1986b) continues to develop some of these characteristics. *Compatibility* with existing needs, values, and practices is an attribute of advanced manufacturing technologies that contributes to their adoption. When felt needs are not met, the innovation is likely to be adopted. These felt needs interact, however, with other personal characteristics of potential adopter, such as values toward change. *Complexity* affects the adoption and implementation of advanced manufacturing technologies in two ways. First, the likelihood of adoption is reduced because these new forms of automation are believed to be too complex for any but the largest and most sophisticated firms. They are not purchased because they are not understood. Second, they are adopted for inappropriate reasons. They are purchased but fail to satisfy the unrealistic expectations of the adoption organization.

Observability of the new technology also plays a role in its adoption and implementation. The production and human resource outcomes of advanced manufacturing technologies are observable to all over time. Workers either like them or do not like them; productivity goes up or down. The high observability of positive outcomes of advanced manufacturing technologies should be related positively to their adoption. Uncertainty regarding outcomes, however, will reduce the likelihood of adoption.

Davis (1986b) concludes that the ability to experiment with these new technologies on a trial basis is problematical. It is difficult to try these new systems on a small scale before making the adoption decision. Although firms may automate incrementally by starting with stand-alone equipment, only when this equipment has been integrated into comprehensive systems are the full benefits of computer-integrated manufacturing (CIM) realized. Although robots or computer numerically controlled (CNC) tools can be easily tried out, CIM cannot. The inability to try CIM on a small scale is a barrier to adoption that is difficult to eliminate.

The first step in the implementation process, therefore, is to map the attributes, dimensions, or characteristics of the innovation. The dimensions themselves may produce potential resistance to change effort. The complexity, variety, and uncertainty of new technologies are themselves sources of perceived stress on the implementation process.

Stage 1 (Panel A): Development of a profile of the innovation.

1. Review objectives and functions of the innovation.
2. Establish degree of complexity, specialization, and uncertainty of the innovation.
3. Identify change targets of the innovation.
4. Map the path of change associated with the innovation.
5. Diagnose differential effects on management/organizational behavior produced by the change.
6. Identify areas of congruence between the characteristics of the innovation and those of the adoptive unit.
7. Identify areas of potential conflict between the characteristics of the innovation and those of the adoptive unit.
8. Identify the motivational strategy for facilitating the adoption of the innovation. (The type of change, innovative–incremental, will partially determine the magnitude and targets of the motivational function.)

Stage 2 (Panel B): Map the functions of the manager in the various stages of the change process.

PRINCIPLE FOR MANAGING CHANGE

The effective implementation of an innovation is facilitated by the design of strategies for managing each stage of the change process.

The innovation process involves four stages—organizational diagnosis, initiation, implementation, and routinization (Figure 9.4). The stages are separated by three decision points which mark movement to the next stage Robey (1986). Between organizational diagnosis and initiation comes the decision to innovate, as one possible response to the need. Between initiation and implementation comes the decision to adopt a particular innovation generated in the initiation stage. Finally, between the implementation and routinization stages comes the decision to design the management infrastructure—for example, programs, operating procedures, a management information system, and control and feedback systems—to support the innovation and to facilitate its integration with management systems. While the stages are conceptually distinct, they are not so clearly separated in practice. Often, initiation uncovers recognition of more needs, and problems encountered during implementation may require more work in initiation. In other words, most successful innovations involve team or reciprocal interdependence rather than sequential or pooled interdependence. We will briefly review the activities generally characteristic of each stage and then consider the role and functions of the manager in the context of these stages.

The dynamic nature of the implementation stage of new technologies is noted by Hetzner et al. (1986). Implementation is not a single decision; it is, rather, a set of decisions made at many different times and levels in the organization. Although decisions concerning the purchase of an advanced manufacturing system may be made at high levels of management, decisions concerning the implementation and use of these systems, of necessity, involve individuals and functions down to the shop floor. These decisions are often unrecognized but are crucial to the success of the implementation process.

Hetzner et al. (1986) further note that technological change moves through stages from initial awareness to evaluation and decision, to adoption, and finally to implementation and use. Up to the point of commitment of resources, which initiates implementation, knowledge dissemination is primarily a cognitive and intellectual activity that consists of learning about the innovation. Implementation, by contrast, requires expenditure of financial, material, and human resources and calls for behavioral changes at many levels in the organization for it to succeed. Thus, the implementation of complex technological systems demands on-site adaptation or reinvention a process in which research on complex sociotechnical implementation

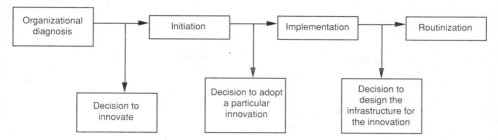

Figure 9.4. Stages in the change process. (Adapted from Robey, 1986. Reprinted with permission of Richard D. Irwin, Inc.)

processes suggests that participation at all levels is critical. Some programs have been significantly enhanced by the involvement of lower-level staff in the design and implementation of the advanced systems. More of this participation may occur as management systems are replaced by structures and practices more supportive of manufacturing systems and the lower-level discretion they seem to require (Hetzner et al., 1986).

9.4.3 The Diagnostic Stage of the Change Process (Panel B1)

The change process is preceded by a diagnostic stage. The objective of the diagnostic stage is for management to identify the nature and extent of the problem area(s) before initiating some action. Diagnosis generally precedes action. To aid in the identification of problems, several questions may be asked:

1. What are the specific problems to be corrected?
2. What are the causes and effects of the problems?
3. What must be changed to resolve the problems?
4. What factors will facilitate or inhibit the changes?
5. What goals or objectives are expected from the changes and what criteria will be designed to measure them?

Answers to these questions will involve the design of methods for scanning the internal and external environment for information. A manager may use the data base of the information systems for engineering, marketing, finance, accounting, and so on, or conduct attitude surveys, job diagnostic surveys, or interviews to identify the problem. From these diagnostic data, he can then formulate the objectives of any technical change. To improve the performance of the organization, the objectives of the change may involve change in organizational structure, technology, market strategies, job design, product portfolios, or the information system in order to cope with its environment, or changes in the behavior of employees. The change in behavior is critical if the organization is to respond to its environment. Every organization has its patterns of decision-making behaviors. These patterns are influenced by the values of top management, the leadership practices of managers, the reward system used to motivate employees, work group norms, and job design. Any change in an organization, whether it is introduced through a new structural design or a technical design, is an attempt to get employees to change their behavior.

During this stage, the organization may determine that the goals are not being met as adequately as in prior periods or as projected by forecasts. In any event, perception of a performance gap between desired and actual performance initiates the process of organizational change. Most organizations have ongoing technical surveillance, monitoring procedures as well as programs for gathering information in nonroutine situations. This performance gap analysis is a routine activity. At this stage, the consensus about the need for change to bridge the performance gap is problematic. The different perceptions of a gap and the alternatives that must be designed to bridge the gap may create conflict and hence inhibit change. Another dilemma facing management during the early part of the change process is whether to engage in large-scale or modest change effort. The type of change may be incremental (minor changes, as in creation of a new position or a manual), or it may be strategic (major changes which may be reflected in changes of the work flow, job design, structural and technical design, or even the goals of the organization). The magnitude of the change will depend on the nature of the problem as indicated in the performance gap analysis, the propensity of the elite to engage in risk-taking behavior and uncertainty, and the culture of the organization (i.e., norms, attitudes, values and motivation of its members to initiate and facilitate change).

Diagnostic Stage: Managerial Strategy

1. Develop diagnostic data on
   a. The adoption–diffusion pattern of the innovation in other adoptive units.
   b. Profile of the innovation regarding the internal consistency among its attributes.
   c. Distinctive attributes of the innovation or change program.
   d. Performance gap analysis and role of the innovation in bridging the gap.
   e. Feedback of data on performance gap analysis to the key decision makers.
2. Develop diagnostic data on
   a. Range of alternative strategies to bridge the gap.
   b. Objectives of the innovation.

c. The change targets of the innovation.
   d. Methods for managing change.
   e. Potential effects of the change (on organizational behavior, structure, processes) produced by the innovation.

Some Functions of New Technologies. With reference to the strategic purposes of new technology, Child (1984) makes the following observations. The strategic intentions will vary in emphasis according to the priorities and purposes of the organization and the problems and prospects it faces. The following purposes are usually prominent: (1) reduction in operating costs, (2) increased flexibility, (3) improvement in the quality of product or service, and (4) increased control and integration. There is some interdependency among these strategic intentions, and they are all concerned with enhancing the organization's ability to absorb the risks posed by external competition or by other threats to the organization's standing and survival.

In the field of new technologies, the availability of relatively inexpensive microelectronics is permitting a synthesis between areas that were previously segregated because each (e.g., CAD/CAM), was tied to costly large computers, namely computer graphical design, numerical-control machine tools, production schedules, and industrial robotics. The new integration between these areas greatly reduces the cost of product change and significantly improves the economics of flexibility.

New technology, Child (1984) further observers, is also facilitating the unification of fragmented control systems. In manufacturing, through flexible manufacturing systems (FMS), the objective is of a unified control system complementary to integrated computer-aided design, manufacture, and production systems. Such a system would combine controls in regard to physical movements, condition of plant, stocks, wastage, energy consumption, unit costs, and deployment of personnel. Integrated control of this kind can optimize the overall balance of production activities and maximize flexibility of adjustment.

The diagnostic phase must determine the functions of the new technologies. These functions, in turn, will indicate the change targets, the path of change, and the criteria for evaluating change. The four stages of the change process for implementing new technologies or any innovation are based on Hage and Aiken's research on change.

9.4.4 The Initiation Stage of the Change Process (Panel B2)

At the initiation stage, the multiple paths to change, the variety of change scenarios, and competing alternatives are reduced to one alternative solution that is perceived as the most desirable for bridging the performance gap. A strategy change can be initiated more effectively by outsiders because of vested interests, established power, status, influence, networks, or preconceptions that may react against change if initiated by insiders. At the same time, outsiders who may be able to initiate new ideas and directions for change are more likely to create conflict and resistance in the process (Hage and Aiken, 1970).

A stronger effort is required to change powerful, integrated, or cohesive units than weak, fragmented units. However, powerful and cohesive units may serve as models for others. Change may be facilitated because of the prestige of the initiating units. In units with a high degree of interdependence, change tends to be diffused more readily. In some cases, change may be initiated as a pilot project in one unit. The demonstration effect of the change in the pilot unit can reinforce management's strategy to extend the innovation or change program to other units (Ullrich and Wiedland, 1980).

In the choice of a solution or change strategy, the organization faces a dilemma. On the one hand, the organization can decide to be cautious and choose a modest change that does not depart noticeably from the previous product, service, activity, or technical program of the organization. On the other hand, the organization can attempt to solve its problem(s) by a solution that represents a radical departure from previous organizational activities. The disadvantage of the former strategy is that the change may not solve the problem(s), but the disadvantage of the latter strategy is that the risk may be too great and may threaten the continued existence of the organization. The greater the scope of the new program being considered, the more acute these problems become. Hage and Aiken (1970) note that there is an inevitable gamble in the development of any new activity. The elite of an organization may perceive that the technical or organizational change will meet the needs of their customers or clients, but, in fact, it may not. Therefore, the elite must decide how innovative they want to be when selecting a solution for the improvement of organizational performances.

Initiation Stage: Managerial Strategy

1. Develop scenario regarding the need for change in policies, programs, and procedures.
2. Take an inventory of the objective attributes of the innovation.
3. Develop contingency plans for the innovation.
4. Assess organizational culture for facilitating change.
5. Design incentive program for change.
6. Plan performance feedback on change program.

9.4.5 The Implementation Stage of the Change Process (Panel B3)

Hage and Aiken (1970) note that at this stage the disequilibrium of the organization is at its greatest. This is because rules, regulations, procedures, structures, and habits that combine to produce order, predictability, standardization, and stability may have to change. Some changes may also alter the power and status of organization members or simply encroach on the turf of influentials in the organization. The incidence of conflict can be expected to be quite high. The plans designed for changing the system will not anticipate all unintended consequences. The implementation plan is a simple strategy for muddling through, since the effects of the change cannot be predicted fully. Some become evident after the implementation process has stabilized, especially in technical fields or in the process of technology transfer.

Ullrich and Wiedland (1980) observe that management cannot complete plans for change. They lack a science of organizational change and detailed information residing at lower levels of the organization. The responses to the change cannot be predicted. The change process must therefore proceed gradually, using feedback to correct mistakes and accommodate unforeseen contingencies. The changes in rules, resource allocation, and authority for facilitating implementation of a new program may create conflict with other units. Management may either use a unilateral or participative approach in implementing the change. Another dilemma is whether a highly structured or unstructured change effort should be used. Management can plan changes in considerable detail and then proceed to implement its plan systematically. On the other hand, it may decide that planning is unfeasible, in which case change will be implemented and coordinated by feedback. It is argued that the advantage of planning is that contingencies are foreseen, costs projected, and pitfalls perhaps avoided. Management can determine its progress by comparing actual to planned results. Anxieties of those responsible for implementation may be reduced to the extent that the direction of the effort is foreseen. Plans, like rules, may reduce the potential for interpersonal conflict. However, flexibility is lost in the process—executing the plan may become more critical than achieving the objectives of the change program (Hage and Aiken, 1970).

Not matter how much the elite may study the situation, a plan is unlikely to consider all the potential sources of discontinuity between the new program and the existing organizational structure. Organizations can have operations researchers carefully design a technical system for handling a new program, but the human element is seldom adequately considered in the implementation of a new product or service. There will be mistakes that will have to be corrected. Alteration of the existing structure will also create conflicts and tensions among the members of the organization Hage and Aiken (1970).

Organizations and networks of organizations are arenas in which coalitions representing opposing organizational interests meet in conflict. The process of organizational decision making reflects this conflict of interest, and maneuvering for position and power. The basis upon which decisions are likely to be made is not rationality, but rather the reflection of the interests of the dominant coalition. The goal of the organization or organizational network is not monolithic. Goals are multiple and often contradictory, reflecting the dynamics of the conflict of coalitions and the interests of the organizational actors who comprise those coalitions. This process is one of continuous conflict, with goals and programs constantly in the process of negotiation.

Implementation Stage: Managerial Strategy

1. Develop the ability of subordinates to plan and organize strategic resources for implementing change.
2. Diagnose organizational climate that predisposes the organization toward change.
3. Clarify the *path* of the change program.
4. Design systems, procedures, and rules for implementing change.
5. Identify behavioral constraints (attitude, perception, values) on the change program.
6. Identify structural constraints (levels of authority, degree of specialization, procedures, etc.) on the change program.
7. Diagnose the "political" constraints (power, conflict, status) on the change program.
8. Design organizational structure to cope with the change.
9. Manage conflict and stress generated by change.
10. Design change techniques for the "negotiation" of constraints on the change initiative.

9.4.6 The Routinization Stage of the Change Process (Panel B4)

Regardless of the criteria used, at some point the elite must make a decision either to retain or reject the new program. If they decide to keep the innovation, a period of consolidation is begun. What was a new activity becomes integrated into the existing structure. If the innovation is abandoned, the organizational structure may revert to the pattern that existed prior to the initiation stage. If the program is continued, rules and regulations must be developed, which may include not only the writing of a rules manual but perhaps a detailed job description for each of the new positions involved in the new activity. The decision to standardize a program marks the beginning of the routinization stage (Hage and Aiken, 1970).

Closely related to the problem of establishing rules and procedures for the program is the problem of defining a proper role for the new program in the existing structure. The positions associated with the program must be fitted into the existing power structure and reward system. The program must be articulated with other programs, which means establishing definite procedures and policies in order to articulate these activities with other parts of the organization. The stabilization of the power and reward structures tends to result in the reduction of internal conflicts. Every occupant of a position now knows his place in the organizational chart, so that much of the ambiguity of the implementation stage has been resolved (Hage and Aiken, 1970).

Routinization Stage: Managerial Strategy

1. Design information systems for reducing uncertainty and complexity.
2. Establish rules, procedures, and programs.
3. Integrate the innovation with relevant management systems.
4. Design operating procedures and manuals.
5. Establish criteria for evaluating the performance of the innovation.
6. Establish feedback and control systems to monitor the changes produced by the innovation.
7. Conduct culture audit and implement culture change mechanisms.

9.4.7 The Domains of an Organization and Change (Panel C)

The model of the organization involves the behavioral, structural, process, management systems, technical, and environmental domains (Figure 9.5). They all interact to determine organizational behavior and performance. The model also serves as a framework for managing technological change and to monitor the potential elements within each domain of the organization that change may impact on. Behavior can best be explained in terms of a continuous reciprocal interaction among these domains. Particularly in the field of organizational behavior, which is still developing and has a changing theoretical base, such a model is extremely important for structuring study and further development. If the organization is viewed as a system, these domains are essentially subsystems functioning with some degree of interdependence and in unity to achieve the goals of the system. The domains are not mutually exclusive. The field of organizational behavior embraces them as complementary. There is no need to choose one level of analysis and exclude others. Luthans (1986) The managers in business, government, health-care, educa-

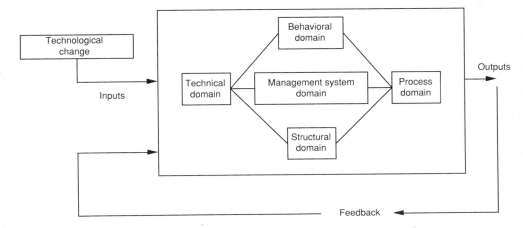

Figure 9.5. The domains of an organization.

tion, and religious organizations have shared problems that require an interdisciplinary approach. An organization's domain is the field of action, internal and external to the organization. There are two segments in the domain, the objective field and perceptual field of action. We will present an overview of each domain of the organization and then discuss in detail some of the major domains in terms of their impact on technological change.

The Environmental Domain (Panel C1). The domain is a set of environmental elements that the organization interacts with to accomplish organization goals. The environment is comprised of several sectors, such as human resources, financial resources, the market, technology, economic conditions, industry, raw materials, government, and the social culture in which the organization functions (Daft, 1989). There are two essential ways in which the environment influences organizations: the need for information and the need for resources.

A critical factor contributing to the crisis of the organization is failure to adapt to changing environments (markets, technology, cultures, etc.). An environmental domain may be stable or changing. The magnitude and rate of change and the abrupt shift of environmental elements can be potential stressors. Environmental complexity refers to heterogeneity or the number of diverse external elements that interact with and influence the organization. These two dimensions of the environment, change and complexity, may produce degrees of uncertainty. Organizations must cope with and manage uncertainty in order to be effective. Uncertainty increases the risk of failure for organizational actions and makes it difficult to compute costs and probabilities associated with decision alternatives, and can therefore be stressful.

The Behavioral Domain (Panel C4). Organizational behavior is unique in recognizing the importance of studying behavior at each of three levels: (1) the individual, (2) the group, and (3) the formal organization. We need to integrate our knowledge about behavior across these levels. The focal point of this domain is the individual and his characteristics that impact on organizational performance. There is also the necessity in this domain to study small-group behavior in terms of norms, leadership, and cohesion. The group's personality is called syntality. Work groups manifest behavior and characteristics apart from and beyond the sum of the personal attributes of the individuals making up the group. The group thinks. The group sets goals. The group behaves. The group acts. It must be studied as an entity, since through cohesiveness, group think, and dysfunctional norms it can generate stress for both the individual and the organization (Luthans, 1986). The culture of the organization is also anchored to the behavioral domain, since it involves patterns of behavior, human values, myths, ideologies and belief systems, and the symbolization of ethical values.

The behavioral domain involves attitudes, personality determinants, perception, leadership behavior, values, and group behavior, among other factors. It is essentially the cluster of behavioral patterns in the organization. While recognizing that behavior is influenced by the structural, process, and management systems domain, we are, in this domain, focusing on the elements of behavior that generate conditions for potential barriers to change; such elements include a lack of consistency in leader behavior, lack of integrity in managerial action, stereotyping of workers, interpersonal conflict episodes, and so on. Cognitive dissonance may be produced by some of the behavioral components.

The Management Systems Domain (Panel C5). Management is a distinct subsystem responsible for directing the other subsystems of the organization, such as production, R&D, maintenance, and boundary spanning. To perform this function effectively, management provides direction, strategy, goals, and policies for the entire organization. It coordinates the information flow, resource flow, and work flow among differentiated units of the organization through management systems. A major system is the management information system (MIS), an organized set of processes that provides information to managers to support the operations and decision making within an organization to manage change. Most management systems involve some sort of transformation process. The processing of information in all management systems implies that they are all goal oriented, regulated, and differentiated in their functions. A related system is the decision support system (DSS), which is an MIS designed exclusively to support management decision making. New forms of technology enable the manager to interact directly with the data base to ask questions and get responses. Personal computers and automated offices represent decision support systems that enable managers to use organizational data bases for personal decision making and for managing change. Another strategic system is the management control system, which is composed of the formal planning, data-gathering, and transmission systems that provide management with information about organizational performance. Control information includes information about targets, activity measurement, and feedback. The main problem with management systems is their potential to become bureaucratic, arbitrary, and irrelevant over time. Because each organizational domain may design systems to management elements of that domain, the management systems domain is pervasive. Even the behavioral domain, which is not as structured and concrete as the other domains, has designed many systems such as performance appraisal systems, career-planning systems, incentive systems, selection systems, manpower planning systems, management by objectives (MBO) systems, and so on. The tendency of systems to be complex, rule oriented, technical, regulatory, specialized, and impersonal may produce potential barriers to change in the organization.

The formal management systems in every organization influence actual behavior patterns by (1) setting the behavior expected of employees, (2) measuring and rewarding expected behavior, and (3) controlling and enforcing expected behavior. Such systems are effective when they facilitate required behavior patterns *and* meet the needs of the people involved; in other words, when there is consistency between the organizational tasks, the design, and the people. If there are major inconsistencies, some combination of changes in management systems is needed to bring these three basic dimensions of task, design, and people back into alignment. This lack of integration can be a potential barrier to change. The interface with computerized management systems can be alienating for the manager because it may fragment his task, abilities, aptitudes, and knowledge. The standardization of work flows, predictability of outcomes, and formalization of work procedures through standard operating procedures can enhance the impersonality of the system and thus serve as another set of potential barrier to change.

The Structural Domain (Panel C2). Organizations have unique characteristics in much the same way as individuals and groups do. Formal organizations can be compared on the basis of specific structural characteristics common to all organizations: hierarchy, rules and procedures, differentiation of organization units, degree of task specialization, span of control, status hierarchy, locus of decision making, degree of centralization, etc. These influence organizational performance. The organization rotates and impacts on all other domains of the organization.

The four principal components in the definition of organizational structure may also be affected by technological change (Luthans, 1986):

1. Organization structure describes the allocation of tasks and responsibilities to individuals and departments throughout the organization.
2. Organization structure designates formal reporting relationships, including the number of levels in the hierarchy and the span of control of managers and supervisors.
3. Organization structure identifies the grouping together of individuals into departments and the grouping of departments into the total organization.
4. Organizational structure includes the design of systems to ensure effective communication, coordination, and integration of effort in both vertical and horizontal directions.

The potential barriers to change in the structural domain include the variety of linking mechanisms to coordinate information flow, resource flow, and work flow and strategies for grouping organizational activities into functional, product, program, and hybrid structures. Other barriers involve, among other factors, role ambiguity, centralization of decision making, ambiguous operating procedures, organizational politics, and status hierarchy.

The Process Domain (Panel C3). This domain focuses on the vital processes that constitute the lifeline of organizations. Such processes as power, conflict, decision making, communications, change, and leadership are related to structure—both resulting from it and leading to it. No manager can expect to direct an organization effectively without an intimate familiarity with these processes. They are far more improtant than structural analysis. Each is a critical component for understanding organizations. For example, power as a process involves patterns of centralization, and conflict is a major contributor to change. Every social relationship can be viewed as involving power, with conflict either manifest or latent in every situation. Conflict may result from the exertion of power, but it is not inevitable. Another critical process is leadership, which is the exercise of power and influence to achieve organizational goals through the decision making process. The communications process, which constitutes the nerve center of the organization, is imperative to effective decision making. The interrelationship among these organizational processes is quite dynamic. Here again, change can be viewed as the consequence of the decision-making process and the leader can be seen as the manager or initiator of that change, a change that may alter the distribution of power within an organization and therefore produce a high incidence of conflict.

Some elements in the communications process, such as inadequate feedback, information overload, centralized communications networks, and information filtration, have the potential to inhibit change. Within the decision-making process, the decision-making style, goal conflicts, uncertainty of goals and decision outcomes, and value conflicts can generate potential problems in managing change. The control process involves bureaucratic control systems, rules and regulations, close supervision, the visibility of power, and limiting of discretion of decision makers. Finally, change as a major organization process can be stressful because of uncertainty, the rate and magnitude of change, change in job routines, and job skills and knowledge. The politics of managing the change, which involves change of power bases and the patterns of power and distribution within the organization, can also be a potential barriers to change. Zero-sum assumption of power in any planned change can create enormous episodes of conflict for managers.

The Technical Domain (Panel C6). The technical domain involves elements of the work flow, job design, and technology. Technology is the knowledge, tools, techniques, and actions used to transform organizational inputs into outputs. Education skills and work procedures are technologies in the transformation process. In large complex organizations, different technologies are used in different parts of the organization. Each department transforms inputs into outputs. Research and development transforms ideas into new products. Marketing transforms inventory into sales. Raw materials flow into and through the organization's production process in a logical sequence, and work activities are performed with a variety of tools, techniques, or machines. Here, the organizational structure must be designed to facilitate internal work flow. The variability of work activities, the degree of mechanization in the transformation process, and the interdependence tasks in the work flow are all examples of technology. The job design strategy, job enlargement, and job enrichment will affect the work flow, work modules, job depth and scope, and horizontal and vertical dimensions of the job and interaction patterns. These are all elements of the technical domain (Sankar, 1988).

Complexity, and variability among elements of the work flow, can produce problems through role ambiguity, conflicting role expectations, and changes in performance expectations. The greater degree of interdependence among elements of the work flow, the greater the potential for conflict. The impact of technology on the work flow and job design can potentially create stress, as experienced in the implementation of computer technology. Job enlargement and job enrichment effected by changing job protocols, routines, skills, knowledge, and work modules can be problem areas in change. Similarly, the fragmentation of work and the degree of job specialization can also produce alienation in the management of technological change (Sankar, 1991).

9.4.8 The Organizational Domains and the Implementation Stage (Panel C)

The Environmental Domain and Change (Panel C1)

PRINCIPLES FOR MANAGING CHANGE

If an innovation requires a change in the integration pattern for an organization that is well matched to its environment, the characteristics of the innovation must be changed to facilitate effective implementation.

Organizational environments have been conceived either as information or as resources. When conceived as information, the environment is easiest defined as streams of signals that are received by an organization about external activities. These signals are first perceived by organizational members who occupy boundary positions. An important role in perceiving new concepts and subsequently adopting innovations is held by individuals occupying boundary roles, known as "technological gatekeepers." They function as receivers of new ideas in the organization. Whether it is an individual who scans the technical and financial press for new developments, or a group responsible for environmental scanning, gatekeepers function to bring potential threats, opportunities, and new developments to the attention of key organizational decision makers. The propensity of organizations to adopt innovations depends on the organization's structure, its personnel, individual characteristics, organizational environment relations, and the type of innovation involved: administrative (organizational structure), goal/value-centered, or technical (product-, process-, client-centered) (Child, 1984).

The interaction between the organization and its environment is crucial to the innovation process. Many technical and organizational changes in an organization are borne as a result of changes in the organization's environment. The organization's environment is important in understanding innovation in three specific ways: as an early signal of performance problems, as a source of technical information, and as a pool of financial resources. Successful innovators must actively sense or scan their environments and recognize the potential performance gaps. Failure in a strategic vision in recognizing external needs and demands is perhaps the primary limitation on a firm's effectiveness in innovation. Organizations vary widely in their abilities to recognize performance problems and exploit environmental resources and technical information.

The environment is a source of constraints, contigencies, problems, and opportunities that affect the internal operations (structure and processes) of the organization. The organization, however, is unlikely to adapt effectively to these without reliable information. If reliable information about business expansion and contraction, technological change, changes in legislation, and the like is not available to the organization, it cannot plan effective responses to them. At any given time, the information coming from the environment is filtered and processed through the mechanisms developed by the organization for this purpose. These may range from technological forecasting through market research activity to tuning in to the business grapevine. The filtered information reaches key decision makers. If it is of acceptable quality and reliability, decisions will be taken that affect the organization's strategy, structure, and information systems. If it is not, the information-gathering and -processing mechanisms may be modified. Thus, how the environment affects the organization depends on two crucial factors: the kind and quality of environmental information reaching decision makers, and the interpretation and use of the information by them. How the information is interpreted and used depends in turn on the goals and attitudes of the decision makers.

Systems Differentiation, Integration, and Environment. The differentiation–integration model (Figure 9.6) incorporates the varying degrees of uncertainty, differentiation, and integration across organizations. The figure suggests a contingency approach to organizational design, emphasizing the organization and the environment. In highly heterogeneous markets and changing technologies, effective organizations tend to become more differentiated than organizations in more homogeneous and stable technological environments. Problems of integrating diverse activities in complex organizations can also vary from high to low, depending on the degree of coordination needed to achieve organizational objectives. By focusing on differentiation in (1) a changing and (2) a stable environment, and examining the modes of integration required by the organization's environment, we can see the

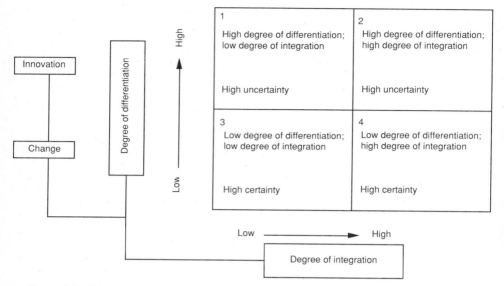

Figure 9.6. Differentiation–integration grid and change. (From Hellrigel et al., 1986. Reprinted with permission from West Publishing Company.)

possible ways in which organizations can vary in terms of these two variables (Hage and Aiken, 1970).

Increasing levels of structural differentiation provide for a greater capacity to generate nonroutine knowledge (Duncan, 1976). The type of knowledge-generating structure adopted should reflect the extent of knowledge demanded by the targeted innovation. Organizations that do not sufficiently differentiate structurally are likely to develop uncreative proposals (Kazanjian and Drazin, 1986). This degree of differentiation also develops linkages among the subsystems so that change can be monitored effectively. While differentiation may create variety in learning styles, frames of reference, perceptual sets, and goal orientations, the potential for conflict is also a condition of differentiation. This conflict may be positive in that it diffuses ideas and information among subsystems, which is conducive to creativity.

We have adapted the differentiation–integration grid to show that the organizational innovation may create changes in the differentiation–integration pattern of the organization. In this case, a number of organizational characteristics or structural features will have to be changed, since the organization will now have to cope with a high degree of perceived uncertainty produced by the change. The innovation may cause the organization to further subdivide or segment its functions, activities, and programs, or to add new programs and tasks, especially if it is a strategic innovation like MIS or a technical innovation. Here, the degree of differentiation for this organization has changed. The problem facing management is to decide on the degree of integration—will it be high or low? If the innovation creates task uncertainty, there will be a need for high information processing, and therefore the type of integration mechanisms will be more flexible, such as lateral relations, mutual adjustment, feedback, and plans. However, if the innovation does not create much task uncertainty, the information processing requirements are low and integration can be achieved through rules, standardization, hierarchy, and other control mechanisms. On the other hand the innovation may simply produce incremental change in the degree of differentiation, an extension of programs or tasks that may require a high or low degree of integration. The certainty here is rather high, so the rules, hierarchy, and plans can integrate the organization's tasks and programs. However, if the innovation is perceived as dysfunctional for the organization that is matched to its environment, then changes in the characteristics of the innovation must be initiated to facilitate the possibility of effective implementation.

The environment is a major determinant of an organization's internal structure. In the differentiation–integration model, as each part of the organization strives to cope with a different part of its external environment, it demands a different mode of integration. Effective organizations operating in unstable environments develop "organic" management systems, whereas effective organizations in stable environments tend to have "mechanistic" management systems. The organic system will facilitate the evaluation and initiation stages of the change process because of the degree of flexibility in organizational rules, regulations, and information flow and control mechanisms. The mechanistic system will facilitate the implementation and routinization stages of the change process because of the emphasis it places on rules, regulations, procedures, control, and the hierarchy of authority.

The Structural Domain and Change (Panel C2)

PRINCIPLE FOR MANAGING CHANGE

The structural characteristics of an organization will have different effects on the stages of the change process. One set of characteristics may facilitate the initiation of change, while others may facilitate the implementation of change.

Because most changes are implemented in bureaucratic structures and because bureaucracy is cited as one of the principal factors that resists change, we will examine the role of the bureaucratic structure in initiating and implementing change.

The impact of advanced technology on organizational structure is noted by Lund and Hansen (1986). They observe that responsibility for operating decisions will be moved to the lowest possible level in the organization. This lowest level will frequently be that of a machine operator or technician who has greater familiarity with the process and is more consistently available to make the choices.

The value and complexity of the manufacturing process will increase. These shifts will give different meaning to the concepts of responsibility and authority. The importance of a position in the organization will no longer be measured in terms of numbers of people supervised. It may be expressed in terms of the value of the investment supervised, the number of decisions that must be made, the value of the output, or the level of complexity of the system being managed. Whatever becomes the new basis for measuring responsibility also becomes the basis for performance evaluation and for reward.

Furthermore, Lund and Hansen (1986) note that since operating decisions will move downward, the upper levels of management will be freed for strategic planning and decision making. Assistance from computer information systems will be available at these levels, making large information-gathering staffs less necessary. Consequently, it is expected that there will be a thinning out of staff groups supporting corporate level executives who have become computer-communications literate. These managers will be able to reach into the communications net for information, bypassing their staffs and line managers.

Design of organizations is open to options made possible by technology. Project teams are not locked into one conceptual scheme, but are free to create structures that fit with their mode of operation and corporate philosophy. Computer-assisted telecommunications aids can be tailored to make a given organizational scheme work. In this respect, the design of organization and the design of jobs have much in common.

Bureaucratic Structure (Cluster A) The Managerial Power Structure and the Initiation Stage. Figure 9.7 speaks to the discussion in the following sections on bureaucratic structure.

PRINCIPLE FOR MANAGING CHANGE

Where the bureaucratic structure (Cluster A) is emphasized, conditions within the organization will interact to inhibit the initiation stage of the change process.

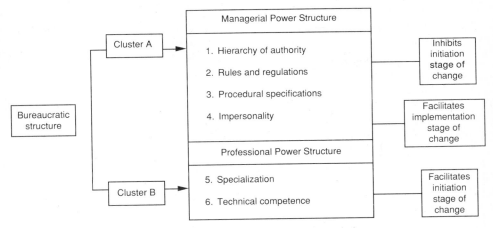

Figure 9.7. Bureaucratic structure, power, and change.

For example, an emphasis on hierarchical authority, procedural specifications, behavioral rules, and impersonality may lead to (1) a crystallized authority system with rigid role expectations, (2) communications systems with limited feedback processes, and (3) standardization of organizational activities. These structural characteristics will constrain the decision-making process and hence the change orientation of the decision outcomes, since change is a consequence of decision making. That is, the manner in which the decision issue is defined, its parameters delimited, and alternatives evaluated will be constrained by an inventory of rules and procedures. This condition becomes very much exaggerated if the culture of the organization emphasizes a zero-sum conception of power, whereby power is perceived as a fixed amount in a system so that any change is viewed as a reduction in power for X and a gain for Y. The sources of dysfunctions that are derived from the bureaucratic structure involve (1) the internalization of rules and procedures, (2) the increased use of categorization as a decision-making technique, (3) the use of management strategies that enhance power visibility, (4) the programming and standardization of organizational activities, and (5) product specifications and pacing rules in the organization.

These structural characteristics influence the problem definition process and its outcome. The actual outcome is determined by the manner in which the decision-making process is constrained, the degree of specific rules and procedures, the hierarchy of authority, and the degree of impersonality in decision making. When these dimensions are highly structured, channels of information and amount of information available within the unit are restricted. When dealing with high environmental uncertainty and change, a very high degree of emphasis on the hierarchy of authority can cause decision-unit members to adhere to the specified channels of communication and selectively to feed back only positive information. Strict emphasis on rigid rules and procedures may hinder the unit from seeking new sources of information when new information inputs are required to adapt to the uncertainty of the environment, which may not have been anticipated when the rules and procedures were initially developed.

Structure should encourage innovation by providing for the pooling of organizational resources and for an appropriate communications and information system. The organizational structure should thus facilitate effective operation of organizational sensors and boundary-spanning units. In particular, the structural linkage between sensors and decision authority centers needs to be such that important information reaches the decision centers so that innovative responses occur.

Bureaucratic Structure (Cluster B) The Professional Power Structure and the Initiation Stage

---

PRINCIPLE FOR MANAGING CHANGE

Where bureaucratic structure (Cluster B) is emphasized, conditions within the organization will interact to facilitate the initiation stage of the change process.

Although the managerial hierarchy diagnoses, legislates, monitors, and evaluates the need for change, it must rationally utilize its resources in the organization for the adoption of innovations involving cognitive, technical, and administrative competencies of subordinates. Technical competence is a potential source of influence and power. Authority resides in expertise. This neutralizes the maximum application of hierarchical authority in organizational activities. Where roles, task structures, technology, means–ends analysis, and information are complex, vague, and nonprogrammed, there is a greater possibility that the professional segment or cluster of a bureaucracy will be emphasized. The innovative requirements of organizations and the nonprogrammed content of strategic decisions also make this strategic suspension of hierarchical authority possible. All of these factors may interact to create sources of discretion or potential bases of power that the expert or professional manager can convert into power assets. The emphasis on specialization and technical competence, elements of the professional power structure, will neutralize the maximum application of hierarchical authority and enable decision making that may be conducive to the initiation of change by professionals in the organization. The existence of parallel problem-solving channels for specialists alongsde the bureaucratic hierarchical communications channels, and a dual ladder for the upward mobility of professionals who do not want to assume positions in the managerial hierarchy and who are keen to maintain their functional authority, also neutralize the strict application of hierarchical authority to professional decision making. While the initiation of change or suggestions regarding the adoption of an innovative program may come from the professional sector, it is the elite in the managerial hierarchy that will decide on the initiation of change.

Managerial values are often ignored in studies of organizational structure. However, recent research shows that the values of an organization's elite can be a more important cause of innovation than organization structure (Robey, 1986). An elite is defined as those members who regularly consult on critical policy decisions and is not restricted to formal boards or top management groups. Values toward change, information, and new programs were positively related to the adoption and use of new programs in sixteen health and welfare agencies; structural variables were also related to innovation, but not as strongly as elite values. Elite values contribute to innovation because top managers have the power to introduce change and this power is a prerequisite of successful innovation (Hage and Aiken 1970). Elites in the managerial hierarchy function as integrators at the top level of an organization. While professionals at lower levels work out the details of an innovation, a committee of top executives has neither the time nor the expertise to scrutinize such matters. However, the elite does resolve basic questions of goals and priorities with respect to innovation. Subgoals of major functional areas can be evaluated in such a setting and basic conflicts settled prior to implementation.

Bureaucratic Structure and the Implementation Stage

PRINCIPLE FOR MANAGING CHANGE

A bureaucratic structure facilitates the implementation of change through the processes of formalization and centralization. Formalization provides the rules, procedures, and controls for implementation, while centralization provides the authority to mobilize resource flow, information flow, and work flow—the management infrastructure for the innovation.

The hierarchy of authority manifests itself in rules and regulations—that is, rules are the bearers of organizational authority, communications, and control strategies. The managerial heirarchy also allocates scarce resources such as conditions for self-actualization, autonomy, discretion, status/prestige, and expertise in decision making. The hierarchy can therefore manipulate motivators, mobilize resources, expertise, information, etc., and initiate a strategic application of rules, regulations, and discretionary powers for facilitating the adoption of innovations that it sponsors or initiates. The managerial hierarchy, by virtue of its allocation of "scarce" resources, creates a dependency orientation and a climate for change facilitation behavior as opposed to change initiation behavior. Strategic change efforts in such a setting and climate reside in the power elite in the hierarchy, while incremental change efforts reside in the lower levels of the organization. Even this type of change effort may be monitored by the hierarchy. The managerial power structure, therefore, can facilitate the implementation of innovations by virtue of its emphasis on the hierarchy of authority, organizational rules and regulations, and control systems. The management infrastructure that implementation requires is controlled by the managerial heirarchy, which can mobilize all power sources within the system to permit the adoption of the innovation. Thus, different structures are appropriate for different stages of innovation, and this presents a basic challenge to the manager designing an organization to be innovative. Conventional organizational studies conclude that bureaucracy inhibits change. This inference can be partially explained by the vague and ambiguous concept of change that these studies employ. The parameters of change are rarely delimited. Even the conclusion that the managerial power structure—the central notions of a bureaucracy—predisposes the organization toward stability or rigidity is erroneous, since there are types and categories of change that are compatible with stability as well as necessary for the maintenance of stability. The success of an innovation is thus due in large part to the structural design that facilitates each phase of the process. Where the professionalization segment of the bureaucratic structure is emphasized, the generation of ideas for innovation will be facilitated. On the other hand, where the hierarchical managerial segment of the bureaucratic structure is emphasized, the implementation phase will be facilitated. Attention to these structural conditions for innovation can increase the organization's capacity for innovation.

Organizational Structure: Managerial Strategy

Identify the structural constraints:
1. Degree of job specialization required for the innovation.
2. Levels of the management hierarchy affected by the innovation, such as chains of command.
3. System procedures, regulations, and rules.
4. Sources of structural stress, such as role ambiguity, integration–differentiation mix, and so on.
5. Power and influence patterns affected by the innovation.
6. Management infrastructure, such as the management information system, organizational programs, and systems and procedures.
7. Degree and type of departmentalization—span of control.
8. Organizational rules and procedures.

9. Type of centralization–decentralization mix.
10. Degree of bureaucracy of management structure.

The Process Domain and Change (Panel C3). There is a dynamic interaction between structure and organizational processes. Structure may be a major determinant of organizational change. A change in structure will alter basic organizational processes such as decision making, communications, and power. In some cases, the change may affect the basic organizational processes rather than the structure of the organization. An innovation will generally require changes in organizational processes such as changes in information flow, resource flow, and work flow. The job design, managerial style, and decision-making strategy may also have to be changed to allow the implementation of the innovation. In such cases, the innovation may be perceived as disruptive or dysfunctional and may produce some degree of resistance. However, the more these changes in organizational processes can be planned and managed in the implementation stage, the more the resistance will be reduced. Planning reduces the degree of uncertainty produced by the change, which is generally a source of stress for organizational members.

PRINCIPLE FOR MANAGING CHANGE

Where the degree of "fit," or congruence, between the innovation and basic organizational processes may facilitate the implementation of the innovation, it may also serve to inhibit more innovative strategies for implementation.

Organizational Process: Managerial Strategy

Identify the process constraints on the innovation:
1. Critical path of the change.
2. Feedback of information on innovation.
3. Information flow required for innovation.
4. Job design and work flow pattern affected.
5. Decision rules and strategies consistent with the innovation.
6. Role expectations and change strategy.
7. Control mechanisms for monitoring change.
8. Methods for implementing change.

Lund and Hansen (1986) observe that the computer-based systems will need just-in-time management responses. Just-in-time management will be made possible by computer integration and analysis of information on a continuing basis. Managers can be brought up to date in a few minutes by accessing information on their computer terminals, thereby avoiding the time-consuming practice of delegating a search for information to a crew of assistants. Computer-based expert systems, teleconferencing, and local area information networks will aid the decision-making process.

These researchers (Lund and Hansen, 1986) further note that machine integration will be paralleled by organizational integration. The interdependencies among materials availability, product quality, delivery schedules, and utilization of human and machine resources are very strong. To obtain fast responses, single individuals or integrated teams will be asked to make the decisions. The need for the professional specialist at the operating level is likely to become a thing of the past.

Organizational Process: Power and Change: The Politics of Implementing Change (Panel C3). Any decision within organizations is reached and implemented through a political process. Politics is about the use of power, and decisions are a formalization of that use, which will have to be reached through negotiation and compromise when power is spread among several parties. When the decision involves a significant change, the political process leading up to it is likely to be the more active. It is not too difficult to identify the probable major sources of resistance to a proposed change if one understands the ideology and perceived interests of the groups concerned and if one can estimate their awareness of their power to influence the change and their propensity to employ it (Child, 1984).

> The area of power in organizations according to Thomas and Bennis (1972), is one where a contingency model has been relatively lacking in the design of strategies for managing change and conflict. In many theories of organizational development, assumptions about the relationship of power to change and conflict appear at best ambivalent, at worse deficient . . . But the capacity to conceive change . . . in terms of the dynamics of power is essential to the effective management of these issues.

An organization can be described as a pattern for the distribution of power and also a pattern for the flow of information. The quickest way to alter the distribution of power in an organization is to change the characteristics of the information flow; that is, to change the structure of the system that regulates the information flow. Power, decision making, and information are distributed as a result of the design of the organization. Where the innovation alters the information flow, it will create changes in other sources of power within the organization. It may do so by creating areas of uncertainty in information processing that then have the potential to be a source of power for the units managing the uncertainty.

The concept that power resides primarily in the hierarchy simply ignores the sources of discretion in organizations as potential power assets (Figure 9.8). It also overlooks the fact that a source of power is knowledge that challenges hierarchical definitions of authority and role. The decision-making

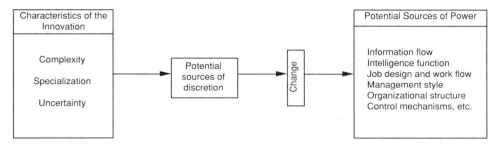

Figure 9.8. Discretion, power, and change.

process involving an innovation and its means–ends analysis are subject to varying degrees of ambiguity, complexity, and specialization. The degree of uncertainty and the expertise required to interpret, apply, and implement the innovation is a source of power for those who manage the uncertainty. The characteristics of the innovation may create change in the resources of the organization: knowledge, intelligence, expertise, and technical facilities— the knowledge infrastructure of the organization. As the uncertainty of the organization's task in managing the innovation increases, coordination increasingly takes place by specifying outputs, goals, or targets. Instead of specifying behaviors to be enacted, the organization undertakes processes to set goals to be achieved and the employees select behaviors that lead to goal accomplishment. Planning reduces the amount of information processing in the hierarchy by increasing the amount of discretion exercised at lower level. The centralization of power in the organizational hierarchy is therefore compromised by this source of discretion in the characteristics of the innovation and its program goals.

The power dimension of discretion is articulated by Crozier (1964), who argues that since it is impossible to eliminate all sources of uncertainty by multiplying impersonal rules and developing centralization, a few areas of uncertainty will remain. Parallel power relationships will develop around these areas. In a system of action where nearly everything is predictable, individuals or groups who control a source of uncertainty have at their disposal a significant amount of power over those whose situations are affected by uncertainty. Moreover, their strategic position is all the stronger because sources of uncertainty are few. Paradoxically, in a bureaucratic system of organization, parallel power increases in direct ratio to its rarity. The position of an expert is much stronger in an organization in which everything is controlled and regulated. This observation is reinforced by the fact that the more sources of uncertainty of contingencies for the organization there are, the more bases there are for power and the larger the number of political positions in the organization. Also, the more a subunit copes with uncertainty, the greater its power within the organization.

Furthermore, if power expands in areas of uncertainty and contracts in areas of routinization, it may be argued that sources of discretion constitute potential power bases in organizations. The expansion of power in an organization is triggered not only by the sources of discretion that prevail in the organization, but also by the type of assumption of "power conditions" that the managerial hierarchy entertains. A "zero-sum" or "positive-sum" assumption of power by the manager will have different influences on the sources of discretion and on the conversion of these sources of discretion into power assets.

PRINCIPLE FOR MANAGING CHANGE

Where the organizational change stimulates innovative decision making, the strategic processes of the organization will be affected and the more difficult the implementation of the innovation. A fundamental strategic resource in the organization is power. The innovation may change the sources of power and the pattern of power distribution in the organization. Hence, the politics of the implementation emerges as a major strategy for managing change.

Since change is a consequence of decisions—that is, organizational decisions may be considered as ways by which organizations adapt to change—it may be argued that the change orientation of an organization (innovative–incremental) will be partially determined by the type of decision strategy adopted by the organization. The structure of the organization and the pattern of power distribution will influence what type of decision strategy is adopted. Depending on the characteristics of an innovation, the decision strategy may trigger incremental or innovative changes. Organizational decision making is never a process in which means are allotted according to a systemic plan or policy; it is rather a process in which a significant role is played by the relative powers of the supporters of various organizational goals and by the power of the decision-making elites of the managerial power structure. Organizational decision making is not merely a cognitive process that balances goals and means, but also a political process that balances various power vectors. Each means–ends connection has, in addition to its relative merits, a different political weight. In short, values, knowledge, and power interact in the making of organizational decisions and in the management of change (Etzioni 1968).

> As Etzioni (1968) argues, it might be noted that power is central to decision making in the form of an information input, and hence, that the power analysis is reducible to an analysis of the knowledge process.

While the objectives of the innovation or change program may be quite logical and consistent with organizational goals, the implementation of the innovation may create changes in the sources of power within the organization, and therefore the logic of the innovation is compromised as members shift their focus on the effects produced by the innovation on their orbit of power and influence. In addition, an innovation may create new sources and bases of power within the organization. It may require new expertise and specialization, it may integrate organizational units, or it may change the information systems of the organization, job design, and uncertainty, which must be managed. The politics of implementing change is primarily the use of power to achieve the objectives of the change program or innovation. The uncertainty and conflict produced by the innovation set the stage for political activities. This is a legitimate domain of activity for the change agent or manager, since power is a basic organizational reality and the power structure can be managed either to encourage or inhibit change. Politics becomes more apparent when the change program is strategic in its implications. Politics is simply an implementation strategy, and a viable one, that must be effectively utilized to achieve the objectives of the change effort. It can be a formidable barrier to implementation, and, unless it is taken into account, it will neutralize the other change techniques—behavioral, logical, technical—that are used to educate or persuade people to accept the innovation.

Organizational Power: Managerial Strategy

Diagnose the politics of implementation:
1. Actual sources of discretion in the organization.
2. Potential sources of discretion created by the innovation.
3. Sources and bases of power in the organization.
4. Sources of power affected by the innovation.
5. Sources of leader power that can be activated for implementation.
6. Zero-sum and positive-sum conditions of power in the organization.
7. Sources of subordinate power.
8. Sources of work design power, such as complexity, variability, and interdependence of work flow.
9. Sources of organizational structural power, such as rules, regulations, and organizational hierarchy.
10. Power acquisition strategies available to the change agent for implementation.

The Behavioral Domain and Change (Panel C4)

PRINCIPLE FOR MANAGING CHANGE

The effective implementation of change is influenced by (1) the attitudes, values, and behavioral patterns of organizational members; and (2) the type of strategies designed by managers to influence changes in these individual characteristics.

One of the principal objectives of organizational change is to achieve changes in the attitudes, styles, and behavioral patterns of individuals within the organization. An organization may not be able to change its adaptation strategy for reacting to its relevant environment unless the members in the organization behave differently in their relationships with one another and to their jobs. Organizations survive, grow, decline, or fail because people make decisions. In this sense, behavioral change is involved in all organizational change efforts. Any organizational change effort not taking into account the necessity for individual behavioral change is likely to prove unnecessarily difficult or, in some cases, fail completely. The role of behavioral attributes such as perception, learning, and personality in the change process is noted by Szilagyi and Wallace (1983).

> Individuals consciously interact with their organizations in a process of mutual change and adjustment. People working in organizations are constantly bombarded by incoming stimuli—some are physical in nature, some are behavioral ones reflected in job titles, status, and power (the reward system of the organization), some are complex information stimuli via MIS, and so on. Perception is defined as the process by which individuals attend to incoming stimuli and organize or interpret such stimuli into a message that in turn influences behavior. The ability to perceive allows people to make continual adjustments to their environments (including work organization) through their behavior. Hence, we characterize perception as an individual process that enables short-term changes in behavior as a result of interaction with organizational environments. In addition to short-term changes, we know that individuals adapt their behavior in rather stable ways over the long run. This process is known as learning and is indicated by the feedback and behavioral change. Note the similarities among the concepts of motives, personality, perception, and learning. First, all take place within the mind; none is tangible or amenable to direct observation by another person. Second, all four processes are determinants of behavior. Finally, perception and learning are proposed as processes that can change both motives and personality. Managers easily explain behavior by referring to these four behavioral processes. In any change effort, the manager must focus on these characteristics.

Motivation and Change. Expectancy theory and motivation provide a more complex model of the human being for managers to work with in the management of change. At the same time, it is a model that holds promise for the more effective motivation of individuals and the more effective design of organizational systems. It implies, however, the need for more thorough diagnosis by the manager to determine (1) relevant forces in the individual and (2) relevant forces in the environment, both of which combine to motivate different kinds of behavior. Following diagnosis, the model implies a need to act—to develop a system of pay, promotion, job assignment, group structures, supervision, etc., to bring about effective motivation by providing different outcomes for different individuals; performance of groups is a critical issue in effectively changing organizations. If a manager is to influence work behavior, attitude, and performance, he must have an understanding of the factors influencing individual and group motivation. Managers can use models (e.g., behavior modification, expectancy theory) to understand the nature of behavior and build more effective organizations in the management of the change process. A manager must understand the nature of the individual's psychological contract with the organization. That is, the individual expects certain rewards in return for meeting the organization's expectations. As with needs, expectations are developed and later met or frustrated through an interaction between the individual and the organization. A manager desiring to maintain a viable psychological contract between the organization and its employees must understand what expectations are created or met by personnel policies, management practices, and organizational arrangements, and how changes in these may affect the fulfillment of these expectations. The expectancy model of motivation will help a manager interpret worker's needs and expectations, their valence for certain outcomes, and the extent to which organizational processes, leadership, job design, and organizational culture affect their valence. These are some of the variables that a manager must interpret to motivate workers to adopt change in organization and management practices.

Organizational Behavior: Managerial Strategy

Develop the motivational function for managing change by
1. Specifying concrete behavioral changes or targets.
2. Recognizing and mobilizing group members' needs for outcomes over which the manager has some control.
3. Interacting with team members to help them to clarify their expectancies, so that they will have a better grasp of the probabilities of particular relations and outcomes ("If I do this and get there, what are the chances that you will (a) recognize it? (b) reward it?").
4. Providing performance feedback to subordinates involved in managing change.
5. Developing path–goal clarification of the change program, that is, design strategies and action plans, to achieve the change objectives.
6. Providing positive reinforcement to attitudes, values, and belief systems congruent with the change objectives.
7. Designing an incentive program that differentiates between active and passive change agents.
8. Giving autonomy, discretion, and authority to subordinates so that they may participate in the management of change.
9. Reducing stress barriers, such as role ambiguity, uncertainty, and complexity, generated by the change.
10. Providing ethical guidelines to manage the change process.

People will resist change if they believe it is detrimental to aspects of their work life and roles that they value. If the change breaks up informal social groups, it is likely to be seen as a threat to their status and power within the organization. A change aimed at simplifying an organizational structure, perhaps by reducing the number of heirarchical levels, will probably be seen by some as a threat to their job security and to their prospects of promotion. A change aimed at enriching the jobs of subordinates may be viewed by a manager as a threat to his authority. A reallocation of functions will be regarded with some anxiety by some senior managers as a diminution in their territorial rights within the organization. The very process of change itself may be seen as an unwelcome disturbance and interference to a well-established routine (Child, 1984).

The Technical Domain and the Management of Change

Principle for Managing Change. The effective implementation of information technology is influenced by job design principles that enhance (1) vertical job loading and horizontal job loading and by (2) strategies to manage technostress produced by the new job design protocols.

Lund and Hansen (1986) observe that the greatest challenge is in the area of job design for new technologies. Not only must manufacturing managers reverse the precepts about the nature of work held by generations of managers, but they will also have to revise their concepts about the relationships between workers and managers. When individual responsibility becomes a key skill requirement for employers, managerial behavior and attitude will have to reflect a realization of this change through organizational realignment, improved communication, and above all, trust. Some of the job design principles that Lund and Hansen recommend are (1) worker participation, (2) information to orient actions, (3) comprehensiveness, (4) social relatedness and (5) opportunity for individual growth.

Zuboff (1982) observes that when information technology reorganizes a job, it fundamentally alters the individual's relation to the task. This is a computer-mediated relationship. This means that a person accomplishes a task through the medium of information system, rather than through direct physical contact with the object of the task. With computer-mediated work, employees get feedback about the task object only as symbols through the medium of the information system. Very often, from the point of view of the worker, the object of the task seems to have disappeared "behind the screen" and into the information system. Workers express frustration in losing a direct experience of their task because it becomes more difficult to exercise judgement over it. In routine jobs, judgement often becomes lodged in the system itself.

On the other hand for more complex jobs, Zuboff notes that by creating a medium of work where imagination instead of experience-based judgement is important, information technology challenges old procedures. Judging a given task in the light of experience thus becomes less important than imagining how the task can be reorganized based on new technical capabilities. While it does seem that those who shift from conventional procedures to computer mediated work feel a degree of stress because of abstract work, it is impossible to forecast what adaptation to the abstraction of work will do to people over the long run.

> Employee involvement and expansion of the scope of jobs are becoming increasingly important, because highly automated plants have a growing amount of integration and coordination at lower levels, pushing decision-making downward . . . . Information, formerly a monopoly of managers, may become available to workers on the shop floor, enabling them to act on their own. The discretionary parts of jobs will increase and employees must diagnose problems in order to prevent system interruptions, which are extremely costly under automation. (Helfgott 1986)

Cougar and Zawacki, (1978) however, point to a phenomenon that has plagued many jobs, namely job alienation. They argue that worker alienation is a major problem in many fields today. Jobs have not kept pace with changes in our society such as in worker attitudes, aspirations, and values. These old signs of job dissatisfaction are emerging in the computer field, too.

> We are emulating the industrial engineers of the 1940s, who kept chipping away at each job on the auto assembly line until it was splintered into the ultimate level of specialization. We seem to be concentrating just as fervently on fragmenting the jobs of analysis and programming. The chief programmer team concept is the latest in the long list of moves for enhancing specialization. The analysis/programming task now is fragmented into the elements performed by the chief programmer, the moderator, the librarian, the recorder, etc. Perhaps jobs should be *enlarged* in scope, rather than reduced.

The Technical Domain: Managerial Audit for Change

Audit changes in the technical domain produced by computer technology.
(1) Monitor changes on the degree of complexity of the workflow.
(2) Specify changes in job depth and job scope.
(3) Determine degree of workflow and task uncertainty.
(4) Identify the degree of change in workflow interdependence.
(5) Determine degree of change in input uncertainty, conversion uncertainty, and output uncertainty.
(6) Identify changes produced by computer technology in core job dimensions:
    (a) skill variety—tasks that challenge the individual's repertoire of skills and abilities.

(b) task identity, that is, completing a whole and identifiable piece of work.
(c) task autonomy, that is, experienced responsibility for outcomes of the job.
(d) feedback, that is, knowledge of results of the job,
(e) task significance, experience meaningfulness on the job.
(7) Assess the need for:
   (a) vertical loading of the job, that is, giving workers more control, autonomy, responsibility for the job,
   (b) horizontal job loading, that is, expansion of task elements through job enlargement.
(8) Make certain that the humanistic imperatives in job design are not compromised by computer technology such as the need for variety and challenge, continuous learning, discretion and autonomy, meaningful social support and opportunity for self actualization.
(9) Determine the feasibility of implementing the following job design action steps to enhance job satisfaction with the new technologies:
   (a) combine related task elements,
   (b) assign work modules to workers,
   (c) allow self-control of work pace,
   (d) allow discretion for work method,
   (e) allow workers to evaluate their performance, and
   (f) create autonomous work groups.

9.4.9 Implementation Dynamics: Participation and Humanism (Panel D)

The involvement of people concerned in the design and implementation of a change will normally offer the best chance of success, according to Child (1984), who makes the following observations on participative management as a strategy for implementing new technologies. One reason is that participation provides an opportunity for the rationale behind the proposed change to be explained and critically examined. This can help lessen people's fears stemming from a lack of kowledge and a feeling of powerlessness. If people contribute actively toward establishing the new development, this helps create among them a degree of commitment to the change and to making it work (Child, 1984).

A second consideration is that a great deal of information required as a basis for planning a change (data on present problems, work activities, decision points, time cycles, etc.) will only be known in detail to the people who are affected. Their participation is therefore necessary if the change is to have a grounding in the realities of the situation. Third, the process of employee participation should assist managers in learning about their employees' attitudes, values, and perceptions, and this learning experience should assist them to plan further necessary changes in ways that provoke less conflict. Just as important, the chance to influence and understand structural change should create employee awareness of the need for frequent reorganization, and perhaps eventually a desire to take the initiative in this field through more far-reaching participative mechanisms, such as planning agreements and management by objectives (Child, 1984).

A participative approach, then, can be appropriate in introducing organizational change, and it offers the best prospects of developing an adaptive learning capacity in organizations. It is also *ethically* the correct procedure for planning changes which affect other people. Participation is a way of confronting the political issues involved in change, not a means of avoiding or smoothing over them. If there is a deep-seated conflict of interest between the parties involved in a proposed change, participation will probably not turn up a mutually acceptable solution (Child, 1984).

Whatever the ethical and ideological attractions of participation, there are only some situations in which it is likely to succeed as a means of implementing organizational change. The conditions are identified by Child (1984):

1. There is no definite time limit on when the reorganization has to be completed—the situation is not urgent and the organization's survival is not at stake.
2. Management anticipates that it will require information from members of the organization to help design the change as well as their commitment in order to make the new organization operate effectively.
3. The need for change is not widely or clearly recognized throughout the organization.
4. The members of the organization expect to be involved in discussions prior to any change—this has become part of the organization's culture.
5. Some resistance to the proposed reorganization is anticipated, but it is not likely to challenge the underlying objectives of this proposal.
6. The power of the initiator of reorganization is limited vis à vis other groups, without being wholly constrained.

This call for a more participative approach to managing technological change is reinforced by Lund and Hansen (1986). They argue that there must be opportunity for individual participation, growth, and advancement concurrent with technological change. This principle relates partly to job design, but it looks beyond the design of individual production systems to the need for a corporate culture in which these values are considered essential to the well-being of the firm.

Early involvement of all employees that will be affected by technological change has been advocated by both management and employee groups. They further observe that involvement means more than a vaguely worded notice in the company newsletter. It means dialogue: sharing in planning and implementation efforts. Such actions remove the mystery surrounding changes that will have been signaled by the grapevine. They serve to motivate employees to get ready for job transitions, and they free employees to take part in job design of the technology itself.

The role of participative management in the implementation of new technologies is described by Hetzner et al. (1986) from a comparative perspective—Japanese versus U.S. implementation processes. Explicit attention to implementation and participation in implementation helps workers and managers avoid surprises. The new technology is always uncertain in its effects, but many principles derived from implementation analysis can guide strategy. In general, the Japanese have been better at applying these principles than have U.S. firms, for reasons that are partly cultural and partly a matter of choice. The Japanese emphasis on consensus building incorporates trials and pauses into the adoption processes, explicitly recognizing that multiple decisions are complex and that all decision makers must to some degree exert rights of "ownership" over the technology. Thus, though adoption is a long and difficult process, when the time comes to implement the technology, the participants are already familiar with it and have accepted it.

The researchers further note that by contrast, U.S. managers are likely to make relatively quick decisions concerning adoption or rejection of new technology, with little understanding of or concern for complexity of the technology and associated implementation processes. If calculations of costs and benefits are made, they tend to be "quick and dirty," without much systematic thought or basis in experience. Furthermore, for those technologies that are adopted, disappointment with the system tends to be interpreted as a failure of the technology rather than as the result of inadequate or incomplete implementation or management because only technical factors have much visibility to the participants in the decision.

Scott et al. (1981) differentiate between two types of participation. There are two modes of increasing consensus through participation in organizational decision making. The first, representative of the U.S. experience, is an attempt to have participation at the task level. Scott et al. (1981) contend that in this frame of reference, participation is a method of increasing productivity and effectiveness. The goal is efficiency and the motive is technical rationality. One area in which people advocate participation is in the organizational change process. Many authors argue that participation is fundamental for effective organizational change. Employees will be more committed to the change and demonstrate greater accetpance of new rules or norms if they have contributed to their formulation. There should be better coordination and less conflict about implementing change.

The second form is participation at an organizational or occupational level. The power base is broader, the conflicts more general, and the underlying rationality is political rather than technical. The redistribution of power is the objective here. The power redistribution position is couched in ethical and political terms, not in technically rational ones. Here, we are involved both with workers' controls as a means of avoiding arbitrary power and with self-management as a major goal. Tannenbaum (1974) contends that this form of participation may help develop trust between management and workers, a sense of responsibility in the plant, and responsiveness to management's influence attempts (which may dilute the power of workers). However, as noted earlier, participation is not only ethical in its assumptions, but also motivational because of the commitment it has the potential to generate. With technological change and the associated degree of uncertainty and potential for stress, participative management is a useful vehicle for managing the organization through a period of strategic change. A contingency approach to participative management is necessary because of the domains of the organization that it impacts on, such as the behavioral domain (a change in leadership style), the structural domain (delegation of decision-making authority), the management systems domain (distributed data bases), the process domain (changes in communications channels and networks), job design (expansion of job depth and scope), and so on. However, it is a humanistic strategy that must be considered because of the potential techno-stress and job alienation associated with technological change.

A Philosophy of Management. What the business enterprise needs is a principle of management that will give full scope to individual motivation and responsibility and at the same time give common direction of vision and effort, establisht teamwork, and harmonize the goals of the individual with the common weal (Beck and Hillmar, 1972).

The only principle that can do this is management by objectives (MBO). It makes the common weal the aim of every manager. It subsitutes for control from outside the stricter, more exacting and more effective control from the inside. It motivates the manager to action not because somebody tells him to do something or talks him into doing it, but because the objective needs of his task demand it.

> The word "philosophy" is tossed around with happy abandon these days in management circles. (I have even seen a dissertation, signed by a vice president, on the "philosophy of handling purchase requisitions"; as far as I could figure out "philosophy" here meant that purchase requisitions had to be in triplicate.) But management by objectives may legitimately be called a "philosophy" of management. It rests on a concept of the job of management; on an analysis of the specific needs of the management group and the obstacles it faces; and on a concept of human action, human behavior, and human motivation. Finally, it applies to every manager, whatever his level and function, and to any business enterprise, large or small. It ensures performance by converting objective needs into personal goals. And this is genuine freedom, freedom under the law (Beck and Hillmar, 1972).

MBO is both a general philosophy of management and a well-defined process. It is a highly rational philosophy and suggests proactive rather than reactive management. Overall goals of the organization and its interrelated parts are jointly developed and specified at different organization levels. It encourages participation and tries to anticipate change. It is an overall philosophy designed to deal with current organizational problems (Scott et al. 1981).

Participative Aspects of MBO (Panel D). The MBO concept requires that all persons involved in achieving objectives have a way of meaningfully influencing those objectives to which they are expected to contribute. Unless a person fully understands the objectives for which he is responsible, he may have difficulty achieving them, or his achievement may not reflect the results desired from the objectives (Beck and Hillmar, 1972).

The individual or team working on an objective must consider the objective both desirable and attainable. It is also important that their participation in the objective-setting process includes an opportunity for synergism. Synergism is a type of group action in which the group result is superior to the sum of the potential of the individual members of the group. It occurs when two or more people work on a problem in a free and open climate in which each person feels free to say whatever he thinks for creative problem solving.

The MBO system must also be controllable by the user and adaptive to his needs, and it must give him feedback on his performance. The system must be one that he understands and can use to improve his performance. Its intention must not be to manipulate more work out of him. The doer is involved in the planning, problem solving, and decision making for his job. He also participates in setting standards that provide him with a measurement tool. This gives him an opportunity to evaluate his actions and make the necessary corrections to improve his results in the future (Beck and Hillmar, 1972).

This type of participatory activity in the organization may necessitate a change in the organizational structure, management style, management systems, and policies and procedures. An organization must develop the ability to generate valid information. Its people should be able to make a free, informed choice so that there is internal commitment to the choices made.

The MBO Process and Technological Change. As a process, MBO is broader in scope than goal setting. It requires not only the setting of clear, concise objectives, but also the development of realistic action plans and the systematic measurement of performance and achievement. Finally, there are built-in corrective measures to deal with problems of goal changes. Figure 9.9 provides an overview of the MBO process. The actual setting of objectives and the goal-setting process can occur at all levels. In terms of technological change, the goals and objectives of the new technology must be discussed at all levels. Throughout the process there needs to be an element of flexibility and negotiation regarding performance targets. Management by directives involves issuing a memo or directive about the performance targets expected unilaterally without consultation. Through MBO, consensus building is

9.4 A MODEL OF THE IMPLEMENTATION PROCESS

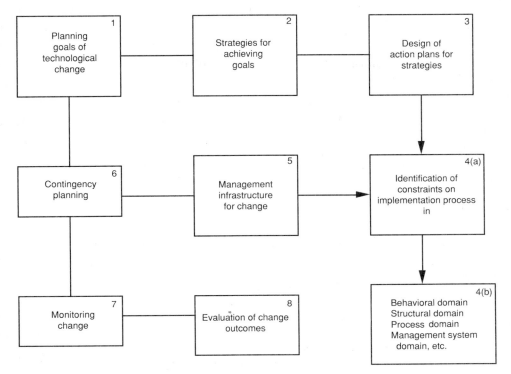

Figure 9.9. Managing change through MBO.

stressed about performance targets, strategies for implementing change, identification of barriers to adoption and implementation of new technologies, contingency planning, incentives to motivate change, the ethical guidelines for managing change, and performance appraisal of the change. Scott et al. (1981) argue that performance appraisal is an integral process of the MBO system. In order to set goals, monitor them, and evaluate goal attainment, we must know what sort of behaviors are important for good performance. Generating the criteria, standards, and yardsticks to evaluate the outcomes of technological change can also be facilitated by a group decision-making process (GDM).

The goals of technological change must be specified in operational terms. The goals are complex sets of value premises to which priorities must be assigned because of the multiple values associated with change. The strategies for achieving these goals must be articulated and the action plans for realizing these strategies designed. The identification of constraints involving the major domains of the organization becomes explicit as action plans are developed. The management infrastructure for iplementing change must be mobilized. Because of the unintended outcomes associated with change, contingency planning must be initiated to neutralize the dependency on crisis management through the period of strategic change. The monitoring of the effects of change through control, information, and feedback systems is also integral to the process. Next follows the evaluation phase, where criteria for assessment of the effects of change are developed through a participative style of management.

Design of the Culture Audit (Panel 4D). Understanding a company's culture is critical to planning, since planning involves a complex set of value premises for change. These value premises are the dynamic elements of corporate culture. Culture can be a positive force when used to reinforce goals and strategy of the organization. Chief executives can influence internal culture to be consistent with corporate strategy. Culture indicates the values employees should adopt to behave in a way consistent with organizational goals. Top executives deal in symbols, ceremonies, and images (Peters, 1978). Managers signal values, beliefs, and goals to employees. Techniques top managers use to convey the appropriate values and beliefs include rites and ceremonies, stories, symbols, and slogans. Symbols, stories, and ceremonies are ways of managing organizational culture that is hard to shape by conventional means. Issuing a written rule or policy, for example, would have almost no impact on the organization's value system (Daft, 1989).

Culture is central to change. Managers must consider culture when they decide to change corporate strategy or the entire organization. If the organization's culture is structured to be flexible, change will take place more easily. However, if the change process conflicts with values, beliefs, attitudes, and other elements of the organization culture, it will be difficult to implement.

There are three prerequisites for changing culture: (1) the strategy and all its elements must explicitly stated and easily understandable; (2) the current culture must be analyzed and made tangible; and (3) strategy must be revealed in the context of the culture to determine where the cultural risks are. This statement deals with the fact that organizations should make changes operational and concise and should attempt to monitor the effects of change on culture. Strategic change will impact on critical elements of corporate culture, including power, status, and prestige, and therefore resistance must be anticipated.

Culture can also inhibit change, according to Lorsch (1986), in two ways. First, it can produce a strategic myopia. Because managers hold a set of beliefs and values, they see events through this prism. Frequently, they miss the significance of the changing external conditions because they are blinded by strongly held beliefs. Second, even when managers can overcome such myopia, they respond to changing events in terms of their culture. Because the beliefs have been effective guides in the past, the natural response is to stick with them.

The area of strategic organizational change is one of the most important to analyze, understand, and manage. Management of an organization's culture should be one of the basic elements of a corporate strategy for an effective business enterprise. Peters and Waterman (1982) refer to excellent companies with strong corporate cultures. Such companies are directed more toward their marketplace and make limited use of policy manuals, charts, rules, and regulations.

REFERENCES

Barton, D. L., and J. Gogan, "Marketing Advanced Manufacturing Processes," in Donald D. Davis and Associates (Eds.), *Managing Technological Innovation,* Jossey-Bass, San Francisco, 1986.

Beck, A. and E. Hillmar, *A Practical Approach to Organizational Development Through MBO,* Addison-Wesley, Reading Mass, 1972.

Child, J., *Organizations: A Guide to Problems and Practice,* Harper & Row, London, 1984.

Couger, J. and Zawacki, R., *What Motivates D.P. Professionals,* Datamation, 1978.

Crozier, M., *The Bureaucratic Phenomenon,* University of Chicago Press, Chicago, 1964.

Daft, R., *Organizational Theory and Design,* West Publishing, St. Paul, Minn., 1989.

Davis, D., "Integrating Technological Manufacturing, Marketing, and Human Resource Strategies," in Donald D. Davis and Associates (Eds.), *Managing Technological Innovation,* Jossey-Bass, San Francisco, 1986a.

Davis, D., "Technological Innovation and Organizational Change," in Donald D. Davis and Associates (Eds.), *Managing Technological Innovation,* Jossey-Bass, San Francisco, 1986b.

Duncan, R., "The Ambidextrous Organization: Designing Dual Structure for Innovation," in R. Kilmann, L. Pandy, and D. Slevin (Eds.), *The Management of Organizational Design: Strategies and Implication,* North Holland Publishing, Amsterdam, 1976.

Ettlie, J. E., "Implementing Manufacturing Technologies: Lessons from Experience," in Donald D. Davis and Associates (Eds.), *Managing Technological Innovation,* Jossey-Bass, San Francisco, 1986.

Etzioni, A., *The Active Society: A Theory of Societal and Political Processes,* Collier Macmillan, London, 1968.

Hackman, J. R., and G. K. Oldham, *Work Redesign,* Addison-Wesley, Reading, Mass., 1980.

Hage, J., and M. Aiken, *Social Change in Complex Organizations,* Random House, New York, 1970.

Helfgott, R., "America's Third Industrial Revolution," *Challenge,* Dec., 1986.

Hellrigel, D., J. Slocum, and R. Woodman, *Organizational Behavior,* West Publishing, St. Paul, Minn., 1986.

Hetzner, W. A., J. D. Eveland, and L. G. Torantzky, "Fostering Innovations: Economic, Technical, and Organizational Issues," in Donald D. Davis and Associates (Eds.), *Managing Technological Innovation,* Jossey-Bass, San Francisco, 1986.

Kazanjian, R., and K. Drazin, "Implementing Manufacturing Innovations: Critical Choices and Staffing Roles," *Human Resource Management,* **25**(3) (1986).

Keen, G and M. Scott Morton, *Decision Support Systems: An Organizational Perspective,* Addison Wesley, Reading, Mass. 1978.

Lorsch, J., "Managing Culture: The Invisible Barrier to Strategic Change," *California Management Review,* 1986.

Lund, R. T. and J. A. Hansen, *Keeping America at Work: Strategies for Employing the New Technologies,* Wiley, New York, 1986.

Luthans, F., *Organizational Behavior,* McGraw-Hill, New York, 1986.

Peters, T. "Symbols, Patterns and Settings An Optimistic Case for Getting Things Done." *Organizational Dynamics,* 1978.

Peters, T., and R. Waterman, *In Search of Excellence,* Harper & Row, New York, 1982.

Robey D., *Designing Organizations,* Irwin, Homewood, Ill., 1986.

Rogers, E., and R. Agarwala-Rogers, *Communication in Organization,* Free Press, New York, 1976.

Sankar, Y., "New Technologies and Corporate Culture," *Journal of Systems Management,* April (1988).

Sankar, Y., *Corporate Culture in Organizational Behavior,* Harcourt Brace Jovanovich, Orlando, Fla., 1991.

Schultz, R., and D. Slevin, *Implementing Operations Research/Management Science,* American Elsevier, New York, 1975.

Scott, W., T. Mitchell, and P. Birnbaum, *Organizational Theory: A Structural and Behavioral Analysis,* Irwin, Homewood, Ill., 1981.

Slocum, J., Jr., and H. Sims, Jr., "A Typology for Integrating Technology, Organization and Job Design," *Human Relations,* **13**(3) (1980).

Sproul, L., and R. Hofmeister, "Thinking About Implementation," *Journal of Management,* **12**(41) (1986).

Szilagyi, A. and M. Wallace, *Organizational Behavior and Performance,* 3rd. ed., Scott, Foresman and Co., Glenview, Illinois, 1983.

Tannenbaum, A., *Hierarchy in Organizations: An International Comparison,* Jossey-Bass, San Francisco, 1974.

Thomas, J., and W. Bennis (Eds.), *Management of Change and Conflict,* Penguin Books, Harmondsworth, England, 1972.

Ullrich, R., and G. Wiedland, *Organization Theory and Design,* Irwin, Homewood, Ill., 1980.

Zuboff, Z., "New Worlds of Computer-Mediated Work," *Harvard Business Review,* Sept. 1982.

10
SUMMARY AND CONCLUSION

10.1 INFORMATION TECHNOLOGIES: DIMENSIONS OF TECHNOLOGICAL CHANGE

The dimensions of work indicate that a humanistic perspective in managing technological change is crucial.

A humanistic perspective on the management of technological change is given major focus in this book. The theme of industrial humanism as a contemporary management creed is noted. The elements of humanism, such as compassion and empathy, identity and integrity, survival and transsurvival needs, the will to meaning, self-actualization, individuality and potentiality, freedom and responsibility, and so on, are identified from major works in the field of humanistic psychology.

The imperatives of humanism conflict with those of technology and have the potential to produce stress. The humanistic perspective argues for actualization of the self and development of value potentialities of the individual. The development is facilitated by freedom and autonomy in the particular work setting. The worker brings to his (meaning his or her throughout this book) job situations certain hopes, expectations, and values. Emotions and perception are important components of the individual complex. The "inner world" of the worker is considered as important as external reality in determining productivity. The worker obtains his sense of identity through interpersonal relationships, and the better they are, the more responsive to the norms of the work group he will be. The employee is responsive to management to the extent that the supervisor can meet his needs for esteem and self-actualization, his will to meaning, and his character development for ethical excellence.

Classical job design based on scientific management and industrial engineering methods involves a set of mechanisms for improving organizational efficiency. While major contributions are made to the goal of efficiency, the design of jobs based on principles of simplification, standardization, and formalization produce a high incidence of job stress and alienation and a dilution of the meaning of work.

The socio-technical systems (STS) design, an alternative job design strategy, provides a more integrated and humanistic perspective to job design. It enhances the meaning of work. This approach, which integrates social, human, and technical elements, can also produce some positive effects, as noted by Davis (1971). We examined the effects of a new technology, robotics, on workers and suggest some sociotechnical design strategies for managing technological change.

Next, we focus on the characteristics of technology, including task uncertainty, work flow uncertainty, interdependence, discretion, and technical uncertainty in terms of input, conversion, and output uncertainty. The impact of these characteristics on job design is reviewed. Some effects of new technologies on job design from the perspective of Lund and Hansen (1986) are also summarized.

The imperative in job design based on the STS design is to link the design to human needs, such as for variety, challenge, learning, autonomy, discretion, recognition, meaning, belonging, and identity. Two models of job design by Chung (1977), a job enlargement model and a job enrichment model, are presented. Some job design action steps for new technologies are suggested. The future scenarios for job design from the viewpoint of Hackman and Oldham (1980) are more positive than negative in terms of the humanistic imperatives.

Occupations and organizations typically build their practices, values, and basic self-image around their underlying technology. If the technology changes substantially, the organization or occupation must not only learn new practices, but must redefine itself in more profound ways that involve deep cultural assumptions (Schein, 1985). Companies must realize that cultural change is only a starting point; it must be followed by the use of all other management tools, such as new leadership styles, structural changes, and changes in reward systems and value premises to promote an innovative organization.

When implementing innovations, ignoring culture invites disaster. Managing around culture may be realistic in the short run; over the long run, a change in culture is imperative (Stonich, 1982). Implementing new technologies in any organization will impact on all elements of culture, such as behavioral patterns, norms, and values, the organizational climate, rules of the game, organizational ideology, and so on.

Three types of corporate cultures are identified: entrepreneurial, bureaucratic, and participative. All three are needed for different stages of technological change; the initiation stage requires an entrepreneurial culture, the implementation stage, a participative culture, and the routinization stage, a bureaucratic culture.

Evaluating an organization's culture is critical when managing an organization through a period of strategic change produced by the adoption and diffusion of new technologies. Various levels of an organization's culture, such as artifacts, creations, values, and basic assumptions, are audited for an assessment of its impact on technological change. From this culture audit comes the conclusion regarding high or low culture/value adjustment for new technologies.

To cope effectively with technological change, organizations must be designed as open systems and must behave as open systems. The systems paradigm provides a set of conceptual tools for the design of an adaptive corporation. The systems approach affords a method of planning and managing technological change.

One approach, which uses the elements of the systems paradigm, is the sociotechnical systems (STS) design. Because STS management focuses on the work processes, organizational culture, job design, environmental domain, subsystems interdependence, feedback loops, etc., rather than on the parts of the system, it is a systems approach. The STS approach in managing technological change is noted by many researchers, including Daft (1989) and Taylor et al. (1986), as quite an effective approach.

The general systems approach and the characteristics of an organizational system are described. These can be used to manage or design the system as an open system in a state of dynamic equilibrium for the new information society.

The role of computer technology in organizational design must be considered. The effectiveness of a management information system is greatly enhanced if the structure of the system conforms to that of the organization of the firm. The geometry of the organization is also affected by computer technology by virtue of its impact on centralization/decentralization, span of control, job design, and decision making.

An information-processing model of organizational design is developed with a call for changes in classical and contingency design principles. The information domain and its effects on the functional, product, and matrix structures are also examined. The information pathologies in the classical and contingency design makes a change in design essential for the information age. The information imperatives for this new age also reinforce this need for a new design. A variety of structural configurations are developed from the information-processing model.

Different structural configurations within a complex organization are needed because of the variety, complexity, and uncertainty of the information domain within it. Changes in the information domain are generally produced by technological change. The systematized modular unit, the discretionary modular unit, the organic modular unit, and the adhocratic modular unit are different structural configurations used for different stages of the change process.

The new information society will be more turbulent and complex than the previous industrial society. Companies will have to adapt to changes made by their competitors in order to retain or improve their share of the market. If companies must change, employees will also have to adapt to these changes. Flexible, well-designed computer-based systems can help workers adapt to change by way of such things as "job aids" built into the systems. Computer technology facilitates the unification of previously fragmented control systems and assists integration through its enhancement of communication. These possibilities present opportunities for change within management with regard to the hierarchical location of decision making, the complexity of coordinative mechanisms, and the size of middle management. The adoption and diffusion of innovations and change can also be made easier by the organization's information system. The management of change, a crucial function for the contemporary manager, must also be linked to computer technology.

A major factor in understanding the innovation process is an understanding of the organization's information system. The strategic objectives of the information technology (e.g., control and integration versus strategic planning and change) will determine the effectiveness and innovation potential of this system. As an instrument for strategic planning, environmental scanning, and the management of change, MIS (management information system) can be effective in the management of technological change.

We must learn how to make existing companies, large companies in particular, capable of innovation. A strategy is needed that will enable businesses first to identify the existing opportunities for innovation and then to give effective leadership for such innovation. It will no longer be sufficient to extend, broaden, modify, or attempt to adapt existing technologies. From now on, the need will be to innovate in the true sense of the word, to create truly new wealth-producing capacity, both technical and social (Drucker, 1985).

The adaptive corporation needs a new kind of leadership. It needs managers of adaptation equipped with a whole set of new, nonlinear skills. Leadership is the force behind every innovation. The efficacy of a leader is linked to his management of the strategic decision-making process, and, since such decision making is an instrument of change and innovation, we consider at length some of its elements.

The first ingredient in reinventing the corporation for the information age is a powerful strategic vision. Basically, the source of a vision is a leader. The systems approach forces him to look upon his organization as an information network. Adapting the organization to the information systems needs for effective strategic planning and the management of technological change is a major function of the leader. Another significant task of the leader who is reinventing the corporation is to select the right corporate structures. Since changes in corporate strategy also dictate changes in structure, the role of the leader in structural change is quite crucial to organizational effectiveness. Three uses of structure by the leader at each level of the organizational hierarchy are necessary in the management of technological change.

We conclude with the observation that a leader's style must reflect the ethical values of the corporate culture and the humanistic imperatives for effectively managing and designing tomorrow's organization.

Managers of large enterprises have urgent sociohumanistic responsibilities to create innovative corporate designs that will emphasize human values over technological imperatives in their design.

The design of organizations is open to options made possible by technology. Firms are not locked into one conceptual scheme, but are free to create structures that fit with their mode of operation and corporate philosophy. Computer and telecommunications aids can be tailored to make a given organizational scheme work (Lund and Hansen, 1986).

We consider a variety of conceptual schemes for organizational design. The need for structural change is indicated by reference to a number of structural problems such as adaptability, role ambiguity, and integration (Duncan, 1976); and environment diversification, growth, and technology (Child, 1981, 1984). A new organizational paradigm (Trist, 1978) is therefore mandated by this evidence of structural problems and the need for structural change. Some new templates for today's organizations are advanced (Drucker, 1974). Attributes of the innovative organization, as developed by Peters and Waterman (1982) and Thompson (1973), are discussed. The characteristics of organic-mechanistic systems noted by Davis (1986) are appropriate for technological change. Another conceptual scheme for organizational design encompasses temporary problem-solving systems (Bennis, 1973). The framework structure and modular units of the adaptive corporation (Toffler, 1985) and collateral organizations (Zand, 1979) also emerge as another alternative design for change.

Any function as complex as the management of change requires some kind of conceptual framework of thinking to guide it. We provide such a conceptual framework for implementing change. In general, the implementation model indicates that an organization's capacity to implement an innovation is contingent on (1) the characteristics of the innovation being adopted; (2) the functions of the managers at each stage of the change process (the evaluation, initiation, implementation and routinization stages); (3) the nature and character of the domains of the organization (the behavioral, structural, process, technical, and management systems domains); and (4) the type of strategies adopted by the manager to modify the characteristics of the innovation and/or organization. The implementation process is a complex contingent process.

The conceptual model for implementing and managing change focuses on a number of principles for managing change. A detailed analysis of each of the four stages of the change process is given with specific operating principles for managing each stage. Next, the five domains of the organization affected in the implementation of an innovation are described. It is argued that the implementation of an innovation may create changes in the behavioral, structural, management systems, technical, and process domains of the adopting organization. The principles for managing each of these domains as they relate to the implementation process are also developed. Managerial strategies associated with these principles of change are also discussed. In addition, the politics of implementing change is considered, since an innovation may create changes in the sources of power within an organization.

The chapter concludes with implementation of dynamics from a participative perspective. Participative management as a strategy for implementing new technologies is recommended because it develops an adaptive learning capacity, it is ethically the correct procedure for planning changes that affect other people, and it is a way of confronting political issues involved in change. The participative aspects of management by objectives (MBO) are reviewed briefly and a model based on MBO is also developed.

10.2 CONCLUDING COMMENT: ON THE INDIVIDUAL IN THE TECHNOLOGICAL SOCIETY

I have argued in this book for a renaissance of humanism for all of mankind on the planet, for the triumph of human values of love, truth, beauty, goodness, compassion, caring, human dignity, non-violence and peace. The logic of technology and production is not the logic of life nor of society. It leads to the devaluation of human beings as either appendages of the machine or as peripherals to such machines. The center of the stage can never be occupied by a machine because of the unique attributes of the human being. No machine will ever possess the human attributes of reason, insight, imagination, creative visualization, love, truth, justice or goodness. The machine and technology must serve authentic human needs and values; the need for self actualization, self esteem, love, identity, growth, beauty and meaning, cannot be supplanted by the imperatives of technology or the logic of production. The beauty and miracle of man is noted in poems, the scriptures, the humanities, art, music, and literature.

Here is Shakespeare's classic exclamation from Hamlet.

> What a piece of work is a man! how noble in reason! how infinite in faculty! in form and moving how express and admirable! in action how like an angel! in apprehension how like a god!

The most powerful current expressions of this poetic image of man is emphasized by Sri Sathya Sai Baba in his discourses.

Man is no mere biped, an animal that struts about on two legs, instead of four. He has the unique destiny of realizing and appreciating beauty, truth, goodness, harmony, melody, and—conferring on himself and others—love, compassion, and sympathy. He can delve not only into the secrets of nature, but also into his own mystery, and discover God who is behind both, nature and himself. The clouds of conceit and ignorance hide from him his destiny. Man is Truth, Goodness, and Beauty. That is why he is drawn by the true, the beautiful, and the good; he is, therefore, an ideal candidate for Divinity, which is his real destiny.

In the design of technology and in the implementation of technology, too often the study is on the object of technology not on man. The effects of science through technology on human behavior, attitudes, and norms indicate that the entire spectrum of human values is affected. The concept of Social Darwinism, which Einstein rejects, is also notable. I personally believe that such values as competition, aggression, selfishness, and malice are traits that belong to our animal ancestry, but as we ascend on the human plane, we must shed these traits and cultivate more humane traits of love, compassion, caring and service. However, because the traits of our animal ancestry are compatible with our economic system, we have created a pseudo-scientific myth that we are genetically programmed to behave in an aggressive, self-centered, competitive, and malicious manner. Similarly, we have rejected religion not because of our "enlightenment" via science and technology, but because we have programmed our society and lives with pagan values such as greed, lust, envy, and malice to our brothers and sisters. As Chesterton (1910) noted "The Christian ideal, it is said, has not been tried and found wanting; it has been found difficult and left untried." A value-free society means everything is negotiable; "we know the price of everything but the value of nothing," according to Oscar Wilde. Far from being an opium of the masses, religion like philosophy, humanism, and the humanities provide values and ideals for an ethical culture which is critical for human excellence.

In the same way that values influence our goals, belief systems, self concept, attitudes, perception, judgements and behavior, technology also has the potential to affect all of these elements of our personality and life. It is therefore imperative that the ethics of technological change be incorporated into all judgements on which to base decisions for implementing new technologies.

Technology makes possible a future of open-ended options that seems to accord well with the prophetic tradition of Judeo–Christian belief system of human hope and responsibility for man becoming and developing his potential. The main task that technological change poses for theology, according to Mesthene, (1977) is that of deliberate religious innovation and symbol reformulation to take specific account of religious needs in a technological age. I believe that while symbol reformulation may be necessary since the medium is the message, the ultimate values are adequate for the technological age. The problem is with the values that are pathological, such as greed, aggressive competition, selfishness and malice. The ultimate values of truth, justice, love, nonviolence, peace, right conduct, service to humanity and goodness do not require redefinition, simply implementation into the managerial apparatus for implementing technological change. Do we need to define love before we love so with justice, goodness, peace, truth as determinants of our behavior and as ideals to which we aspire? If we link love with our ego needs and emphasize reciprocity and an exchange value or a utility value then maybe we have to define it. However, if it is unconditional, then spontaneous expression of this magnificent value is more than possible for any age. It serves a multiplier effect as it generates the necessary conditions for an ecological awareness, harmony and compassion for nature, all species, and the planet.

Technology creates new possibilities for human choice and action but leaves their disposition uncertain. What its effects will be and what ends it will serve are not inherent in the technology, but depend upon what man will do with the technology and what values will be actualized through technology. The use of technology reflects man's values, goals, and priorities. The organization that manages technological change is also created through man's values, and reflects his assumptions about human nature.

With reference to the problem of technological change and human values, Mesthene argues that technological change leads to an increased questioning and reformulation of values. Technology has a direct impact on values by virtue of its capacity for creating new opportunities, options, and questions of choice. As such, technological change falls within the orbit of ethics. Ethics is the science of judging human ends and the relationship of means to those ends. It is also the art of controlling means so they will serve authentic human ends. Viewed in this way, ethics involve the use of human knowledge to tell us something about the relationship among men or about the applicability of available instruments. Additionally, as an art, it comprises techniques of judging and decision making as well as the tools of social control and personal development. Essentially, ethics really is or should be involved in all human activities. Harrison (1975)

One of the functions of the ethics of change is to provide principles for managing change. These principles of change must be based on the humanistic imperatives that were developed in Chapter one—such humanistic imperatives as identity and integrity, self actualization, the will to meaning, character values, the dynamics of caring, individuality and potentiality, responsibility and humanity, freedom and autonomy, survival and transurvival needs, compassion and empathy. These imperatives are simply self evident components of Kant's categorical imperative and Herzberg's dynamics of caring for leadership excellence. It is not the information domain, nor the dominant norm of efficiency, nor the organizational control and coordination that are at the center of the change strategy, but the human element. He/She is not a constraint on systems efficiency but the central factor in the equation or in the formula for change. To conceptualize the individual as a constraint is to reduce him/her to a marginal value that must be negotiated in the change process. Any change program that compromises any of the humanistic imperatines above is problematic for the individual. The change must have as one its objectives the actualization of the values of individuals who interface with the new technologies. The principles of change can be derived from the questions posed as one initiates technological change and innovation. For example, what change in creative values will this technological change program create for workers and managers? What changes in experiential values will be necessary to facilitate technological change? What attitudinal values must be articulated in this program of change that may have the potential to produce stress and alienation? Finally, what ethical values, truth, integrity justice, equity, and so on must I as a manager articulate in my principles for managing change?

The greatest problem for technological ethics is to provide conditions that enable human beings to actualize their potentialities and to reinforce those human values that enhance the search for meaning in our lives.

Technological change involves rational actions of change agents, technocrats, managers, engineers, scientists and so on. They act rationally according to a rational decision-making procedure. Their rational actions affect people and the eco-system. Hence these actions can be evaluated from a moral point of view. The consequences of technological change can be evaluated from a moral and ethical perspective. Some of these consequences are good, bad, or both and therefore can be subject to an ethical assessment. We must now engage in this ethical discourse. The urgency is now as we embark globally on a new contract to save the planet, endangered species, human lives, and the ultimate values that guided our ancestors in their search for a meaningful life.

For the first time in recorded history, science and technology have given man enough resources to ensure for every human being born upon this earth adequate material, intellectual, and spiritual inputs to be able to live a full and rewarding life. If we can creatively utilize the wealth of the world and constructively utilize its resources, no one need go hungry, no child need to be deprived of education, no person need to be without adequate medical care, and the spectre of unemployment would be laid to rest for ever. But this will require not simply a readjustment of our existing perceptions, values, and priorities, it will need a paradigm shift, a break from existing orthodoxies and a bold thrust into a new dimension of awareness (Ophuls 1982).

REFERENCES

Child, J., *Organizations,* Harper & Row, New York, 1984.

Chung, K. H., *Motivational Theories and Practices,* Grid, Inc., Columbus, Ohio, 1977.

Daft, R., *Organizational Theory and Design,* West, St. Paul, Minn., 1989.

Davis, L., "The Coming Crisis for Production Management: Technology and Organization," *Journal of Production Research,* **9** (1971).

Drucker, P., "New Templates for Today's Organization," *Harvard Business Review,* 1974.

Hackman, J. R. and G. K. Oldham, *Work Redesign,* Addison-Wesley, Reading, Mass., 1980.

Harrison, E., *The Managerial Decision-Making Process,* Houghton Mifflin Co., Boston, 1975.

Lund, R. T. and J. A. Hansen, *Keeping America at Work: Strategies for Employing the New Technologies,* Wiley, New York, 1986.

Mesthene E., "The Role of Technology in Society" in *Technology and Man's Future,* A. Teich (Ed.), St. Martin's Press, New York, 1977.

Peters T., & R. Waterman, *In Search of Excellence,* Harper & Row, New York, 1982.

Ophuls, W., "The Scarcity Society" in *Ethical Issues in Business, A Philosophical Approach,* T. Donaldson and P. Werhane (Eds.), Prentice Hall, Englewood Cliffs, New Jersey, 1982.

Sai Baba, Sri Sathya, *Discourses: The Voice of the Avatar,* Prasanthi Nilayam, A. P. India, 1980.

Schein, E., *Organizational Psychology,* Prentice Hall, Englewood Cliffs, New Jersey, 1970.

Toffler, A., *The Adaptive Corporation,* McGraw-Hill, New York, 1985.

Trist, E., "The Socio Technical Perspective" in *Perspectives in Organizational Design,* A. H. Van der Ven and W. Joyce (Eds.), Wiley, New York, 1981.

Zand, D., "Collateral Organization: A New Change Strategy," in W. French et. al., *Organizational Development: Theory, Practice and Research,* Business Publications, Dallas, 1979.

INDEX

Adhocracy, 180–181, 303–305
Alienation, 5, 7
   analytic description of, 13–17
   increase in, 8–9, 61
   work and, 9–12
Attitude, 17–19
Autonomy, 32

Behavior, 330–331, 349–350
Bureaucracy, 210–211, 338–342

Centralization, 164–166
Change design(s), 274–311
   adaptive corporation attributes, 293–295
   adhocracy and, 303–305
   collateral organization and, 295–297
   humanism and, 305–309
   innovative organization attributes, 289–290, 292–293
   matrix organization and, 297–303
   need for new design, 275–281
   organic-mechanistic systems and, 290–291
   overview of, 274–275
   paradigm for, 286–287
   problem-solving systems and, 291–292
   reorganization and, 283–286
   symptoms of problems, 281–283
   template for, 288–289
Change implementation model, 312–360
   change process concepts, 315–316
   diagnostic stage in, 323–325
   environmental domain and, 334–337
   implementation dynamics and, 352–359
   implementation stage in, 326–328
   initiation stage in, 325–326
   management and, 338–352
   organizational domains and, 329–334
   overview of, 312–315, 316–318
   routinization stage in, 328–329
   stages in, 318–319
   structural domain and, 337–338
   technical innovation dimensions, 319–323
Character values, 29–30
Cognition, 223–225
Communications, *see also* Information-processing model
   information-processing model and, 154–155
   systems approach and, 138–140
Compassion, 24
Computer, *see* Information-processing model; Information technology; Integrated data base
Corporate culture, 75–114
   change and, 91–93
   change implementation model and, 358–359
   cultural audit and, 83–84
   culture types and, 81–83
   ethics and, 102–103, 105–106

371

INDEX

Corporate culture (*Continued*)
  industrial engineering ethics and, 107–112
  leadership and, 250–253
  levels of, 84–89
  management strategies for, 83
  merging of, 103–105
  model of, 89–91
  organizational culture perspective, 79–81
  overview of, 75–76
  resistance of, to change, 93
  technology and, 76–79, 106–107
  values and, 94–102
Creativity, 181–185
Culture, *see* Corporate culture
Cybernetics:
  humanism and, 145–147
  information-processing models and, 159–161, 167–169
  management and, 140–142
  systems approach and, 135–138

Data base, *see* Integrated data base
Decision making:
  information technology and, 202–206
  leadership and, 239–242
  values and, 268–272
Differentiation, 129–130
Discretionary modular unit, 179
Diversification, 284

Ecology, 226–231. *See also* Environment
Empathy, 24
Entropy, 130–131
Environment:
  change designs and, 283–284
  change implementation model and, 330, 334–337
  information technology and, 197–198
  technological impact on, 226–231
Equilibrium, 134–135
Ethics, 38–40. *See also* Values
  corporate culture and, 95, 101
  cultural change and, 102–103
  industrial engineering and, 107–112
  leadership and, 266–267, 270–272
  technology and, 72–73, 105–106

Feedback:
  information technology and, 201
  systems approach and, 136–137
Freedom, 32

Growth, 284

Homeostasis, 134–135
Humanism, 1–42
  alienation and, 9–17
  change design and, 305–309
  change implementation model and, 352–359
  cybernetics and, 145–147
  importance of perspective of, 361–366
  industrial humanism, 23–24
  job design and, 61–63
  leadership and, 266
  management and, 33–40
  perspectives on, 24–33
  stress and, 19–22
  technology and, 17–19, 22–23
  work and, 3–7
Humanity, 33
Human nature, 9–11

Identity, 24–25
Individual, 280–281, 366–369
Individuality, 32–33
Industrial humanism, 23–24. *See also* Humanism
Information-processing model, 150–186
  centralization and, 164–166
  change implementation model and, 343–344
  classical/contingency design principle changes and, 169–171
  classical organization design and, 152–154
  computer-based systems and, 163–164
  computer technology and, 161–163
  creativity and, 181–185
  description of, 167–169
  design principles and, 156–157
  organizational design and, 166–167
  organizational design changes and, 158–161
  organizational design imperatives and, 174–176
  overview of, 150–152
  pathologies and organizational design, 173–174
  power relations and, 155–156
  structural configurations design for, 176–181
  structure and, 154–155
  structure/systems matching and, 171–173
  systems approach and, 137–138
Information technology, 187–233. *See also* Technology
  bureaucratic framework and, 210–211

change management and, 206–210
change planning and, 196–197
control/integration considerations and, 193–194
decision making and, 202–206
ecology and, 226–231
environmental scanning and, 197–198
feedback and, 201
human considerations and, 221–225
importance of, 188–191
integrated data bases and, 211–214
limits to, 218–220
management process and, 195–196
operations control and, 198–200
organizational change and, 215–217
overview of, 187–188
stability considerations and, 194
strategic objectives of, 191–193
technology assessment and, 231
uncertainty reduction and, 194–195
Innovation, 129–130
Integrated data base, 211–214, 218
Integration, 193–194, 209
Integrity, 25–26
Interdependence, 133–134
Interpersonal relationships, 19

Job design, 43–74
ethics and, 72–73
future scenario for, 70–72
humanism and, 61–63
models for, 63–69
overview of, 43–46
robotics and, 50–52
sociotechnical design, 52–53
technology and, 46–49, 49–50, 53–61, 69–70

Leadership, 234–273
adaptive coping role of leader, 238–239
humanistic values and, 266
innovation and, 236–238
Northern Telecom example of, 242–249
organizational culture and, 250–253
organizational structure and, 260–265
overview of, 234–236
strategic decision making and, 239–242
strategic planning and, 239
strategic vision and, 253–254
systems approach and, 143–145
transformational leader role, 255–260
values and, 266–272

Machine, 4. *See also* Technology
Management:
change implementation model and, 331–332, 338–352, 355–358
corporate culture and, 83
cybernetic principles for, 140–142
humanism and, 33–40
information technology and, 195–196, 206–210
systems approach and, 125–126, 142–143
technology and, 7
Management by objectives (MBO), 313, 355–357. *See also* Change implementation model
Management information system (MIS), *see* Information technology
Matrix organization, 297–303
Meaning, 27–29
Middle management, 215–216
Motivation, 18, 348–349

Negative entropy, *see* Entropy

Organic modular unit, 179–181
Organizational culture, *see* Corporate culture
Organizational systems, *see* Systems approach

Pathology, 173–174
Perception, 33
Planning, *see* Leadership; Management
Potentiality, 32–33
Power relations, 5
change implementation model and, 340–341, 344–346, 347
information-processing model and, 155–156
information technology and, 216–217
Progressive segregation, 132–133

Relationships, *see* Interpersonal relationships
Responsibility, 33
Robopathic behavior, 15–16
Robotics, 50–52

Scientific management, 47–48
Self-actualization, 30–32
Self-esteem, 267–268

Sociotechnical design, 52–53
Sociotechnical systems (STS) design, 48–49, 52–53. *See also* Systems approach
Strategic decision making, *see* Decision making
Strategic planning, *see* Leadership
Stress, 19–22
Survival needs, 26–27
Systematized modular unit, 178–179
Systems approach, 115–149
   change implementation model and, 332–334, 337–338
   communication and, 138–140
   cybernetics and, 135–138, 140–142, 145–147
   differentiation and, 129–130
   equilibrium and change, 134–135
   general systems concept and, 120–121
   information imperatives and, 174–176
   information-processing model and, 166–167
   information systems matching and, 171–173
   information technology and, 152–154, 156–157, 215–217
   interdependence and change, 133–134
   law of requisite variety and change, 131–132
   leadership and, 143–145, 260–265
   management implications of, 142–143
   manager and, 125–126
   negative entropy/change and, 130–131
   open organizational systems characteristics, 121–125
   organization framework and, 126–129
   organization of a system, 121
   overview of, 115–118
   progressive segregation and change, 132–133
   sociotechnical systems theory, 118–120

Technology, *see also* Information technology
   change designs and, 285–286
   change implementation model and, 333–334, 350–352
   corporate culture and, 76–79
   ethics and, 72–73, 105–106
   future functions of, 325
   humanism and, 17–19, 22–23, 33–35
   individual and, 366–369
   job design and, 46–49, 53–61, 69–70
   job design improvement and, 49–50
   leadership and, 242
   stress and, 19–22
   values and, 106–107
Transformational leadership role, 255–260
Transsurvival needs, 26–27

Values, *see also* Ethics
   character values, 29–30
   corporate culture and, 94–102
   decision making and, 268–272
   industrial engineering and, 107–112
   leadership and, 266–272
   technology and, 105–106, 106–107
Vision, 253–254

Work:
   alienation and, 9–12
   humanistic perspective on, 3–7